Fish & Fisheries Series

Volume 41

Series Editor

Kai Lorenzen, University of Florida, Gainesville, FL, USA

The volumes in the Fish & Fisheries series will cover topics ranging from the biology of individual species or groups of fishes, to broader concepts in fisheries science, conservation and management. The series is directed to professionals and researchers in fish biology, senior undergraduate and postgraduate students, and those concerned with commercial production or harvest of fishes.

Prospective authors and/or editors should consult the Publishing Editor for more details. Comments or suggestions for future volumes are welcome:

Eva Loerinczi
Publishing Editor
eva.loerinczi@springer.com

More information about this series at http://www.springer.com/series/5973

John Clay Bruner • Robin L. DeBruyne
Editors

Yellow Perch, Walleye, and Sauger: Aspects of Ecology, Management, and Culture

Springer

Editors
John Clay Bruner
Department of Biological Sciences
University of Alberta
Edmonton, Alberta, Canada

Robin L. DeBruyne
Lake Erie Center
University of Toledo
Toledo, Ohio, USA

Fish & Fisheries Series
ISBN 978-3-030-80677-4 ISBN 978-3-030-80678-1 (eBook)
https://doi.org/10.1007/978-3-030-80678-1

This Springer imprint is published by the registered company Springer Nature Switzerland AG.
The registered company address is: Gewerbestrasse 11, 6330 Cham, Switzerland

Preface

This book is a snapshot of a symposium originally planned for the 150th American Fisheries Society annual meeting in Columbus, Ohio, September 14–25, 2020. The symposium was entitled, "Biology, Management, and Culture of Walleye, Sauger, and Yellow Perch: Status and Needs" and 39 presenting groups, representing 165 percid researchers, were scheduled to give talks. The worldwide COVID-19 pandemic forced a switch to a virtual meeting causing 13 of the presenting groups to drop out of the symposium, leaving 26 talks and 83 people still involved. Some of the presenting groups enquired about the possibility of producing a symposium proceeding. A book proposal by John Bruner and Robin DeBruyne to Springer Fisheries Series was accepted. Twelve presenting groups of the remaining 26 decided to participate in the book proposal.

Walleye, one of the most sought-after species of freshwater sport fishes in North America, and its "sister" species, the Sauger, have demonstrated appreciable declines in their numbers from their original populations since the beginning of the twentieth century. Similarly, Yellow Perch, once the most commonly caught freshwater sport fish in North America and an important commercial species, have also shown declines. Yet, some Western states and provinces are trying to extirpate Walleye and Yellow Perch where they have been introduced outside of their native range. The purpose of this book is to present up-to-date information on the biology and management of Walleye, Sauger, and Yellow Perch since the 2011 publication of the AFS book, *Biology, Management, and Culture of Walleye and Sauger*, the book *Biology and Culture of Percid Fishes Principles and Practices* (Springer Press, 2015), and pertinent review papers in *Biology of Perch* (CRC Press, 2016).

Part I Yellow Perch *Perca flavescens*

The first four chapters provide new information on the effects of water level fluctuations on spawning, best practices of aquaculture techniques for increasing production of fingerlings, evaluating statewide regulations for bag limits meeting management objectives, and ecological factors influencing the early life history of larval Yellow Perch distribution in Lake St. Clair. The first chapter by Matt, Welsh, and Smith, evaluates water level fluctuations of a Central Appalachian hydropower reservoir on the spawning characteristics of Yellow Perch. This data will inform decisions regarding management of fish populations and lake level drawdown regulations. The second chapter by Doyle et al. documents 41 years of a western Ohio fish hatchery and what changes were made to eventually increase production of fingerlings from 13 ± 4 to 53 ± 6 fish·m^{-2} (mean ± SE). In the third chapter, Clapp et al. detail how a Michigan daily bag limit was obtained by evaluating law enforcement and biological concerns, interjurisdictional management issues, and ongoing comments from agency staff and the public. DeBruyne et al. in the fourth chapter discuss Yellow Perch larval distribution and abundance in Lakes St. Clair, a source of Yellow Perch for Lake Erie, and the ecological factors influencing their early life history.

Part II Walleye *Stizostedion vitreum*

The next set of chapters discuss methods to manage, raise, treat, and track Walleye. The fifth chapter by Euclide et al. used new genomic tools of reduced representation sequencing (RAD-seq), or RAD-capture (Rapture) and GT-seq (genotyping-in-thousands) panels for managing Walleye in the Great Lakes. Johnson, Kelsey, and Summerfelt, in the sixth chapter compare and contrast Walleye larviculture in innovative reuse aquaculture systems (RAS) at two-state fish culture facilities in Vermont and Iowa. The seventh chapter, by Eroh et al., reports on the efficacy of hydrogen peroxide treatments in RAS to control infections of *Saprolegnia* spp. on Walleye eggs and improve hatching success. Smith, Welsh, and Hilling, in the eighth chapter, used telemetry to understand the ecology and spatial distribution of Walleye in a hydropower reservoir in northern West Virginia, USA. The tenth chapter, by Shultz et al. evaluated acoustic tag types, designed a gridded array of receivers, and evaluated the effect of the implanted tags on Walleye for a telemetry pilot project in Mille Lacs Lake, Minnesota. The purpose of the telemetry project was to understand the movements, distribution, and interactions of juvenile and adult Walleye in the lake.

In the ninth chapter, Klimah discusses what challenges a prospective fisheries biologist might expect working for a tribal fishery or natural resource department and suggestions on how to make a positive impact within tribal communities and workplaces.

Part III Sauger *Stizostedion canadense*

In the eleventh chapter, Brewer et al. discuss the success of a stocking program of the New York State Department of Environmental Conservation to establish a self-sustaining population of Sauger in the upper Allegheny River watershed, beginning in 2014, and expected to continue through 2023.

Part IV *Perca* and *Stizostedion* Management, Research, and Culture Progress in North America and Europe

In the twelfth chapter, DeBruyne and Roseman bring together the work presented in this book, together with other recently published *Perca* and *Stizostedion* research and management progress, to discuss the different approaches to enhance the percid fisheries in North America and Europe.

This book demonstrates how advances in technology and management can intersect to improve fishery management and culture outcomes, move toward broad-scale management considerations, and highlight collaboration. As technology (e.g., genomic, telemetry, and aquaculture) continues to improve, and with increased information from user groups, agencies may be able to reduce uncertainty associated with decisions affecting resource use. Unfortunately, the effects of other factors, such as invasive species and climate change, on percid habitat and populations continue to remain unclear, requiring continued knowledge sharing and cooperation to maintain successful percid fisheries around the world.

Edmonton, AB, Canada — John Clay Bruner
Oregon, OH — Robin L. DeBruyne

Chapter Reviewers

The editors are grateful to the following individuals and other anonymous reviewers who served as peer reviewers of the chapters in this book. Without their guidance and expertise, this book would not have been possible.

Dr. L. Zoe Almeida, Ohio State University, 230 Research Center, 1314 Kinnear Rd., Columbus, OH 43212-1156 USA

David F. Clapp, Research Station Manager, Michigan Department of Natural Resources, Fisheries Division, Charlevoix Fisheries Research Station, 96 Grant Street, Charlevoix, MI 49720 USA

Dr. Mrinal K. Das, MacEwan University, 10700 104 Ave NW, Edmonton, AB, T5J 4S2 Canada

Dr. Dan Dembkowski, Wisconsin Cooperative Fishery Research Unit, UWSP College of Natural Resources Stevens Point, WI 54481 USA

Kevin Donner, Little Traverse Bay Bands of Odawa Indians, Natural Resource Department, 7500 Odawa Circle, Harbor Springs, MI 49740 USA

Cathleen Marie Doyle, Department of Evolution, Ecology, and Organismal Biology, The Ohio State University, Columbus, OH, 43212-1803 USA

Guy Darrell Eroh, Warnell School of Forestry and Natural Resources and USGS Georgia Cooperative Fish and Wildlife Unit, University of Georgia, 180 E Green St, Athens, GA 30602 USA

Maxwell A. Field, Little Traverse Bay Bands of Odawa Indians, Natural Resource Department, 7500 Odawa Circle, Harbor Springs, MI 49740 USA

Dr. Corbin D. Hilling, U.S. Fish and Wildlife Service, Maine Field Office, 306 Hatchery Road, East Orland, ME 04431 USA

Dr. James R. Jackson, Cornell Biological Field Station, Department of Natural Resources, Cornell University, Bridgeport, NY 13030 USA

Kevin Kelsey, Vermont Fish and Wildlife Department, Ed Weed Fish Culture Station, 14 Bell Hill Rd, Grand Isle, VT 05458 USA

Carl Klimah, Mille Lacs Band of Ojibwe, 43408 Oodena Dr, Onamia, MN 56359-2236 USA

Dave Lucchesi, Area Fisheries Supervisor, South Dakota Game, Fish and Parks, 4500 S. Oxbow Ave, Sioux Falls, SD 57106 USA

Dr. W. C. (Bill) Mackay, Professor Emeritus, Department of Biological Sciences, University of Alberta, Edmonton, AB, T6G 2E9 Canada

Dr. Christine Mayer, Professor, Lake Erie Center, Department of Environmental Sciences, University of Toledo, Oregon, OH 43616 USA

Dr. Kevin Neves, College of Arts and Sciences Biological Sciences, Bowling Green State University, 328 Life Sciences Building, Bowling Green, OH, 43403 USA

Jeff Reed, Fisheries Research Scientist, Minnesota Department of Natural Resources, 23070 North Lakeshore Drive. Glenwood, MN, 56334 USA

Dr. Aaron Dean Shultz, Inland Fisheries Biologist, Great Lakes Indian Fish and Wildlife Commission, 72682 Maple St, Odanah, WI 54861 USA

Elizabeth Slaughter, MA, 245 Raven Hills Rd, Colorado Springs, CO, 80919-1323 USA

Dr. Dustin M. Smith, West Virginia Department of Natural Resources Wildlife Resources Section, PO Box 99, 1110 Railroad St., Farmington, WV 26571 USA

Stephen Spencer, PhD, Senior Fisheries Biologist, Resource Stewardship, Environment and Parks, Box 4240, 250 Diamond Avenue Spruce Grove, AB T7X 3B4 Canada

Dr. Carol A. Stepien, National Oceanic and Atmospheric Administration, Pacific Marine Environmental Laboratory, 7600 Sand Point Way NE, Seattle, WA 98115-6349 USA

Dr. Bill Tonn, Professor Emeritus, Department of Biological Sciences, University of Alberta, Edmonton, AB T6G 2E9 Canada

Dr. Amy Welsh, Associate Professor, Division of Forestry and Natural Resources, West Virginia University, 311B Percival Hall, Morgantown, WV 26506-6108 USA

Dr. Stuart Welsh, U.S. Geological Survey, West Virginia Cooperative Fish and Wildlife Research Unit, Room 322, Percival Hall, Morgantown, WV 26506-6108 USA

Dr. Melissa Wuellner, Department of Biology Bruner Hall of Science, University of Nebraska at Kearney, Kearney, NE 68849-1140 USA

Contents

Part I
Yellow Perch (*Perca flavescens*)

Spawning Characteristics of Yellow Perch During Periods of Water Level Fluctuations in a Hydropower Reservoir

Kyle J. Matt, Stuart A. Welsh, and Dustin M. Smith

Abstract Water level fluctuations alter reservoir ecosystems, causing direct and indirect effects on fish populations. The dewatering of eggs, a direct impact of lake level drawdowns, can affect reproductive success of species that spawn in littoral zones, such as Yellow Perch (*Perca flavescens*). We examined relationships between water level fluctuations and spawning characteristics of Yellow Perch in a Central Appalachian hydropower reservoir where water levels were permitted to be drawn down to 4 and 2.1 m below the full pool elevation in March and April, respectively. Daily presences of egg masses were recorded on artificial spawning structures at two sites for the spring spawning seasons of 2019 and 2020. Spawning structures were placed at different distances from the shoreline, spanning water depths with and without the potential for dewatering based on the lowest permitted levels for lake elevation drawdowns. Generalized estimation equations were used to analyze egg mass presence and six covariates: Secchi disk depth, distance to the shore, water temperature, water depth, lunar illumination, and lake level fluctuation. We also examined the proportion of egg masses in potential dewatering zones based on the minimum lake elevation drawdowns permitted for March and April. Data supported an additive effects model of year + water depth + lunar illumination + water temperature. The predicted probability of egg mass presence was negatively associated with water depth and lunar illumination and positively associated with water temperature. A year effect, in part, reflected a between-year difference in the timing

K. J. Matt
West Virginia Cooperative Fish and Wildlife Research Unit, West Virginia University, Morgantown, WV, USA
e-mail: kjmatt@mix.wvu.edu

S. A. Welsh (✉)
U.S. Geological Survey, West Virginia Cooperative Fish and Wildlife Research Unit, West Virginia University, Morgantown, WV, USA
e-mail: swelsh@wvu.edu

D. M. Smith
West Virginia Division of Natural Resources, Farmington, WV, USA
e-mail: Dustin.M.Smith@wv.gov

J. C. Bruner, R. L. DeBruyne (eds.), *Yellow Perch, Walleye, and Sauger: Aspects of Ecology, Management, and Culture*, Fish & Fisheries Series 41,
https://doi.org/10.1007/978-3-030-80678-1_1

of spawning, where the number of egg masses during April exceeded that of March in 2019, a relationship that was reversed in 2020. During the 27-day spawning period in 2019, 52% (54 of 104) of egg masses had the potential to be dewatered, whereas 70% (30 of 43) had the potential to be dewatered in the 22-day spawning period of 2020. Our results could have direct implications for fishery and hydropower management, as data on the characteristics and timing of spawning of Yellow Perch relative to water level fluctuations may help inform decisions regarding management of fish populations and lake level drawdown regulations.

Keywords *Perca* · Eggs · Spawning habitat · Reservoir · Water fluctuation · Hydropower

1 Introduction

Water level fluctuations in reservoirs, which can vary in amplitude, frequency, duration, and timing, result from climate-induced controls on regional precipitation or from planned drawdowns, such as in hydropower reservoirs (Leira and Cantonati 2008; Hirsch et al. 2017). Drawdowns dewater near-shore littoral areas, reducing available habitat complexity such as riparian-contributed woody structure and aquatic vegetation (Gaboury and Patalas 1984; Zohary and Ostrovsky 2011; Gaeta et al. 2014). This loss of near-shore structure homogenizes habitat, forcing fish to find foraging and resting habitats elsewhere (Logez et al. 2016). Drawdown-induced habitat losses also have trophic level consequences that indirectly impact fish populations. For example, drawdowns reduce available forage by decreasing vegetation and primary production (Ploskey 1986; Wilcox and Meeker 1991; Hill et al. 1998) and by reducing invertebrate population sizes (Ploskey 1986; Aroviita and Hämäläinen 2008; McEwen and Butler 2010; White et al. 2011). Moreover, a common management concern is the direct effect of drawdowns on fish reproductive success (Clark et al. 2008; de Lima et al. 2017). During periods of lake level drawdowns, fishes that spawn in littoral zones may experience loss of spawning habitat or post-spawn dewatering of eggs (Walburg 1976; Ploskey 1983; Gaboury and Patalas 1984; Gasith and Gafny 1990; Hirsch et al. 2017). At the fish assemblage level, reduced spawning success of one or more fish species leads to lower numbers of fish larvae and young-of-year fish, resulting in a reduced forage base for piscivorous fishes (Forney 1974; Pierce et al. 2006).

Cheat Lake, a 700-hectare hydropower reservoir on the Cheat River in northern West Virginia, experiences water level fluctuations resulting in part from the storage and release of water for electric power production. Lake water surface elevation at full pool is 265.2 m above sea level (asl). Three regulation periods are in place to limit the extent of lake drawdown below the full pool elevation. From May through October, lake elevation is relatively constant with a permitted fluctuation of 0.6 m between full pool (265.2 m asl) and 264.6 m asl. Lake elevation is permitted to be lowered 4.0 m below full pool (265.2 m asl) to the minimum level of 261.2 m asl from November through March. During April, lake elevation can be drawn down 2.1 m below full pool (265.2 m asl) to a level of 263 m asl. Water level fluctuations during spring months may result in egg dewatering and spawning failure for individuals of some species, such as Yellow Perch (*Perca flavescens*). Cheat Lake currently supports a Yellow Perch fishery of regional and economic importance, so it is relevant from a management perspective to understand the potential of population impacts owing to water level fluctuations (Taylor 2013; Smith and Welsh 2015; Hilling et al. 2018).

The reproductive ecology of Yellow Perch is generally well-understood. Egg masses are long, transparent, gelatinous, ribbon-like, and accordion-shaped. A gravid female may have from 2000 to 157,600 eggs depending on body size and age (Brazo et al. 1975; Hardy 1978), but estimates of average numbers of eggs within an egg mass range from 23,000 to 35,400 (Herman et al. 1959; Hardy 1978; Hanchin et al. 2003; Weber and Les 1982). Spawning periods have been reported to range from 7 to 22 days (Weber and Les 1982) to >9 weeks (Fitzgerald et al. 2001). Yellow Perch typically spawn in the shallow waters of near-shore littoral zones where egg masses are draped across vegetation or woody debris (Echo 1955; Muncy 1962; Scott and Crossman 1973; Nelson and Walburg 1977; Becker 1983). In the absence of spawning structures, egg masses are deposited onto lake bottom substrates (Noble 1970; Smith 1986; Robillard and Marsden 2001). Water depths at spawning locations range from 0.4 to 2.1 m (Weber and Les 1982), 1.5 to 3.0 m (Herman et al. 1959), 1.0–3.7 m (Krieger et al. 1983), and 2.0 to 3.0 m (Forney 1971). However, spawning depths may exceed 5 m in lakes with low levels of dissolved organic carbon where ultraviolet radiation may damage eggs in shallower waters (Williamson et al. 1997; Huff et al. 2004). The length of the egg incubation period, which may be extended by colder water temperatures (Hardy 1978), has been reported as 6–17 days (Powles and Warlen 1988), 8–10 days (Herman et al. 1959), 10–20 days (Whiteside et al. 1985), 14–20 days (Weber and Les 1982), and 25–27 days (Mansueti 1964). Although egg masses are not protected by parental care, egg predation is thought to be rare (Newsome and Tompkins 1985; Almeida et al. 2017).

The timing, duration, and habitat characteristics of Yellow Perch reproduction in Cheat Lake may differ from those reported elsewhere. Most studies on Yellow Perch reproduction are from the Midwestern and Northern United States, and similar information is sparse from the Central Appalachians. Currently, we have little information on Yellow Perch spawning characteristics in Cheat Lake, so information on timing and duration of spawning periods, as well as data on spawning water depths and distances from the shoreline, are needed to understand the potential for egg dewatering during periods of lake level drawdown. The primary objectives of this study were to (1) determine the timing of Yellow Perch spawning; (2) examine variables that potentially influence spawning habitat characteristics; and (3) evaluate water level fluctuation as a variable of influence on the timing of spawning and the potential for egg dewatering.

2 Methods

During spring 2019 and 2020, 40 artificial spawning structures were submerged at two sites on Cheat Lake: 20 structures at Crammys Run and 20 at Canyon Bend (Fig. 1). Lake bottom contours of near-shore areas of Crammys Run were mostly of gradual slope (3% on average), whereas those of Canyon Bend were comparatively steeper (9% on average). Each spawning structure was comprised of a 2.4-m piece of 51-mm diameter PVC pipe (Schedule 40), 10 sections of 1.8-m strands of artificial aquatic plants (reelweeds by LaDredge Outdoors; https://www.reelweeds.com/), and two, 2.4-m pieces of 13-mm diameter rebar. These parts were assembled into a 1.8-m tall by 2.4-m long structure (Fig. 2). The ends of the PVC pipe were sealed with caps

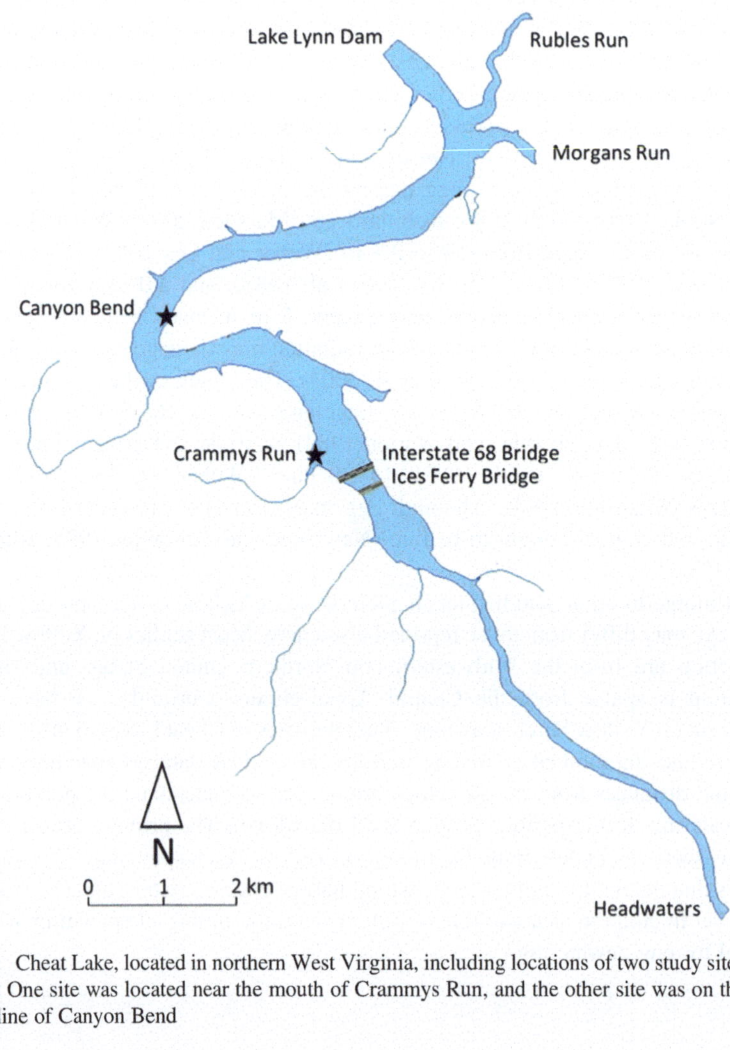

Fig. 1 Cheat Lake, located in northern West Virginia, including locations of two study sites (black stars). One site was located near the mouth of Crammys Run, and the other site was on the inside shoreline of Canyon Bend

Fig. 2 Artificial spawning habitat structures used in a study of Yellow Perch on Cheat Lake, West Virginia

so that the pipe served as a float. Zip ties were used to attach the tops of the artificial plant strands to the PVC float and bottoms of the strands to the rebar. When deployed, the rebar end of the spawning structures rested on the lake bottom, and the structure maintained a vertical position in the water column, owing to the floatation of the PVC pipe. If the water was <1.8 m deep at the deployment site, then the 2.4-m piece of PVC pipe floated on the water's surface (Fig. 3). When deployed at locations with steep bottom contours, the spawning habitat unit was oriented parallel to the shoreline to reduce water depth variation along the unit's length. A harness of 550 paracord attached at each end of the PVC pipe was connected to a longer strand of 550 paracord terminating in an attached location buoy (Fig. 2). Each buoy was labeled with a unique number for identification. When the spawning structure was deployed, the tethered buoy floated on the water's surface, providing a way to find and retrieve the structure. We attempted to position the 20 habitat units at each site so that 10 were in the potential dewatering zone and 10 were in deeper areas that were outside of this zone.

The 40 spawning structures were checked daily for the presence of egg masses during the expected spring spawning period by removing the structures from the water. We removed egg masses from the structures on a daily basis to prevent the double counting of egg masses on consecutive sampling days. We recorded presence/absence of egg masses on spawning structures and counted the number of egg masses per structure. Egg masses were removed from the structures, placed in a bucket of lake water, and relocated to nearby areas with submerged tree habitat and deep water (i.e., areas with a low chance for egg mass dewatering).

Fig. 3 Study sites at Crammys Run (top) and Canyon Bend (bottom). White buoys mark the locations of the spawning habitat units. When the water depth was less than or equal to 1.83 m (6 ft), then the white PVC floats of the spawning habitat units were on top of the water (see bottom right). An organization contact and phone number was printed on each white buoy. One large buoy at each site (see bottom left) was used to alert boaters and provide information about the research project

A total of ten egg masses were preserved in 50% ethanol for estimation of the average number of eggs per egg mass. Egg masses were removed from the ethanol, strained, and measured for length. The gravimetric method was used to determine fecundity (Ganias et al. 2014). Each egg mass was weighed (g) on an Ohaus digital scale. A subsample was removed from the middle of each egg mass and weighed. Each subsample contained greater than 600 eggs. Partitions of the subsample, consisting of 10–30 eggs, were placed onto a gridded dish, and the eggs were

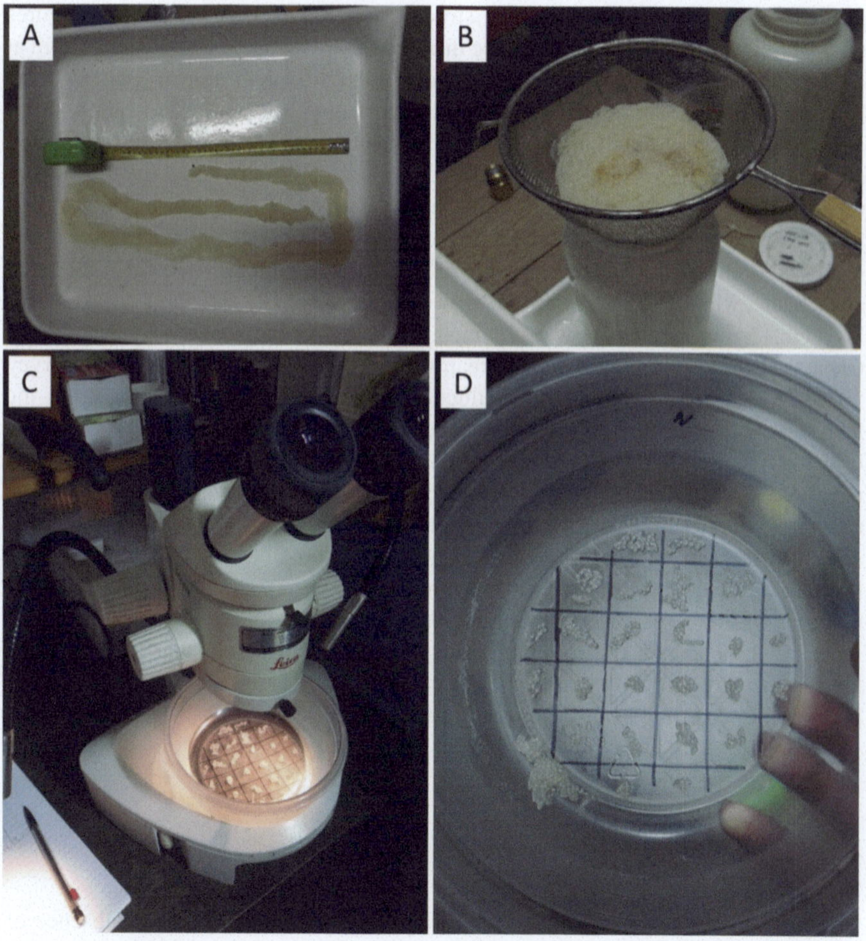

Fig. 4 Photographs of the egg counting process, including an ethanol-preserved Yellow Perch egg mass (**a**), ethanol strained from a preserved egg mass (**b**), and partitioned subsamples of eggs (**c, d**)

counted under a microscope (Fig. 4). Fecundity was estimated with the formula $N = Wn/W^1$, where N = the number of eggs in the egg mass, W = the weight of the egg mass, n = the number of eggs counted in the subsample, and W^1 = the weight of the subsample (Ganias et al. 2014). An average value was calculated from eight fecundity estimates. The estimate of the average number of eggs per egg mass was used to calculate total egg numbers by site and year.

A suite of habitat covariates was determined based on literature review and our understanding of Yellow Perch reproduction. Given that Yellow Perch often spawn at night (Raney 1959; Scott and Crossman 1973), we considered that levels of darkness associated with lunar phase may influence spawning. Lunar illumination was recorded as a fraction or percentage of the moon face, a value ranging from 0 (new moon) to 1 (full moon). Other habitat covariates were recorded daily, primarily at the time when spawning structures were checked. Water temperatures (°C) were measured at the lake surface in a near-shore area and at the lake bottom at or near the deepest habitat unit with either a Marcum LX-9 unit or a Hobo tidbit logger. The mean value of the two water temperatures was used as a water temperature covariate. The depth of water at each spawning structure was recorded during deployment and retrieval using a handheld sonar unit. We measured the distance of the structure to the nearest shoreline's high-water mark (i.e., full pool elevation level) using a laser range finder. We also recorded the distance of the structure to the nearest shoreline's current water level. A Secchi disk depth (cm) was also recorded at each site, which provided an index of water turbidity. The elevation of lake water levels in 15-min increments was obtained from a US Geological Survey gage at the Cheat Lake hydrostation (USGS streamgage 03071590, U.S Geological Survey 2020). A covariate for water level fluctuation was calculated by subtracting the lake elevation at the time of the structure retrieval from the lake elevation at the time of deployment on the previous day. The water level fluctuation covariate was either negative or positive depending on the direction of change of water level during the time period between daily sampling events. A caveat with this approach is that the actual time of the spawning event is unknown. It is possible that a change in water level elevation could occur after a spawning event. For example, consider a habitat unit that was deployed at 11:00 am and then retrieved with the presence of an egg mass at 11:00 am on the following day. We could document that a water level increase occurred from 4:00 am to 10:00 am on the day of retrieval, but we would not know if the spawning event occurred before or after 4:00 am.

2.1 Data Analysis

Generalized estimation equations (GEE) for binary response (presence/absence of egg masses) with a logit link were used to analyze data of Yellow Perch egg masses on artificial habitat units and associated covariates. The GEE analysis was equivalent to a logistic regression analysis but allowed for the use of a correlation matrix structure to properly address spatial clustering of data. In our study, 20 habitat units were clustered together on each sampling day. To select an appropriate working correlation structure, we fit models with autoregressive AR(1), compound symmetry, and independent working correlation matrices to our global model and used the correlation information criterion (CIC) to select a working correlation structure (Hin and Wang 2009).

Before analysis, habitat variables were examined with Pearson correlation coefficients, which supported near collinearity ($r = 0.98$) between two distance measures: distances of habitat units to (1) the full pool level on the shoreline and to (2) the current water level at the time of sampling. The distance to the full pool level was retained for analysis and hereafter referred to as "distance to the shore." Near collinearity was not observed between other variables, resulting in the use of six covariates: water temperature, water depth, lunar illumination, Secchi disk depth, distance to the shore, and lake level fluctuation.

A set of 35 candidate models were fit to the data using GEE analyses with a binomial distribution, a logit link function, and an AR(1) correlation structure (Statistical Analysis System, SAS 9.4; PROC GENMOD) (Table 1). Based on the recommendation of Burnham and Anderson (2002), our candidate models (1) were constructed a priori, (2) did not include all possible models, and (3) were structured with variables supported by our current understanding of Yellow Perch reproduction based on peer-reviewed literature. Twelve of the candidate models included six single covariate models with a year effect and six single covariate models with a site effect. An additional 20 candidate models of two-variable or three-variable additive effects of covariates included 10 with a year effect and 10 with a site effect. Three models included all six covariates: one with a year effect, one with a site effect, and a global model with both a year effect and a site effect.

We used an information-theoretic approach for model selection and inference. The best model (or suite of competing models) was selected with the Quasi-likelihood Information Criterion (QIC_u) of Pan (2001). We also estimated QIC distances among models (ΔQIC_u) and QIC_u model weights (w_i) following methods of Burnham and Anderson (2002). Models, which represented alternative hypotheses, were considered to be supported by the data if ΔQIC_u values were less than 2.0 (Burnham and Anderson 2002). Predicted probability plots (i.e., effect plots) of covariates provided a visual aid for interpretation of model selection results. Further, descriptive statistics of covariates (means and standard errors), histogram plots, and time series plots aided interpretation of modeling results.

The number of spawning peaks was examined using a mixture model-based approach. For this analysis, we used a time series histogram of total daily counts of Yellow Perch egg masses. Daily counts of egg masses from the two study sites were combined, representing the total daily egg mass count from 40 artificial habitat units. The 2019 and 2020 datasets were examined separately. First, a normal model was fit to the histogram data, which represented a hypothesis of a unimodal peak. Next, we fit four normal mixture models (2–5 mixtures) representing hypotheses for a range of multimodal distributions (JMP, version 12.0.1 SAS Institute Inc. 2015). We used AIC-model selection with small sample size correction (AIC_c) to determine the best approximating model (Burnham and Anderson 2002). Using this analysis approach, we determined whether one or more modes or peaks in spawning were present during the 2019 and 2020 spawning periods.

A main focus of this research was on the relationship between fluctuations in water levels of Cheat Lake and the potential for dewatering of egg masses. Daily water level elevation changes of the lake were plotted from data downloaded from the US Geological Survey (USGS 2020). For analysis, we estimated the proportion of egg masses located in potential dewatering zones and included estimates of 95% profile likelihood confidence intervals. Analyses were based on two scenarios. First, we assumed that egg masses were deposited on the lake bottom. In the second scenario, the assumption was that egg masses were deposited onto a structure at a position of 1 m above the bottom. The range of 0.0–1.0 m was based on observations

Table 1 Model selection statistics for 35 candidate models (i.e., alternative hypotheses) fit to egg mass presence/absence data from Cheat Lake, West Virginia. Models included a year effect (2019 and 2020) or site effect (Crammys Run and Canyon Bend). Covariates were water temperature (Temp), water depth (Depth), lunar illumination (Lunar), distance to shoreline (Distance), lake level fluctuation (LLF), and Secchi disk depth (Secchi)

Model	QIC_u	Delta	Model L	W_t
Year + Temp + Depth + Lunar	842.4	0.0	1.0	1.0
Year + Temp + Depth + Distance + LLF + Lunar + Secchi	850.0	7.6	0.0	0.0
Site + Temp + Depth + Lunar	853.0	10.6	0.0	0.0
Global	853.1	10.6	0.0	0.0
Site + Temp + Depth + Distance + LLF + Lunar + Secchi	854.0	11.5	0.0	0.0
Year + Depth + Lunar	855.8	13.3	0.0	0.0
Year + Depth + Lunar + Secchi	857.7	15.3	0.0	0.0
Year + Temp + Depth	860.4	18.0	0.0	0.0
Site + Depth + Lunar + Secchi	864.6	22.2	0.0	0.0
Site + Depth + Lunar	866.0	23.6	0.0	0.0
Year + Lunar	869.9	27.5	0.0	0.0
Year + Temp	870.8	28.4	0.0	0.0
Year + Depth + LLF	870.9	28.5	0.0	0.0
Year + Lunar + Secchi	871.8	29.4	0.0	0.0
Year + Depth	872.8	30.3	0.0	0.0
Year + Depth + Distance + LLF	874.3	31.9	0.0	0.0
Year + Depth + Secchi	874.5	32.1	0.0	0.0
Year + Depth + Distance	875.2	32.8	0.0	0.0
Site + Temp + Depth	877.2	34.8	0.0	0.0
Site + Lunar + Secchi	878.9	36.5	0.0	0.0
Site + Lunar	880.1	37.6	0.0	0.0
Year + Distance + LLF	880.3	37.9	0.0	0.0
Year + Distance	880.4	38.0	0.0	0.0
Site + Depth + Secchi	882.2	39.8	0.0	0.0
Site + Depth + LLF	883.4	40.9	0.0	0.0
Year + LLF	884.9	42.5	0.0	0.0
Site + Depth	885.4	43.0	0.0	0.0
Site + Depth + Distance + LLF	886.5	44.1	0.0	0.0
Site + Temp	887.0	44.6	0.0	0.0
Year + Secchi	887.5	45.0	0.0	0.0
Site + Depth + Distance	887.6	45.2	0.0	0.0
Site + Distance + LLF	890.9	48.5	0.0	0.0
Site + Distance	891.1	48.7	0.0	0.0
Site + Secchi	894.4	52.0	0.0	0.0
Site + LLF	896.3	53.8	0.0	0.0

of egg mass positions on natural structures in near-shore habitats of Cheat Lake (S. Welsh, Personal observation).

3 Results

Artificial spawning structures were deployed at the Crammys Run and Canyon Bend study sites for 51 days in 2019 (11 March to 30 April) and 40 days in 2020 (11 March to 19 April). The time periods of egg mass presence on spawning structures in 2019 and 2020, which we refer to as spawning periods, were documented during a 27-day (21 March to 16 April) and 22-day period (21 March to 11 April) in 2019 and 2020, respectively (Fig. 5). Presences of egg masses were documented 46 and 35 times in 2019 and 13 and 26 times in 2020 on spawning structures at Crammys Run and Canyon Bend, respectively. Usually, a single egg mass was present on a spawning structure, but multiple egg masses were found occasionally on a single spawning structure. In 2019, for 46 instances of egg mass presence on structures at Crammys Run, 36 were of single egg masses, 7 included 2 egg masses, and 3 included 3 egg masses. Thus, a total of 59 egg masses were found on structures at Crammys Run. For 35 instances of egg mass presence on structures at Canyon Bend, 28 were single egg masses, 5 represented 2 egg masses, and single occurrences were found for 3 and 4 egg masses (45 egg masses in total). In 2020, only 13 single egg masses were found on structures at Crammys Run. For 26 instances of egg mass presence on structures at Canyon Bend, 22 were single egg masses, and 4 represented 2 egg masses (i.e., 30 egg masses in total). Egg masses were generally attached to the spawning structures in two ways: spiraled around a single artificial vegetation strand or draped over one or more strands (Fig. 6).

Two of the ten egg masses preserved poorly in ethanol, so eight egg masses were examined. The number of eggs per mass varied between 10,538 and 84,570 (38,237 $\pm$ 17,865, 95% confidence interval). Total estimated eggs produced at Crammys Run and Canyon Bend were higher in 2019 compared to 2020 (2,255,983 vs. 479,081 eggs at Crammys Run and 1,720,665 vs. 1,147,110 eggs at Canyon Bend).

3.1 GEE Analysis and Model Selection

For the GEE analysis, a three-variable additive effect model with a year effect was the only model supported by the data (Table 1). The QIC_u-selected model was year + water temperature + water depth + lunar illumination. The GEE parameter estimates for this model (with confidence intervals and p-values) were year (0.60, 0.15–1.05, $p = 0.0087$), water temperature (0.14, 0.0259–0.2475, $p = 0.0156$), water depth (-0.08, -0.15 – -0.0044, $p = 0.0378$), and lunar illumination (-1.2, -1.9002 – -0.4836, $p = 0.0010$).

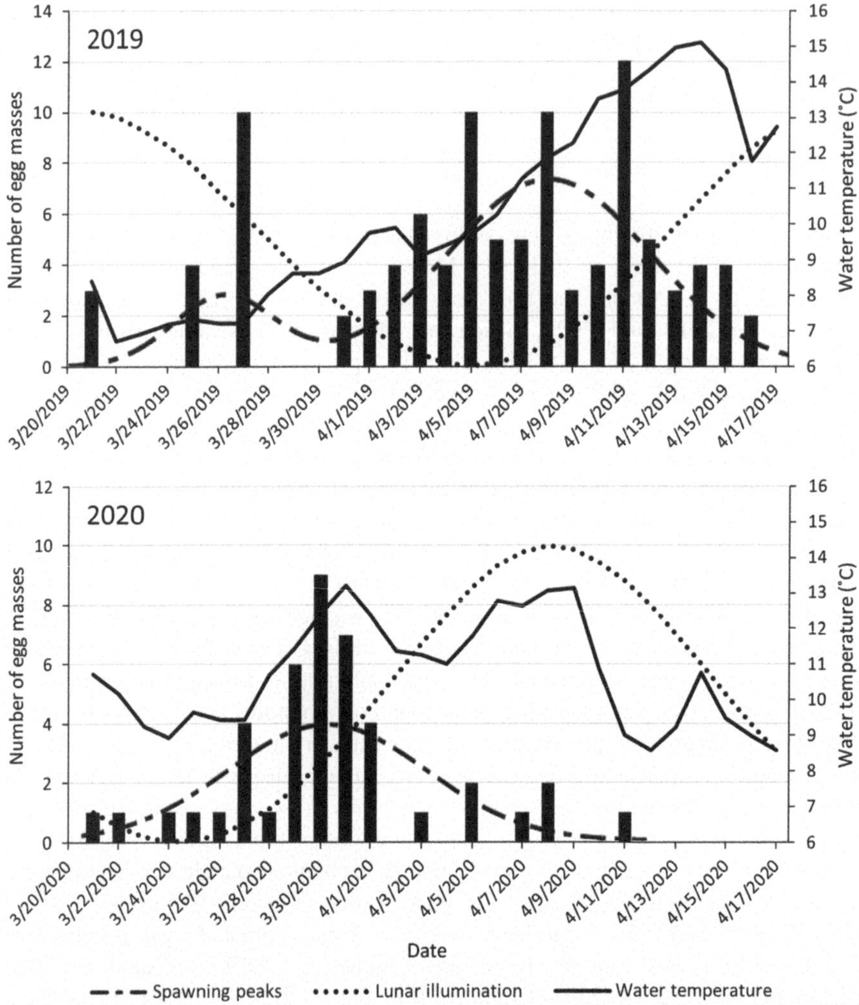

Fig. 5 Time series of daily counts of Yellow Perch egg masses on 40 artificial habitat units. Water temperature and lunar illumination are plotted for the spawning periods, which ranged from 21 March to 16 April in 2019 and 21 March to 11 April in 2020. Spawning peaks were determined by fitting a set of candidate mixture models to daily egg mass counts, where AIC model selection supported a 2-mixture model for 2019 and a unimodal model for 2020

Plots of predicted probability for presence from the GEE analysis as well as time series plots and descriptive statistics of raw data aided interpretation of the QIC_u-selected model and its GEE parameter estimates. The predicted probability of egg mass presence was positively associated with water temperature (Fig. 7). This relationship is visually supported by an overlay plot of the water temperature time series and the dominant peaks in daily egg mass counts for April 2019 and March 2020 (Fig. 5). The predicted probability of egg mass presence was negatively

Fig. 6 Yellow Perch egg masses spiraled (left) or draped (right) around artificial spawning habitat structures

associated with water depth and lunar illumination (Fig. 7). The mean values of water depths for habitat units with the presence of egg masses in 2019 (2.7 m) and 2020 (2.6 m) were less than those of all habitat units in 2019 (3.5 m) and 2020 (3.6 m; Table 2, Fig. 8). The mean values of lunar illumination for habitat units with the presence of egg masses in 2019 (0.24) and 2020 (0.35) were less than those of all habitat units in 2019 (0.38) and 2020 (0.41; Table 2). An overlay plot provided visual support for the association of daily egg mass counts with low levels of lunar illumination for the 2019 and 2020 spawning periods (Fig. 5).

Several patterns are worth noting relative to the three covariates (distance to shore, lake level fluctuation, and Secchi disk depth) not supported by the QIC_u-selected model. Egg masses were generally not present where distances from the shoreline exceeded 45 m (Fig. 9). The average distances to the shore of structures with egg masses in 2019 (23.1 m) and 2020 (18.3 m) were less than those of all habitat units in 2019 (24.9 m) and 2020 (25.2 m; Table 2). The numbers of egg masses associated with increasing lake levels ($n = 68$) exceeded those of decreasing lake levels ($n = 46$; Fig. 10). The predicted probability of egg mass presence in 2019 was positively associated with Secchi disk depth, but the opposite pattern occurred in 2020 (Fig. 7). The mean values of Secchi disk depths at Crammys Run and Canyon Bend in 2020 (171 and 156 cm) were lower than those of 2019 (208 and 228 cm; Table 2).

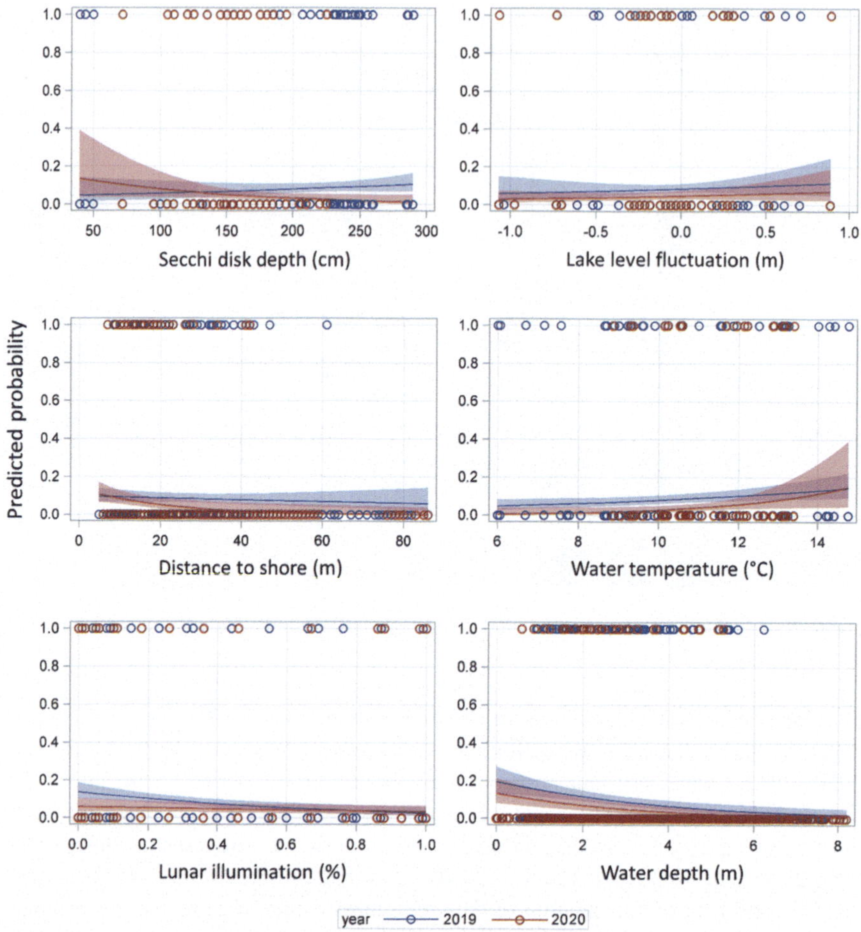

Fig. 7 Predicted probability of egg mass presence on artificial spawning habitat based on analyses using generalized estimating equations (GEE). Plots depict relationships from 2019 and 2020 of single model covariates; Secchi disk depth, lake level fluctuation, distance to shore, water temperature, lunar illumination, and water depth

3.2 Spawning Peaks

A mixture modeling approach supported two spawning peaks during the spawning period of 2019 and a single spawning peak in 2020 (Fig. 5). The 2019 spawning peaks occurred in March and April, although the March peak was of lower magnitude than that of April. Spawning in 2020 peaked in March. Further summary of the timing of spawning peaks and associated egg presence on spawning structures between March and April are relevant, given that the minimum lake elevation level changes from 261.2 m asl in March to 263 m asl in April. In 2019, a total of

Table 2 Summary statistics of habitat variables for all spawning habitat units and for those units with presence of Yellow Perch egg masses (N = sample size, *SE* standard error, *min* minimum value, and *max* maximum value)

	Crammys Run					Canyon Bend					Sites combined				
Variable	*N*	Mean	SE	min	max	*N*	Mean	SE	min	max	*N*	Mean	SE	min	max
All habitat units in 2019															
Secchi depth (cm)	24	208	11.8	44	285	24	228.2	11.1	40	290	48	218	8.1	40	290
Water depth (m)	465	3.2	0.12	0.31	6.1	470	3.8	0.08	0.46	7.6	935	3.5	0.05	0.3	7.6
Distance from shore (m)	465	27.2	0.79	5.0	86	470	22.5	0.45	7.0	47	935	24.9	0.46	5.0	86
Water temperature (°C)	24	10	0.55	6.0	14.8	24	10.2	0.51	7.1	14.4	48	10.7	0.37	6.0	14.8
Lunar illumination	24	0.38	0.07	0.0	1.0	24	0.39	0.07	0.0	1.0	48	0.38	0.05	0.0	1.0
All habitat units in 2020															
Secchi depth (cm)	20	171.1	8.7	105	225	20	155.6	8.0	72	225	40	163.4	6.0	72	225
Water depth (m)	400	3.4	0.08	0.0	6.0	400	3.8	0.1	0.0	8.2	800	3.6	0.06	0.0	8.2
Distance from shore (m)	400	27.1	0.79	6.0	86	400	23.3	0.56	6.0	56	800	25.2	0.49	6.0	86
Water temperature (°C)	20	10.8	0.26	8.9	12.5	20	11.7	0.29	10	13.4	40	12.2	0.28	8.9	13.4
Lunar illumination	20	0.41	0.08	0.0	1.0	20	0.41	0.08	0.0	1.0	40	0.41	0.06	0.0	1.0
Habitat units with egg presence 2019															
Secchi depth (cm)	46	211.6	8.3	44	285	35	248.7	7.4	40	290	81	227.6	6.0	40	290
Water depth (m)	46	2.4	0.16	0.91	5.4	35	3.1	0.24	1.1	6.2	81	2.7	0.14	0.91	6.2
Distance from shore (m)	46	25.3	1.7	8.5	61	35	20.2	1.5	9.0	47	81	23.1	1.2	8.5	61
Water temperature (°C)	46	10.9	0.37	6.0	14.8	35	11.1	0.36	7.1	14.3	81	11.0	0.26	6.0	14.8
Lunar illumination	46	0.27	0.04	0.0	0.99	35	0.2	0.04	0.0	0.76	81	0.24	0.03	0.0	0.99
Habitat units with egg presence 2020															
Secchi depth (cm)	13	161.9	10.2	105	225	26	145.1	5.4	72	195	39	150.7	5.0	72	225
Water depth (m)	13	1.7	0.15	0.58	2.8	26	3.0	0.25	0.9	5.3	39	2.6	0.2	0.58	5.3
Distance from shore (m)	13	14.9	1.3	7.0	26	26	20	2.0	9.0	42	39	18.3	1.4	7.0	42
Water temperature (°C)	13	10.7	0.34	8.9	12.2	26	12.3	0.19	10.1	13.4	39	11.8	0.21	8.9	13.4
Lunar illumination	13	0.27	0.1	0.0	1.0	26	0.39	0.05	0.01	0.98	39	0.35	0.05	0.0	1.0

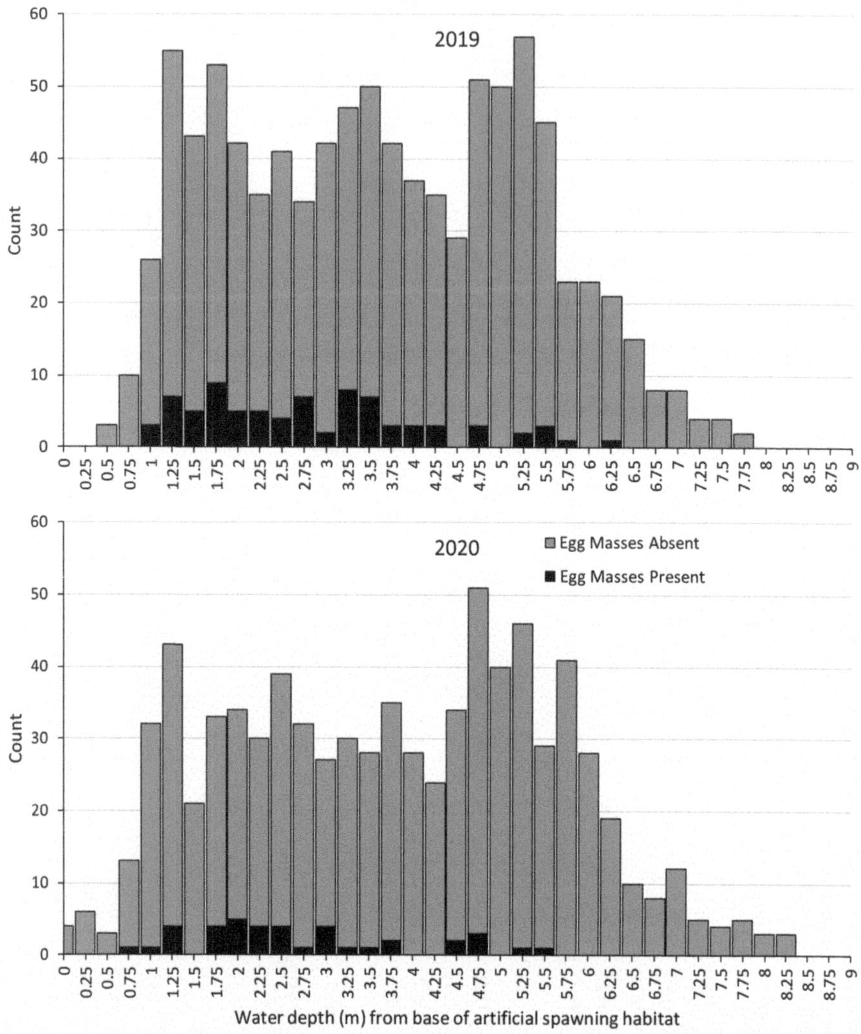

Fig. 8 Water depths of artificial spawning habitat units during sampling in 2019 and 2020 with and without the presence of egg masses

19 egg masses were found on 9 structures during 21–31 March, whereas 85 egg masses were found on 72 structures during 1–16 April. In 2020, a total of 32 egg masses were found on 29 structures during 21–31 March, and a total of 11 egg masses were found on 11 structures during 1–11 April. Based on an average estimate of 38,237 eggs per egg mass, the calculated numbers of eggs per time period were 726,503 (March 2019), 3,250,145 (April 2019), 1,223,584 (March 2020), and 420,607 (April 2020). Thus, the number of eggs during April exceeded that of March in 2019, but this relationship was reversed in 2020.

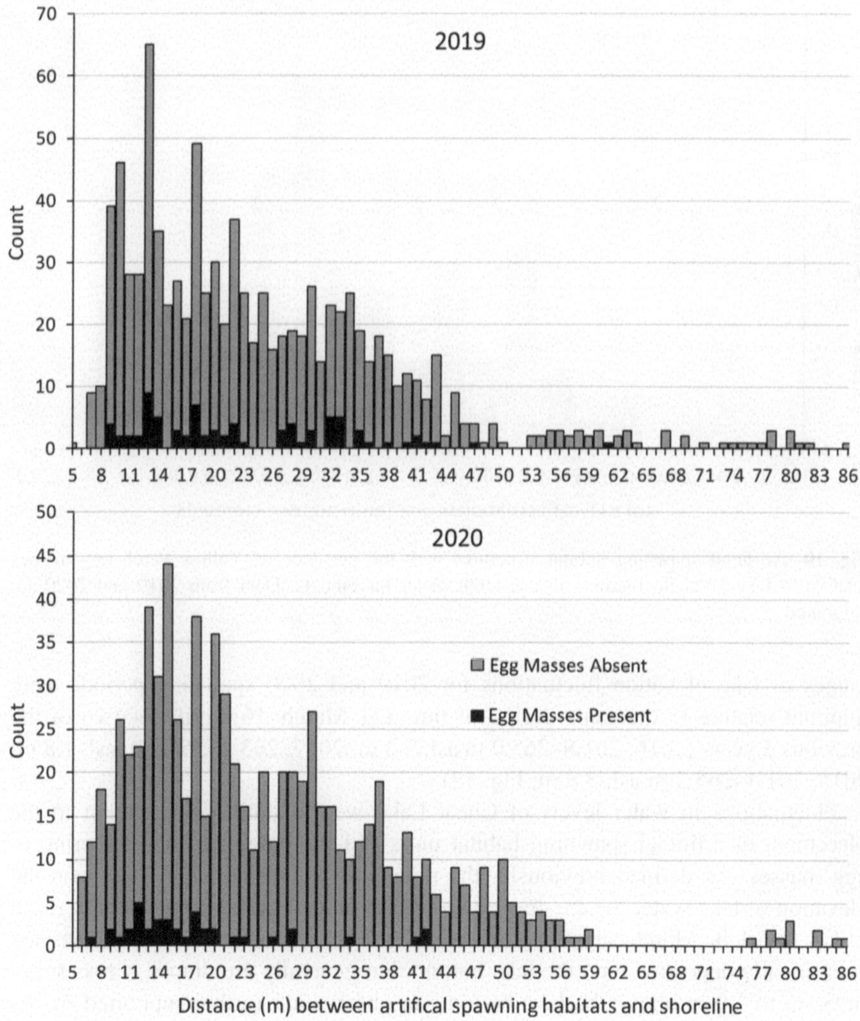

Fig. 9 Distances from shoreline of artificial spawning habitat units with and without the presence of egg masses for 2019 and 2020. The *y*-axis is a count of habitat units. Distances were measured from the water surface (directly above submerged habitat units) to the full pool water mark on the nearest shoreline

3.3 *Lake Level Fluctuation*

The 2019 and 2020 fluctuations in water levels during spawning periods were similar but differed from those of some years prior to our study. During the spawning period of 21 March–16 April 2019, water level elevations of Cheat Lake varied between 263.4 and 265.1 m asl (Fig. 11). Similar variabilities were noted during the spawning period of 21 March–11 April 2020 (263.5–265.1 m asl; Fig. 11). The

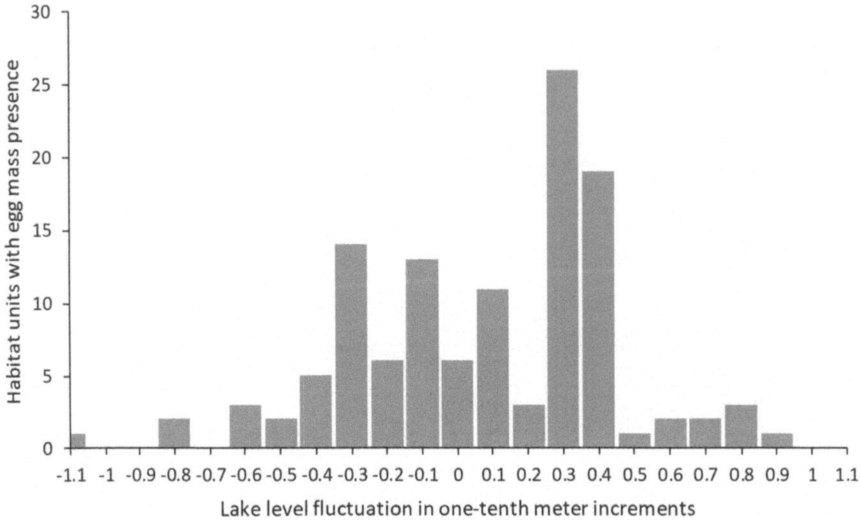

Fig. 10 Artificial spawning habitat structures with the presence of Yellow Perch egg masses relative to lake level fluctuations in one-tenth meter increments. Data from 2019 and 2020 are combined

ranges of lake elevation fluctuations for 2019 and 2020 spawning periods were minimal relative to the same period of time (21 March–16 April) for two of the previous 3 years (2016, 261.8–265.0 m asl, 3.2 m; 2017, 263.4–265.2 m asl, 1.8 m; 2018, 261.4–265.2 m asl, 3.8 m; Fig. 12).

Fluctuations in water levels of Cheat Lake were examined in relation to the placement of artificial spawning habitat units and the potential for dewatering of egg masses. As defined previously, the potential for dewatering is based on the elevation of lake water, where drawdown of lake elevations could potentially reach 261.2 m asl in March and 263 m asl in April. In Cheat Lake, egg masses are occasionally deposited onto the lake bottom, but generally are draped across structures up to 1.0 m above the lake bottom, an observation further supported by the locations of egg masses on our artificial structures. We attempted to place half (0.5) of the artificial spawning structures in areas with the potential for dewatering and half (0.5) in areas outside of the potential for dewatering. However, the proportion of habitat units placed in areas of potential dewatering of the lake bottom ranged from 17 to 29% (Table 3). Additionally, for these structures in potential dewatered areas, the proportion of egg masses located 1.0 m above the lake bottom ranged from 37 to 44% (Table 3).

Based on the maximum range of water level fluctuations during the spawning periods of 2019 and 2020, we estimated the proportion of egg masses located in potential dewatered areas (Table 3, Fig. 13). For 2019, if all egg masses were deposited onto the lake bottom, then 36% of egg masses (21 of 59) were in potential dewatered areas at Crammys Run, whereas 9% of egg masses (4 of 45) were in potential dewatered areas at Canyon Bend. With the two sites combined, 24% of egg masses (25 of 104) were in potential dewatered areas. If egg masses were deposited

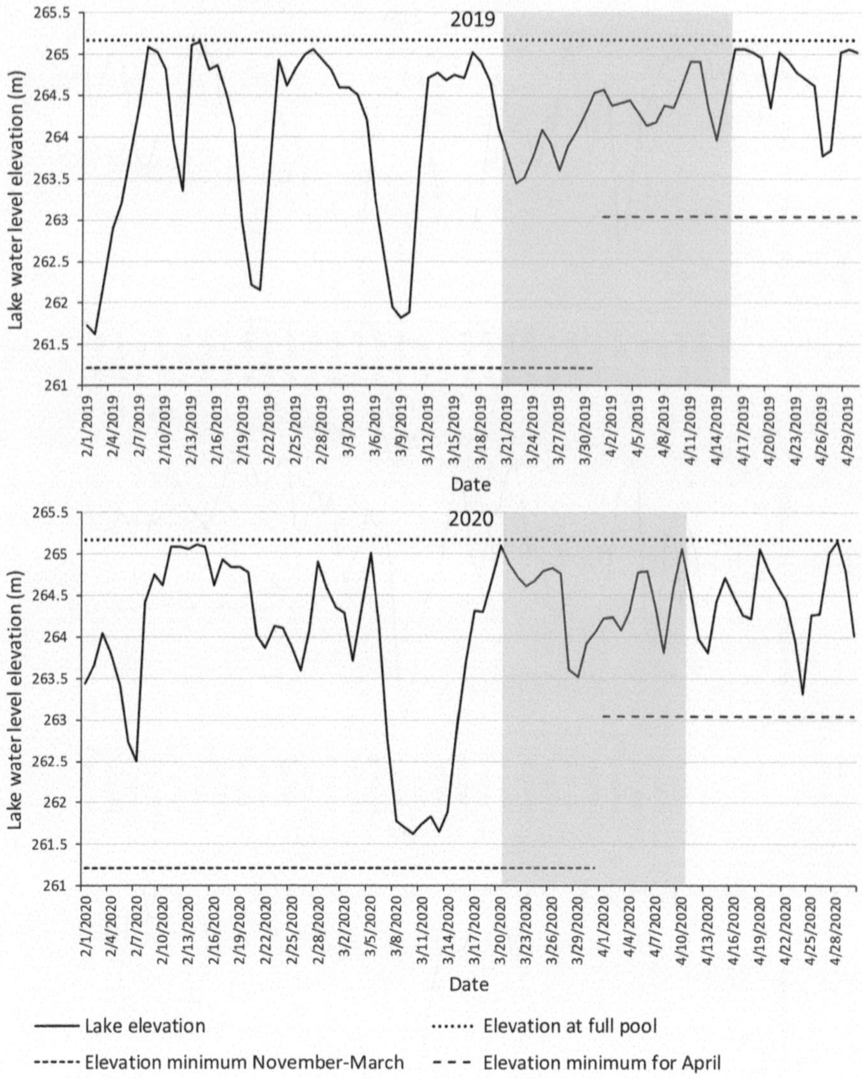

Fig. 11 Amplitude, frequency, duration, and timing of fluctuations in surface elevation of Cheat Lake during February–April of 2019 and 2020. Elevation at full pool is 265.2 m. The minimum permitted drawdown elevation is shown for February–March (261.2 m) and April (263 m). Gray zones represent spawning periods of Yellow Perch

onto structures at 1.0 m above the lake bottom, then estimates of egg placement in potential dewatered areas were 64% (38 of 59), 36% (16 of 45), and 52% (54 of 104) for Crammys Run, Canyon Bend, and the two sites combined, respectively. For 2020, if all egg masses were on the lake bottom, then 85% (11 of 13) at Crammys Run, 43% (13 of 30) at Canyon Bend, and 56% (24 of 43) at both sites combined

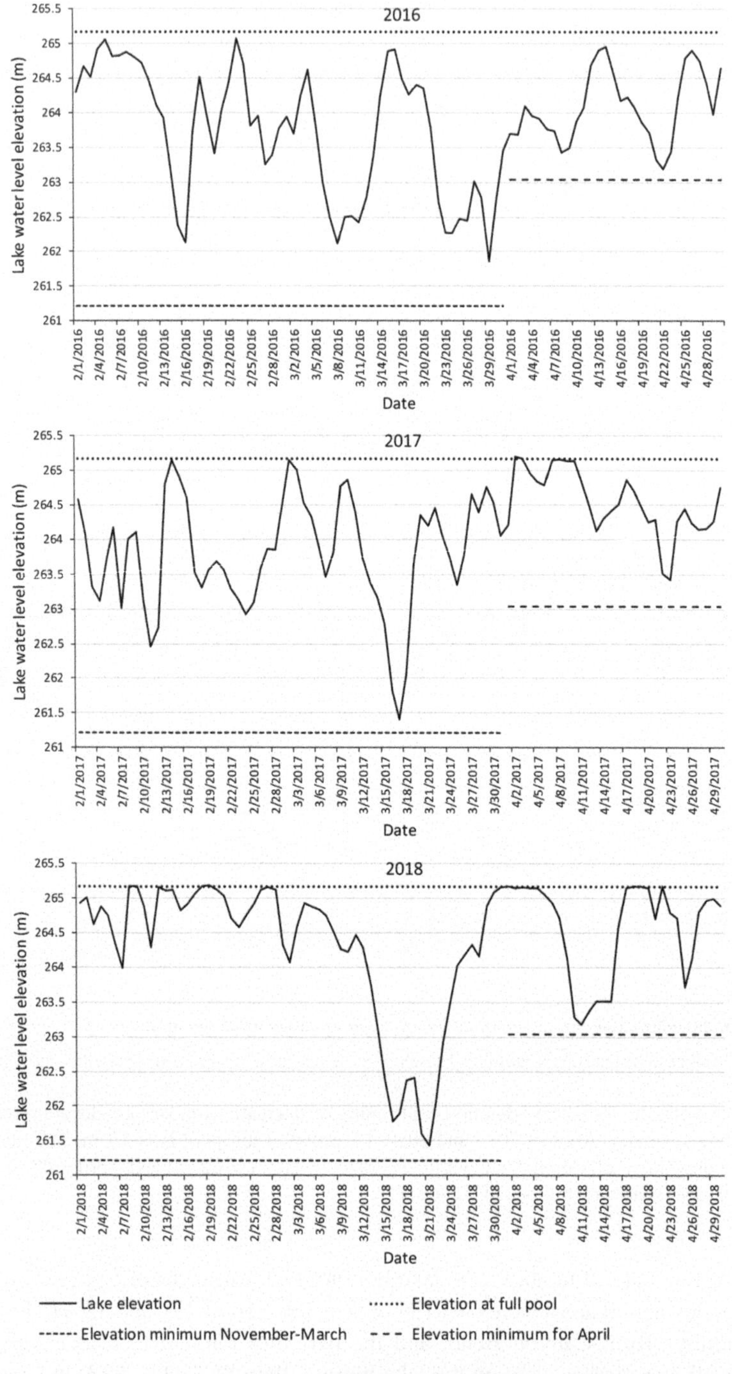

Fig. 12 Amplitude, frequency, duration, and timing of fluctuations in surface elevation of Cheat Lake during February–April of 2016–2018. Elevation at full pool is 265.2 m. The minimum permitted drawdown elevation is shown for February–March (261.2 m) and April (263 m)

Table 3 Proportion of artificial spawning habitat units with and without egg masses located in areas of potential dewatering zones, as defined by minimum lake drawdown regulations. An elevated egg mass is located on structures at 1.0 m above the lake bottom, and a bottom egg mass is located on the lake bottom. Proportions (Estimate) are provided with lower (LCI) and upper (UCI) 95% profile likelihood confidence intervals

	Egg mass	Dewatering zone					
Site	Location	Outside	Inside	Total	Estimate	LCI	UCI
All habitat units 2019							
Crammys	Elevated	268	197	465	0.42	0.38	0.47
Crammys	Bottom	353	112	465	0.24	0.20	0.28
Canyon	Elevated	298	172	470	0.37	0.32	0.41
Canyon	Bottom	390	80	470	0.17	0.14	0.21
All habitat units 2020							
Crammys	Elevated	228	172	400	0.43	0.38	0.48
Crammys	Bottom	287	113	400	0.28	0.24	0.33
Canyon	Elevated	223	177	400	0.44	0.39	0.49
Canyon	Bottom	285	115	400	0.29	0.24	0.33
Habitat units with egg presence 2019							
Crammys	Elevated	21	38	59	0.64	0.52	0.76
Crammys	Bottom	38	21	59	0.36	0.24	0.48
Canyon	Elevated	29	16	45	0.36	0.23	0.50
Canyon	Bottom	41	4	45	0.09	0.03	0.19
Combined	Elevated	50	54	104	0.52	0.42	0.61
Combined	Bottom	79	25	104	0.24	0.17	0.33
Habitat units with egg presence 2020							
Crammys	Elevated	2	11	13	0.85	0.60	0.97
Crammys	Bottom	2	11	13	0.85	0.60	0.97
Canyon	Elevated	11	19	30	0.63	0.46	0.79
Canyon	Bottom	17	13	30	0.43	0.27	0.61
Combined	Elevated	13	30	43	0.70	0.55	0.82
Combined	Bottom	19	24	43	0.56	0.41	0.70

were in the dewatering zone. If egg masses were deposited onto structures at 1.0 m above the lake bottom, then estimates of egg masses in potential dewatered areas were 85% (11 of 13), 63% (19 of 30), and 70% (30 of 43) at Crammys Run, Canyon Bend, and the two sites combined, respectively (Fig. 13).

4 Discussion

The use of artificial spawning structures provided an opportunity to (1) model the relationship of Yellow Perch egg mass presence with a suite of six covariates across two spawning periods and (2) determine the potential for egg mass dewatering based

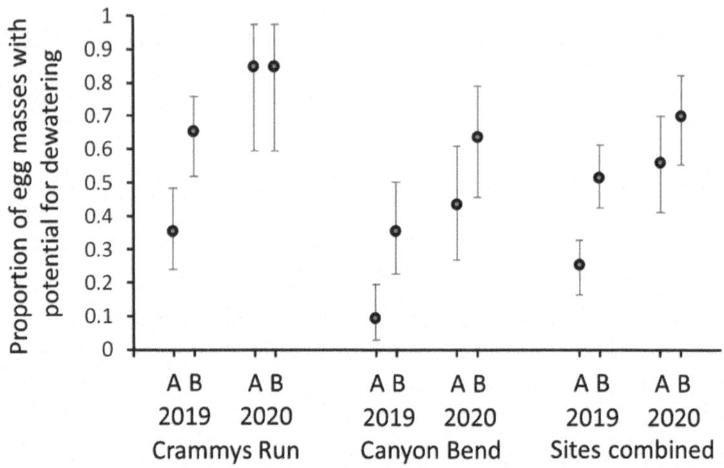

Fig. 13 Proportion of egg masses in 2019 and 2020 with potential for dewatering at Crammys Run and Canyon Bend, Cheat Lake, West Virginia. Estimates are based on two scenarios, where egg masses are deposited directly onto the lake bottom (A) or egg masses are deposited onto structures at 1.0 m above the lake bottom (B). Error bars are 95% profile likelihood confidence intervals

on water level fluctuation regulation periods. Year-to-year variation in Yellow Perch spawning characteristics was documented, a finding consistent with that reported by others (Weber and Les 1982; Sztramko and Teleki 1997). Also, models supported an association of egg mass presence with water temperature, water depth, and lunar illumination. We expected water temperature to influence the onset of spawning (Dabrowski et al. 1996; Feiner and Höök 2015), but our study demonstrated that water temperature fluctuations influence daily spawning activity during the spawning period, a relationship that has also been reported elsewhere (Starzynski and Lauer 2015). Yellow Perch spawned more often on habitat structures in shallower water, a finding that has long been supported by other studies, but spawning depths up to 6.2 m were also documented, exceeding typical depths in the range of 0.4–3.7 m as reported elsewhere (Herman et al. 1959; Forney 1971; Krieger et al. 1983; Weber and Les 1982). The relationship of egg mass presence with lunar illumination, where spawning typically occurred near the new moon, may be an artifact of our short-term study. Although our modeling approach did not support an influence of water level fluctuations on the timing of spawning during a spawning period, a larger issue is the potential for water level fluctuations, specifically lake level drawdowns, to dewater egg masses. We found year-to-year and between-site variation in estimates of the potential for egg dewatering, which was influenced by the timing of drawdowns, water depths at spawning locations, and bathymetric differences between sites.

The interaction of photoperiod and water temperature likely influences the onset and peak of spawning in Yellow Perch (Dabrowski et al. 1996, Ciereszko et al. 1997, Kolkovski and Dabrowski 1998, Feiner and Höök 2015, but see Kayes and Calbert 1979). Also, fluctuations of water temperature following the onset of spawning may influence the length of the spawning period, as well as daily spawning activity and peak periods of spawning (Starzynski and Lauer 2015). In our study, spawning began on March 21 in both years, a consistency that could reflect a photoperiod influence. However, water temperatures at the onset of spawning differed from 6.1 °C in 2019 to 10.6 °C in 2020. Spawning temperatures of our 2-year study ranged from 6.0–14.8 °C in 2019 and 8.9–13.4 °C in 2020, values similar to those reported by Herman et al. (1959; 7.2–11.1 °C), Mansueti (1964; 8.5–12 °C), Hardy (1978; 5–12.8 °C), Krieger et al. (1983; 7–13 °C), and Starzynski and Lauer (2015; 11–13 °C). A range of water temperatures for peak spawning (8.5–10.0 °C) was documented by Tsai and Gibson (1971). Based on our GEE analysis of data collected during spawning periods, we found that the predicted probability of egg mass presence was positively associated with water temperature. This relationship with daily egg mass presence and water temperature will likely influence the distribution of spawning efforts between March and April. Water temperature variation may also influence the spawning effort distribution between deeper and shallower water, as water temperature of shallower water general exceeds that of deeper water.

Our study documented several characteristics useful for understanding where Yellow Perch spawn within Cheat Lake, particularly regarding water depth and distance to the shore. Water depth and distance to the shore are often correlated, especially when lake bottom gradients have moderate to steep slopes, but shallow mud flats do not generally follow this pattern. In our study, Crammys Run had mostly shallow mud flats with some areas of steep bottom slopes, and Canyon Bend had mostly steep slopes with one shallow mud flat. We realize that our placement of habitat units may have influenced the results. Shallow mud flats (distant from the shore) and deeper habitats generally did not contain many natural spawning structures. Fish may have spawned in these areas because of the presence of our artificial habitat structures and may have otherwise spawned in near-shore areas in the absence of structures. Our finding that Yellow Perch will spawn in deep water supports an option for placement of spawning structures in water depths outside of the potential dewatered zone.

Our data supported a relationship between egg mass presence and lunar illumination. Most egg masses were present during periods near the new moon, and the dominant spawning peaks in 2019 and 2020 occurred during a waxing crescent. Our review of the literature did not find a reference to a relationship between lunar phase and spawning of Yellow Perch. Lunar synchronization of fish reproduction is not unusual, but it is often associated with marine fishes as a tide-related or reef-related phenomenon (Taylor 1984). Lunar synchronization of reproduction in freshwater fishes, however, has been reported for cichlids (Schwanck 1987; Watanabe 2000) and sturgeon (Forsythe et al. 2012). Yellow Perch behavior has been associated with levels of ambient light. Helfman (1979) found that individuals of Yellow Perch increased activity levels during dusk and dawn periods. Yellow Perch generally spawn at night (Raney 1959; Scott and Crossman 1973), although some studies have reported day spawning (Harrington 1947; Hergenrader 1969). The relationship between egg mass presence and lunar illumination is possibly a coincidental artifact in our study and could be better understood with a longer time series from additional years of study or by modeling covariates of cloud cover or sky brightness.

Our modeling results did not support an association between water level fluctuations and egg mass presence. However, egg mass presence was more commonly associated with an increase in lake level than with lake level drawdown. A positive relationship between spawning and increased water levels has been reported for Yellow Perch (Henderson 1985; Kallemeyn 1987), as well as other reservoir species within the littoral zone (Ebel 1979; Ozen and Noble 2002). It is also possible that there is a lag effect associated with lake level fluctuation, where changes in lake levels in days previous may influence the timing of spawning, but we did not address this in our models. Lake level fluctuation at or near the time of spawning, however, may not be the main concern. A larger issue is that eggs in the dewatered zone are vulnerable to post-spawn drawdowns during their 6–27 d incubation period (Mansueti 1964; Whiteside et al. 1985; Weber and Les 1982; Powles and Warlen 1988).

We are uncertain as to why the number of egg masses on our artificial spawning structures in 2020 was less than that of 2019. However, year-to-year variation in spawning characteristics of Yellow Perch populations is not uncommon (Weber and Les 1982). The between-year difference may be explained in part by a longer spawning season in 2019 relative to that of 2020. Also, an extended period of lake level drawdown for the dredging of a boat launch area at a local marina occurred during the first half of March 2020, which may have led to Yellow Perch leaving the shallow Crammys Run area to spawn elsewhere. The higher levels of turbidity during the 2020 spawning season possibly reduced the use of artificial spawning structures. It is also possible our counts were biased owing to egg mass detachment from the artificial structures during retrieval, particularly in water deeper than 3 m. Egg masses detached during retrieval of artificial structures on a few occasions for shallow sets (<3 m), but floated upward with the lifting of the structure owing to their near-neutral buoyancy. Thus, egg masses that detached during shallow-water structure retrieval were observed and counted. For deeper water (>3 m), it is possible that some egg masses detached during retrieval of artificial structures and may have gone unnoticed and uncounted.

We were particularly interested in water level fluctuations relative to (1) the potential for egg dewatering and (2) the duration and effort of spawning between March and April, because a 4-m lake level drawdown is permitted during March and a 2.1-m drawdown is permitted during April. Thus, the dewatering of Yellow Perch egg masses would likely be less if the majority of the spawning period and spawning effort occurred during April than in March. During our 2-year study, the spawning periods were similar in timing and duration, where spawning occurred from 21 March to 16 April in 2019 and from 21 March to 11 April in 2020. The effort of spawning, however, differed between years, where peak spawning and the majority of egg masses in 2019 were found on structures in April and most egg masses (peak spawning) in 2020 were documented during March. During each year, the length of time of peak spawning was considerably less than the duration of the spawning period. Short periods of peak spawning may increase vulnerability of

Fig. 14 Examples of lake level drawdown of Cheat Lake, West Virginia (**a**, **b**), egg masses associated with near-shore natural structure (**c**, **d**), and a dewatered egg mass on a natural structure (**d**)

Yellow Perch reproductive efforts to catastrophic failure (Isermann and Willis 2008). Under current lake level regulations, egg losses from dewatering may be increased during years when Yellow Perch spawning efforts during March exceed those of April. Single or consecutive years when most of the spawning effort occurs during a short time period in March could result in reduced recruitment to the adult Yellow Perch population.

In addition to egg masses on our artificial structures, we also observed many egg masses on near-shore natural structures, including submerged and dewatered eggs (Fig. 14). Considering that egg masses were present on natural structures at our study sites, as well as expected along near-shore habitats outside of our study sites, then it is reasonable to assume that the number of eggs with dewatering potential is much larger than the 5.6 million eggs documented in this 2-year study. Bathymetric characteristics of littoral areas, such as depth, slope, and topography, will likely influence the potential for eggs to be dewatered (Henderson 1985; Zohary and Ostrovsky 2011). Walburg (1976) noted that species spawning in shallower water, including Yellow Perch, have much lower rates of spawning success when water levels are lowered. This was also demonstrated by our data, as the proportion of egg masses that were susceptible to dewatering was lower at Canyon Bend than that at Crammys Run. Near-shore areas at Crammys Run are typically shallower with lower slopes than those at Canyon Bend, resulting in a higher dewatering potential of egg masses at Crammys Run. Based on data from our study, future modeling efforts using bathymetry data could provide insights into the potential for dewatering of eggs at a lake level scale. A potential caveat of our study is that the difference in egg dewatering potential between sites, as well as the overall estimates of egg dewatering potential, may be biased by the depths of placement location of our artificial structures. Structures placed in deeper water, which in some areas correspond with farther distances from the shore, may have influenced which spawning locations were selected. It is possible that near-shore and shallower areas would have been used in the absence of these deep-water artificial spawning structures. A higher proportion of spawning events in shallower water would have resulted in a higher estimate of egg dewatering potential. The finding that Yellow Perch will spawn on structures in deeper water outside of dewatering zones provides an option for using deep-water artificial habitats to reduce egg dewatering potential. However, further studies would be needed to address differences in egg-to-dispersal survival between deep-water and shallow-water spawning events.

The dewatering of Yellow Perch eggs may result in fewer larvae and fewer young-of-year. Reductions in the number of young Yellow Perch can impact recruitment to the adult fishery and influence other predators in the community. Larvae and young-of-year Yellow Perch provide a substantial forage base for Walleye (*Stizostedion vitreum*) in Cheat Lake (Smith 2018). Predator/prey relationships between Walleye and young Yellow Perch have been reported from other systems (Maloney and Johnson 1957; Forney 1974; Hartman and Margraf 1993; Hansen et al. 1998; Meerbeek et al. 2002; Pierce et al. 2006). Thus, dewatering and associated egg losses may not only impact the Yellow Perch population, but may also have a bottom up effect on the Walleye population and possibly on other fish populations of Cheat Lake. Although this study focused on one aspect of Yellow Perch life history, additional studies are needed to address potential impacts of water level fluctuations on population dynamics of larval and juveniles (see Henderson 1985; Kallemeyn 1987; Dembkowski et al. 2016). Furthermore, a broader ecosystem level study would be useful toward addressing trophic level effects of water level fluctuations (Ploskey 1983; Leira and Cantonati 2008).

5 Conclusions

Susceptibility of eggs to dewatering from lake level drawdowns has long been a management concern for fish populations in hydropower reservoirs. In Cheat Lake, the potential for dewatering of Yellow Perch eggs exceeded 50% when considering data from both sites and both years of our study. Thus, based on our results, hydropower drawdown has the potential to reduce egg survival of the Cheat Lake Yellow Perch population by more than half. Susceptibility of eggs to dewatering differs across the March–April spawning period owing to water level fluctuation regulations; 4-m and 2.1-m drawdowns are permitted in March and April, respectively. The 2-year study demonstrated year-to-year variation as to whether peak spawning occurred in March or April. Under the current lake level drawdown regulations, the largest egg losses could occur when Yellow Perch focus their spawning efforts in March as opposed to April. Susceptibility of eggs to dewatering is also influenced by depth of spawning location. Our study demonstrated that Yellow Perch in Cheat Lake spawn in shallow near-shore areas, but will also spawn in a wide range of depths and distances from the shoreline. Spawning in deeper water reduces the potential for dewatering of eggs during lake level drawdowns but may be inhibited by a lack of spawning structures. Our study demonstrated that artificial spawning habitat, placed in appropriate locations, may mitigate these losses by providing spawning structure in deeper water.

Acknowledgments We thank Jacob Harrell, Chris Matt, Tamara Matt, Tatiana Matt, Zane Sine, Corey Tucker, and David Wellman for assistance with field work. Funding and support were provided by Cube Hydro, LLC, and the West Virginia Division of Natural Resources. This study was performed under the auspices of West Virginia University IACUC protocol #1901020543. Any use of trade, firm, or product names is for descriptive purposes only and does not imply endorsement by the US Government.

References

Almeida LZ, Guffey SC, Krieg TA, Höök TO (2017) Predators reject Yellow Perch egg skeins. Trans Am Fish Soc 146:173–180

Aroviita J, Hämäläinen H (2008) The impact of water-level regulation on littoral macroinvertebrate assemblages in boreal lakes. Hydrobiologia 613:45–56

Becker GC (1983) Freshwater fishes of Wisconsin. University of Wisconsin Press, Madison

Brazo DC, Tack PI, Liston CR (1975) Age, growth, and fecundity of yellow perch, *Perca flavescens*, in Lake Michigan near Ludington, Michigan. Trans Am Fish Soc 104:726–730

Burnham KP, Anderson DR (2002) Model selection and multimodel inference: a practical information-theoretic approach. Springer, New York

Ciereszko RE, Dabrowskl K, Ciereszko A, Ebeling J, Ottobre JS (1997) Effects of temperature and photoperiod on reproduction of female yellow perch *Perca flavescens*: plasma concentrations of steroid hormones, spontaneous and induced ovulation, and quality of eggs. J World Aquacult Soc 28:344–356

Clark ME, Rose KA, Chandler JA, Richter TJ, Orth DJ, Van Winkle W (2008) Water-level fluctuation effects on centrarchid reproductive success in reservoirs: a modeling analysis. N Am J Fish Manag 28:1138–1156

Dabrowski K, Ciereszko RE, Ciereszko A, Toth GP, Christ SA, El-Saidy D, Ottobre JS (1996) Reproductive physiology of yellow perch (*Perca flavescens*): environmental and endocrinological cues. J Appl Ichthyol 12:139–148

De Lima FT, Reynalte-Tataje DA, Zaniboni-Filho E (2017) Effects of reservoirs water level variations on fish recruitment. Neotrop Ichthyol 15(3):e160084

Dembkowski DJ, Willis DW, Wuellner MR (2016) Synchrony in larval yellow perch abundance: the influence of the Moran Effect during early life history. Can J Fish Aquat Sci 73:1567–1574

Ebel W (1979) Effects of hydroelectric projects on fish populations. In: Major DC, Hanchey JR, Willis EJ Jr, Fratus J (eds) Hydropower: a national energy resource: 1979 engineering foundation conference. Tidewater Inn, Easton, MD, pp 170–176

Echo JB (1955) Some ecological relationships between yellow perch and trout in Thompson Lake. Montana Trans Am Fish Soc 84:239–248

Feiner ZS, Höök TO (2015) Environmental biology of percid fishes. In: Kestemont P, Dabrowski K, Summerfelt RC (eds) Biology and culture of percid fishes principles and practices. Springer, New York

Fitzgerald DG, Dale AR, Thomas MV, Sale PF (2001) Application of otolith analyses to investigate broad size distributions of young yellow perch in temperate lakes. J Fish Biol 58:248–263

Forney JL (1971) Development of dominant year classes in a yellow perch population. Trans Am Fish Soc 100:739–749

Forney JL (1974) Interactions between yellow perch abundance, walleye predation, and survival of alternate prey in Oneida Lake, New York. Trans Am Fish Soc 103:15–24

Forsythe PS, Scribner KT, Crossman JA, Ragavendran A, Baker EA, Davis C, Smith KK (2012) Environmental and lunar cues are predictive of the timing of river entry and spawning-site arrival in lake sturgeon *Acipenser fulvescens*. J Fish Biol 81:35–53

Gaboury MN, Patalas JW (1984) Influence of water level drawdown on the fish populations of Cross Lake, Manitoba. Can J Fish Aquat Sci 41:118–125

Gaeta JW, Sass GG, Carpenter SR (2014) Drought-driven lake level decline: effects on coarse woody habitat and fishes. Can J Fish Aquat Sci 71:315–325

Ganias K, Murua H, Claramunt G, Dominguez-Petit R, Gonçalves P, Juanes F, Keneddy J, Klibansky N, Korta M, Kurita Y, Lowerre-Barbieri S, Macchi G, Matsuyama M, Medina A, Nunes C, Plaza G, Rideout R, Somarakis S, Thorsen A, Uriarte A, Yoneda M (2014) Chapter 4: Egg production. In: Domínguez-Petit R, Murua H, Saborido-Rey F, Trippel E (eds) Handbook of applied fisheries reproductive biology for stock assessment and management. Digital CSIC, Vigo. http://hdl.handle.net/10261/87768

Gasith A, Gafny S (1990) Effects of water level fluctuation on the structure and function of the littoral zone. In: Tilzer MM, Serruya C (eds) Large Lakes, ecological structure and function. Springer, Berlin, pp 156–171

Hanchin PA, Willis DW, St Sauver TR (2003) Influence of introduced spawning habitat on yellow perch reproduction in Lake Madison, South Dakota. J Freshw Ecol 18:291–297

Hansen MJ, Bozek MA, Newby JR, Newman SP, Staggs MD (1998) Factors affecting recruitment of walleyes in Escanaba Lake, Wisconsin, 1958–1996. N Am J Fish Manage 18:764–774

Hardy Jr JD (1978) Development of fishes of the Mid-Atlantic Bight. An atlas of egg, larval, and juvenile stages, Volume III US Fish and Wildlife Service, Biological Services Program FWS-OBS-78/12

Harrington RW Jr (1947) Observations on the breeding habits of the yellow perch, *Perca flavescens* (Mitchill). Copeia 1947:199–200

Hartman KJ, Margraf FJ (1993) Evidence of predatory control of yellow perch (*Perca flavescens*) recruitment in Lake Erie, U.S.A. J Fish Biol 43:109–119

Helfman GS (1979) Twilight activities of yellow perch, *Perca flavescens*. J Fish Res Board Can 36: 173–179

Henderson BA (1985) Factors affecting growth and recruitment of yellow perch, *Perca flavescens* Mitchill, in South Bay, Lake Huron. J Fish Biol 26:449–458

Hergenrader GL (1969) Spawning behavior of *Perca flavescens* in aquaria. Copeia 1969:839–841

Herman E, Wisby W, Wiegert L, Burdick M (1959) The yellow perch, its life history, ecology, and management. Wisconsin Conservation Department Publication 228, Madison

Hill NM, Keddy PA, Wisheu IC (1998) A hydrological model for predicting the effects of dams on the shoreline vegetation of lakes and reservoirs. Environ Manag 22:723–736

Hilling CD, Taylor ND, Welsh SA, Smith DM (2018) Population characteristics of Yellow Perch in a central Appalachia hydropower reservoir. J Fish Wildl Manag 9:486–495

Hin L-Y, Wang Y-G (2009) Working-correlation-structure identification in generalized estimating equations. Stat Med 28:642–658

Hirsch PE, Eloranta AP, Amundsen PA, Brabrand Å, Charmasson J, Helland IP, Power M, Sánchez-Hernández J, Sandlund OT, Sauterleute JF, Skoglund S (2017) Effects of water level regulation in alpine hydropower reservoirs: an ecosystem perspective with a special emphasis on fish. Hydrobiologia 794:287–301

Huff DD, Grad G, Williamson CE (2004) Environmental constraints on spawning depth of yellow perch: the roles of low temperature and high solar ultraviolet radiation. Trans Am Fish Soc 133: 718–726

Isermann DA, Willis DW (2008) Emergence of larval yellow perch, *Perca flavescens*, in South Dakota lakes: potential implications for recruitment. Fish Manag Ecol 15:259–271

Kallemeyn LW (1987) Correlations of regulated lake levels and climatic factors with abundance of young-of-the-year walleye and yellow perch in four lakes in Voyageurs National Park. N Am J Fish Manage 7:513–521

Kayes TB, Calbert HE (1979) Effects of photoperiod and temperature on the spawning of Yellow Perch *Perca flavescens*. Proc World Maricul Soc 10:306–316

Kolkovski S, Dabrowski K (1998) Off-season spawning of yellow perch. Progress Fish Cult 60: 133–136

Krieger DA, Terrell JW, Nelson PC (1983) Habitat suitability information: yellow perch. Western Energy and Land Use Team, Division of Biological Services, Research and Development, U.S. Fish and Wildlife Service

Leira M, Cantonati M (2008) Effects of water-level fluctuations on lakes: an annotated bibliography. Hydrobiologia 613:171–184

Logez M, Roy R, Tissot L, Argillier C (2016) Effects of water-level fluctuations on the environmental characteristics and fish-environment relationships in the littoral zone of a reservoir. Fund Appl Limnol 189:37–49

Maloney JE, Johnson FH (1957) Life histories and interrelationships of walleye and yellow perch, especially during their first summer, in two Minnesota lakes. Trans Am Fish Soc 85:191–202

Mansueti AJ (1964) Early development of the Yellow Perch, *Perca flavescens*. Chesap Sci 5:46–66

McEwen DC, Butler MG (2010) The effects of water-level manipulation on the benthic invertebrates of a managed reservoir. Freshw Biol 55:1086–1101

Meerbeek JR, Isermann DA, Willis DW (2002) Influence of age-0 yellow perch abundance on walleye populations in two eastern South Dakota lakes. Proc S Dak Acad Sci 81:93–99

Muncy RJ (1962) Life history of the yellow perch, *Perca flavescens*, in estuarine waters of the Severn River, a tributary of Chesapeake Bay. Chesap Sci 3:142–159

Nelson WR, Walburg CH (1977) Population dynamics of Yellow Perch (*Perca flavescens*), Sauger (*Stizostedion canadense*), and Walleye (*S. vitreum vitreum*) in four main stem Missouri River reservoirs. J Fish Res Board Can 34:1748–1763

Newsome GE, Tompkins J (1985) Yellow perch egg masses deter predators. Can J Zool 63:2882–2884

Noble RL (1970) Evaluation of the Miller high-speed sampler for sampling yellow perch and walleye fry. J Fish Res Board Can 27:1033–1044

Ozen O, Noble RL (2002) Relationship between water level fluctuations and largemouth bass spawning in a Puerto Rico reservoir. Am Fish Soc Symp 31:213–220

Pan W (2001) Akaike's information criterion in generalized estimating equations. Biometrics 57: 120–125

Pierce RB, Tomcko CM, Negus MT (2006) Interactions between stocked walleyes and native yellow perch in Lake Thirteen, Minnesota: a case history of percid community dynamics. N Am J Fish Manage 26:97–107

Ploskey GR (1983) A review of the effects of water-level changes on reservoir fisheries and recommendations for improved management. Technical Report E-83-3, U.S. Army Engineer Waterways Experiment Station, Vicksburg, Mississippi

Ploskey GR (1986) Effect of water-level changes on reservoir ecosystems, with implications for fisheries management. In: Hall GE, Van Den Avyle MJ (eds) Reservoir Fisheries Management: strategies for the 80's. American Fisheries Society, Bethesda, MD, pp 86–97

Powles PM, Warlen SM (1988) Estimation of hatch periods for yellow perch, based on otolith readings from juveniles (age-0). Am Fish Soc Symp 5:60–67

Raney EC (1959) Some young fresh-water fishes of New York. NY State Conserv 14:22–28
Robillard SR, Marsden JE (2001) Spawning substrate preferences of yellow perch along a sand–cobble shoreline in southwestern Lake Michigan. N Am J Fish Manag 21:208–215
Schwanck E (1987) Lunar periodicity in the spawning of *Tilapia mariae* in the Ethiop River, Nigeria. J Fish Biol 30:533–537
Scott WB, Crossman EJ (1973) Freshwater Fishes of Canada. Fisheries Research Board of Canada Bulletin 184, Ottawa
Smith CL (1986) Inland fishes of New York State. New York Department of Environmental Conservation, Albany
Smith D (2018) Evaluation of a re-established Walleye population within a hydropower reservoir recovering from acidification PhD Dissertation. West Virginia University, Morgantown, WV
Smith D, Welsh S (2015) Biological monitoring of aquatic communities of Cheat Lake, and Cheat River downstream of the Lake Lynn Hydro Station, 2011-2015 FERC project number 2459
Starzynski D, Lauer TE (2015) How temperature affects timing and duration of yellow perch spawning in the Indiana waters of Lake Michigan. J Freshw Ecol 30:445–453
Sztramko L, Teleki GC (1997) Annual variations in the fecundity of yellow perch from Long Point Bay, Lake Erie. Trans Am Fish Soc 106:578–582
Taylor MH (1984) Lunar synchronization of fish reproduction. Trans Am Fish Soc 113:484–493
Taylor N (2013) Assessment of age, diet, and growth of yellow perch (*Perca flavescens*) in Cheat Lake, West Virginia. MS Thesis, West Virginia University, Morgantown, WV
Tsai C-F, Gibson GR Jr (1971) Fecundity of the yellow perch, *Perca flavescens* Mitchill, in the Patuxent River, Maryland. Chesap Sci 12:270–274
USGS, US Geological Survey (2020) National Water Information System data: Cheat Lake near Stewartstown, WV, accessed 15 July 2020 at https://waterdata.usgs.gov/usa/nwis/uv?site_no=03071590
Walburg CH (1976) Changes in the fish population of Lewis and Clark Lake, 1956–74, and their relation to water management and the environment. Research Report 79, U.S. Fish and Wildlife Service, Washington, DC
Watanabe T (2000) Lunar cyclic spawning of a mouthbrooding cichlid, *Cyprichromis leptosoma*, in Lake Tanganyika. Ichthyol Res 47:307–310
Weber JJ, Les BL (1982) Spawning and early life history of yellow perch in the Lake Winnebago system. Technical Report 130, Wisconsin Department of Natural Resources, Madison, WI
White MS, Xenopoulos MA, Metcalfe RA, Somers KM (2011) Water level thresholds of benthic macroinvertebrate richness, structure, and function of boreal lake stony littoral habitats. Can J Fish Aquat Sci 68:1695–1704
Whiteside MC, Swindoll CM, Doolittle WL (1985) Factors affecting the early life history of yellow perch, *Perca flavescens*. Environ Biol Fish 12:47–56
Wilcox DA, Meeker JE (1991) Disturbance effects on aquatic vegetation in regulated and unregulated lakes in northern Minnesota. Can J Bot 69:1542–1551
Williamson CE, Metgar SL, Lovera PA, Moeller RE (1997) Solar ultraviolet radiation and the spawning habitat of yellow perch, *Perca flavescens*. Ecol Appl 7:1017–1023
Zohary T, Ostrovsky I (2011) Ecological impacts of excessive water level fluctuations in stratified freshwater lakes. Inland Waters 1:47–59

A Comparison of Aquaculture Production Methods for Optimizing Production of Fingerling Yellow Perch (*Perca flavescens*)

Cathleen M. Doyle, David A. Culver, Morton E. Pugh, and Jesse E. Filbrun

Abstract Yellow Perch aquaculture has increased since the 1980s to reverse declines in wild populations and meet increased demands by anglers. Over the past 41 years, staff at the St. Marys State Fish Hatchery (SFH) in western Ohio used different methods to obtain Yellow Perch eggs, support embryonic development and hatch eggs, and rear the fry in ponds to the fingerling stage for stocking. We used hatchery records from 1977 through 2017 to statistically compare production outcomes among various rearing methods including (1) natural vs manual spawning, (2) embryo hatching methods, (3) organic vs inorganic pond fertilization, and (4) fry residence time in ponds before harvest. We found that the most reliable production of Yellow Perch fingerlings consisted of placing hormone-induced females in tanks with males, hatching embryos in Heath trays, and stocking fry in ponds fertilized using liquid inorganic fertilizers. While our study is retrospective, and thus precludes assigning causality to any observed improvements in yield with method changes, adopting these methods at St. Marys SFH has increased harvest density of fingerlings produced from 13 ± 4 to 53 ± 6 fish m^{-2} (mean $\pm$ SE).

Keywords Pond fertilization · Fingerling production · Pond management · Yellow Perch

C. M. Doyle (✉) · D. A. Culver
Department of Evolution, Ecology, and Organismal Biology, The Ohio State University, Columbus, OH, USA
e-mail: doyle.108@osu.edu; culver.3@osu.edu

M. E. Pugh
Division of Wildlife, Ohio Department of Natural Resources, St. Marys State Fish, Hatchery, OH, USA

J. E. Filbrun
Department of Biology, Southern Arkansas University, Magnolia, AR, USA
e-mail: jessefilbrun@saumag.edu

J. C. Bruner, R. L. DeBruyne (eds.), *Yellow Perch, Walleye, and Sauger: Aspects of Ecology, Management, and Culture*, Fish & Fisheries Series 41,
https://doi.org/10.1007/978-3-030-80678-1_2

1 Introduction

Yellow Perch (*Perca flavescens* Mitchill) belong to one of the largest families of fishes (Percidae) in North America (Page and Burr 1991). Except for darters, percids represent important commercial and sport fisheries in the Great Lakes region and are subject to intense harvest pressure (Malison 2003). Since the mid-twentieth century, commercial harvest of Yellow Perch populations in the Great Lakes drastically declined, and recruitment remains low (Lesser and Vilstrup 1979; Craig 2000; Baldwin et al. 2018), yet commercial and recreational demand remains high (Riepe 1998). Therefore, Yellow Perch aquaculture has become increasingly important to enhance wild populations experiencing natural declines in recruitment, establish new populations (Fox 1989; Ellison and Franzin 1992; Mitzner 2002), and/or subsidize the market demand for food fish (Malison 2000).

Although percid aquaculture has been refined for Walleye (*Stizostedion vitreum*) and saugeye (Walleye ♀ × Sauger, *S. canadense*, ♂) (Briland et al. 2015), production results have not been compared for the many available Yellow Perch production methods. Previous studies in Ohio have shown that the timing of pond filling, pond fertilization regimens, and source water quality impact Walleye and saugeye production (reviewed in Briland et al. 2015). Managing ecological parameters associated with plankton dynamics in earthen ponds can increase production and survival of these percids (Tew et al. 2006; Jacob and Culver 2010; Briland et al. 2015).

Yellow Perch egg production and hatching methods differ from other percids. Female Yellow Perch require a "chill period" of a few months during winter for egg development. Spawning occurs in the spring, and fecundity is negatively affected by an insufficient number of cold days during ova development (Hokanson 1977; Farmer et al. 2015). Yellow Perch spawn in the spring when water temperatures reach 8–13 °C, with peak spawning at 10 °C (Nelson and Walburg 1977). As with Walleye and Sauger, Yellow Perch are synchronous spawners, producing one batch of eggs annually, and offer no parental care. Yellow Perch eggs are extruded in skeins of a gelatinous matrix which females then attach to submerged aquatic vegetation, rocks, or woody debris, such as fallen trees (Thorpe 1977; Craig 1987).

Broodstock (parents) for culture can be obtained either from the wild, just before spawning, and held at the hatchery until ready to spawn or purchased from a commercial producer. Broodstock may also be held at the hatchery in outdoor production ponds or indoor tanks. Spawning adults may be kept for multiple years, or new parental fish can be chosen each year, depending on the desired genetic composition of the hatchery recruits. Eggs are obtained from either natural spawning occurring in ponds or manual spawning, which requires egg ribbons to be obtained from females and milt obtained from males. The eggs and milt are dry mixed in a bowl for fertilization to occur (Hart et al. 2006). Hatching is temperature-dependent, and the incubation period ranges from 10 to 20 days (Hinshaw 2006). Hatching generally occurs over 2 weeks (Nelson and Walburg 1977; Hart et al. 2006).

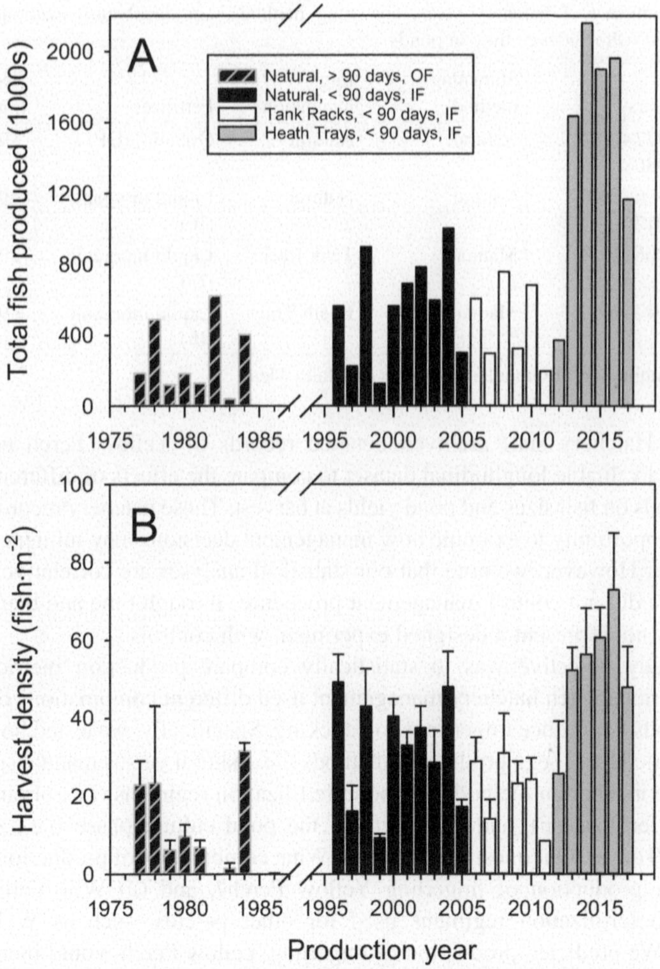

Fig. 1 (**a**) Comparisons of total Yellow Perch fingerlings harvested from ponds at St. Marys State Fish Hatchery from 1977 through 2017. Values are harvest totals summed across all ponds with each year. (**b**) Comparisons of Yellow Perch pond yields through time. Values represent mean + SE of pond yields for each year. Production methods compared are natural spawning versus manual spawning (Tank Racks and Heath Trays), culture duration (<90 days and >90 days), and fertilization regimens (*OF* organic fertilization and *IF* inorganic fertilization)

In this study, we provide a retrospective analysis of historical hatchery data, for Yellow Perch fry and fingerling culture methods by statistically comparing production results for a 41-year dataset (1977–2017) at St. Marys State Fish Hatchery (SFH), Ohio, USA. St. Marys SFH is one of three warmwater hatcheries operated by the Ohio Department of Natural Resources, Division of Wildlife (ODNR-DOW). Yellow Perch culture did not occur every year during the time series, but only when the ODNR-DOW requested fish, and hence production was variable through time

Table 1 Summary of hatchery years, spawning methods, egg incubation methods, fertilizer regimens, and fish residence time in ponds

Period	Years	Spawning method	Egg incubation	Fertilizer	Residence time
I	1977–1984, 1986	Natural	Natural	Organic[a] (OF)	>90 days
II	1996–2005, 2012	Natural	Natural	Liquid inorganic (IF)	<90 days
III	2006–2011	Manual	Tank Racks	Liquid inorganic (IF)	<90 days
IV	2012–2017	Manual	Heath Trays	Liquid inorganic (IF)	<90 days

[a]OF is a combination of Granular Inorganic + Alfalfa Meal

(Fig. 1a). Hatchery staff archived detailed records of Yellow Perch production providing a valuable longitudinal dataset to compare the effects of different production methods on fish sizes and pond yields at harvest. These hatchery records provide a unique opportunity to examine how management decisions may influence production results. However, we note that our statistical analyses are correlative in nature because we did not control management procedures through time and therefore this study does not represent a designed experiment with controls.

Our study objective was to statistically compare production metrics among distinct periods when hatchery management used different combinations of production methods to produce fingerlings for stocking. Specifically, we tested how harvest metrics varied by (1) egg production methods (i.e., natural versus manual spawning), (2) gamete incubation methods, (3) pond fertilization regimens (i.e., organic versus inorganic fertilization), and (4) length of the pond culture phase (i.e., <90 days versus >90 days). Our questions were: (1) What combination of production methods is best for production of fingerling Yellow Perch?, and (2) Will Yellow Perch respond to fertilization regimens used for other percids, such as Walleye and saugeye? We predicted production of fingerling Yellow Perch would increase with controlled inorganic fertilization in ponds, manual spawning, and a shorter duration of rearing in ponds (Table 1).

2 Methods

2.1 Study Site

St. Marys SFH is located in western Ohio, USA (WGS84: 40.527, −84.418). The hatchery was built in 1913 by the Western Ohio Fish and Game Association as a warmwater hatchery and was dedicated as an Ohio SFH in 1936. It historically produced Largemouth Bass (*Micropterus salmoides*), White Crappie (*Pomoxis annularis*), Common Carp (*Cyprinus carpio*), and other species in 51 ponds of various sizes, ranging from 0.11 to 0.85 ha. During 1995–1996, extensive renovations were completed to ponds and the hatchery plumbing system. Thereafter, the hatchery contained 26 ponds of more uniform sizes. Most production ponds are 0.35 ha (0.86 acres) with a mean depth of 1.3 m (pond volume = 4515 m^3). After the renovations, the hatchery produced Walleye, saugeye, Yellow Perch, Fathead Minnow (*Pimephales promelas*), Channel Catfish (*Ictalurus punctatus*), and Blue Catfish (*I. furcatus*).

The hatchery draws hypereutrophic water from the bottom of the adjacent shallow (mean depth = 1.6 m) Grand Lake St. Marys reservoir to gravity-fill all grow-out ponds. The water is filtered through 0.5-mm screens to prevent undesired fish eggs and larvae from entering the ponds. Zooplankton and phytoplankton, however, pass easily through the screens. These screens have been used by all Ohio SFHs for many years, allowing for the regular development of zooplankton without the introduction of wild fish. Originally, the hatchery raised Yellow Perch from eggs to fingerlings exclusively in the outdoor ponds, but the construction of 3 m^3 indoor tanks fed by flow from the wells enabled managers to produce gametes and fry indoors, allowing them to stock known numbers of fry in ponds. The hatchery now contains indoor tanks for producing and hatching eggs of Walleye, saugeye, and Yellow Perch, plus the outdoor ponds for rearing fry to harvestable fingerlings. Over 2010, 2011, and 2014, four wells were drilled (maximum flow of 2.65 m^3 min^{-1} per well) to provide flow-through water for incubating gametes in the indoor tanks at 12 °C, stabilizing hatching dates and providing high-quality water for incubation.

2.2 Production Dataset

St. Marys SFH produced Yellow Perch in 34 of 41 years from 1977 through 2017 (1977–1984; 1986; 1990–1992; 1996–2017). Researchers at The Ohio State University performed experiments with Yellow Perch ponds during 1990–1992 for other projects, so we excluded those years for all analyses, leaving 31 production years in the dataset. In each year, staff maintained records of all production metrics, including pond areas and volumes, numbers of adult pairs stocked into ponds (for years with natural spawning), numbers of fry stocked into ponds (for years with manual spawning), fish residence time in ponds, numbers of fingerlings harvested from ponds, total fish weight at harvest, and average fish total length at harvest. We also calculated average fish weight at harvest (g $fish^{-1}$), fish harvest yields (fish m^{-2} and kg ha^{-1}), and percent survival to harvest (for years with manual spawning).

2.3 Broodstock Management

Yellow Perch broodstock originating from Lake Erie were maintained in two or three winter holding ponds accompanied by advanced yearling perch (future broodstock). Adult broodstock were typically used for several years and were supplemented by adults obtained from commercial breeders when wild-caught broodstock did not meet the needed numbers of adults used for spawning. A formal plan to minimize the effects of inbreeding depression on fish production characteristics has never been used at St. Marys SFH for Yellow Perch production. However, new broodstock were regularly introduced through time to produce embryos. Moreover, hatching success and fry production increased, not decreased, through time, suggesting there were likely no genetic limitations related to fish production. In October, all potential brood fish were placed in tanks containing pond water and 0.5% NaCl and treated with formalin at a rate of 100 ml m^{-3} for 0.5 to 1.0 h, as recommended, to shed any parasites (Hart et al. 2006). Tanks were provided elemental oxygen and mixed during formalin treatments. Fathead Minnows were stocked in each pond as forage for Yellow Perch (510–770 kg ha^{-1}) after first receiving the same prophylactic treatments. Oxygen levels in ponds were monitored weekly to assure they remained normoxic (at or above 8 mg L^{-1}). If they became hypoxic or if ice formed on the ponds, air was injected at the bottom of each pond using compressors to maintain adequate oxygen concentrations.

2.4 Gamete Collection and Incubation

Staff at St. Marys SFH used different methods to obtain gametes over the years. Initially, natural spawning was used for obtaining fertilized eggs, which were produced in ponds (1975–2005). Different densities of mature parental fish were stocked into ponds to spawn, with slight variations year to year due to broodstock availability and differences in pond sizes. As outlined in Briland et al. (2015), an even number of male and female Yellow Perch were stocked at either a low density ($<$60 parental fish ha^{-1}), a medium density (60–100 parental fish ha^{-1}), or a high density ($>$100 parental fish ha^{-1}), as early as 10 March, to as late as 13 April (depending on the pond water temperatures). Ponds were filled with water from Grand Lake St. Marys one day before stocking parents, and dead conifer trees (burned to remove needles) were added to ponds for females to attach egg skeins. Spawning is influenced by both temperature and photoperiod (Hart et al. 2006). Parents were removed by seining once egg skeins were observed to avoid predation on fry. Although the natural spawning method involves a relatively small amount of labor by the staff prior to frequent pond fertilization to support zooplankton to feed the fry, there is no control of when fish spawn, when eggs hatch, or the number of fry produced. Weather influences pond temperatures, altering the time from spawning to fry hatching dates. Thus, there is no control over the number of fingerlings collected at harvest, and percent survival of fry to harvested fingerlings cannot be calculated.

In 2006, St. Marys SFH managers switched to manual spawning and incubation indoors based on methods developed by researchers at the University of Wisconsin-Madison and Michigan State University (Hart et al. 2006), which improved the control of timing of hatching and allowed the estimation of the numbers of fry stocked in ponds. Broodstock were removed from winter holding ponds in mid-March to prevent spawning in the ponds and moved indoors wherein 200 males and 200 females were placed into holding tanks (3.0–3.8 m^3). Adults were reliably sexed by examining features of the anus and urogenital opening (Malison et al. 2011). Tanks were supplied with well water and maintained at a constant temperature (12 °C). Dissolved oxygen concentrations were maintained at or above 9 mg L^{-1} via gravity filtration through a packed column prior to iron filtration. The water was then passed through sand and gravel filters to remove iron precipitates from the oxygenated water. To synchronize spawning, female Yellow Perch are commonly injected with human chorionic gonadotropin (hCG) (Dabrowski et al. 1996; Hart et al. 2006). At St. Marys SFH, all female brood fish were injected with a constant dose of 0.05 ml Chorulon hCG (50 IU), regardless of size, to minimize stress from handling time. All tanks were covered to prevent fish escape, and egg maturation occurred 3–8 days after injection. No spawning substrate was placed into these tanks, and no air was added because air causes egg skeins to float to the water surface. Egg skeins were deposited on the bottom of the tanks, typically at night. Tanks were checked twice daily, and fertilized spawn skeins were retrieved by coaxing them off tank bottoms with a stick.

To prevent fungal outbreaks and parasite infections while spawning fish are held in tanks, broodstock were provided a salt bath at 7 g NaCl L^{-1} and a Terramycin HCl treatment (20 mg L^{-1}) for 4 h every 2–3 days. During treatment, the water volume in the tank was drawn down to half full, and the water supply was turned off. Aeration was kept running while the treatments were administered, and fish were periodically checked for stress. All spent male and female fish were given this treatment before returning them to holding ponds over the spring and summer.

Two methods of indoor incubation in aerated well water were used at St. Marys SFH. The first method, which we call "Tank Racks," incubated eggs in 3 m^3 tanks with egg skein support racks constructed of polyvinylchloride (PVC) pipe frames and zinc-coated 2.5 cm diameter mesh (chicken wire) (Fig. 2a). Incubating egg skeins with Tank Racks involves keeping the egg ribbons submerged in the water and not allowing the ribbons to touch one another, avoiding fungal infestation, suffocation, and eventual death of the eggs. Each Tank Rack held eight egg ribbons, with about 15,000–20,000 eggs per ribbon. Water flow was set at 80% turnover per hour to prevent metabolite buildup.

The second method, which we call "Heath Trays," used stacks of shallow trays common to salmonid aquaculture (Fig. 2b, c, d). Each stack contained eight 61 × 65 cm Heath Trays that received filtered well water initially at about 8 L min^{-1}, followed by 11 L min^{-1} after embryos reach the eyed stage. Water traveled through the tray system via gravity so that all trays continuously receive water, flowing from

Fig. 2 Comparison of the Tank Racks and Heath Trays used by St. Marys SFH for egg incubation. (**a**) Tank Racks are made with 0.5-inch PVC pipe and wire mesh and are set at a 60° angle with egg ribbons anchored at the top, set between the wire mesh to remain submerged (Photo credit: Mark Pummell). (**b**) Heath Tray stacks at the St. Marys SFH (Photo credit: David A. Culver). (**c**) Each Heath Tray unit consists of eight trays with screens covering the egg ribbons to prevent loss as water flows via gravity from the top tray to the bottom tray (Photo credit: David A. Culver). (**d**) Egg ribbons incubating inside a single Heath Tray (Photo credit: David A. Culver)

the top tray through each consecutive tray to the bottom. Egg ribbons were removed from spawning tanks and staff measured ribbons in a volumetric cylinder, placing 750 ml into each tray. Staff estimated 75 eggs ml^{-1} by counting the numbers of eggs in 10 mL subsamples settled in graduated cylinders, which equates to about 56,250 eggs $tray^{-1}$. During incubation, all white, moribund eggs were removed daily and counted.

Regardless of the manual spawning incubation technique used, once all eggs have been taken and water-hardened for 24 h, a daily prophylactic treatment of Formalin (Parasite-S) is administered via a calibrated peristaltic pump at a rate of 1000 ppm for 15 min to guard against fungal proliferation. These treatments ceased once eyes become visible in the developing embryos.

As the egg skein degrades and the eyes become fully pigmented, the eggs were ready for forced hatching. This is commonly assessed by testing if a small sample of eggs can be forcibly hatched by aggressive finger swirling them in a beaker (Hart et al. 2006). For the St. Marys SFH incubation regimen at 12 °C, the eggs hatched in 11 or 12 days. If the Tank Racks were used, five to eight skeins of the same age were siphoned into a 20 L bucket full of oxygenated well water and vigorously stirred for 20 s using a paint mixer attached to a power drill. The bucket of water containing eggs was weighed before and after the eggs were added to quantify total egg volume. While stirring, a 10 ml subsample was collected, and fry were counted to determine the number of fry mL^{-1} and hence to estimate the number of fry in the bucket. The fry were then stocked into grow-out ponds. If eggs were incubated using Heath Trays, a 1 L volume of eggs (~75,000 eggs) was poured into a 20 L bucket and stirred as above to hatch eggs for stocking into ponds. Egg and fry numbers were estimated consistently by counting the numbers of each in settled volumes of random samples.

2.5 *Fertilization Regimens*

Yellow Perch pond fertilization regimens can be divided into two distinct periods (Table 1) that were separated by the period of hatchery pond renovations. From 1975 through 1986, a combination of inorganic and organic agricultural fertilizers was added to ponds. Herein, we call this method "organic" fertilization. Staff applied 6:10:4 (6% N, 10% P_2O_5: 4% K_2O) granular Vigoro® fertilizer at a rate of 168 kg ha^{-1} and alfalfa meal at a rate of 112 kg ha^{-1} to each pond weekly. The Vigoro® fertilizer application contained 10.09 kg N ha^{-1} and 7.34 kg P ha^{-1}. The alfalfa meal, at 2.9% N and 0.24% P, contained 3.25 kg N ha^{-1} and 0.27 kg P ha^{-1}. Combined (assuming all the fertilizer and alfalfa meal dissolved), they provided an addition of 13.33 kg N ha^{-1} and 7.61 kg P ha^{-1}, N:P = 1.8 by mass, equivalent to 1034 μg N L^{-1} and 590 μg P L^{-1}, to each pond each week in addition to whatever was already present in the water. This high phosphorus addition to the already high N and P content of Grand Lake St. Marys water promoted the proliferation of toxic cyanobacteria, including *Anabaena*, *Aphanizomenon*, *Microcystis*, and *Planktothrix*, and nuisance filamentous green algae, such as *Hydrodictyon* (Filbrun et al. 2013a). Moreover, using organic fertilization, there were few "edible" algae species available for zooplankton grazing (Helal and Culver 1991; Culver et al. 1993). Organic fertilization also caused high free ammonia and low dissolved oxygen concentrations (Helal and Culver 1991). Free ammonia is highly toxic to Yellow Perch fry and fingerlings (Espey 2003).

During 1996 through 2017, hatchery managers applied liquid inorganic fertilizers of N and P to ponds weekly to restore N:P to levels that promoted growth of a desirable phytoplankton community (Jacob and Culver 2010). Herein, we call this method "inorganic" fertilization. Briefly, inorganic N and P concentrations were measured in each pond weekly. Using these measured concentrations and known volumes of individual ponds, liquid inorganic N (NH_4NO_3 + Urea) and P (H_3PO_4) of measured concentrations were diluted with pond water and sprayed over pond surfaces to restore each pond concentration to 600 μg N L^{-1} and 30 μg P L^{-1} (20:1 N:P by mass). This method decreased the production of cyanobacteria and filamentous green algae and helped increase the concentrations of "edible" algae needed for zooplankton production (Helal and Culver 1991; Tew et al. 2006; Briland et al. 2015). Inorganic fertilization resulted in zooplankton communities dominated by small-bodied cladocerans (e.g., *Bosmina* and *Chydorus*), cyclopoid copepods (e.g., *Diacyclops thomasi* and *Mesocyclops edax*), and rotifers (Briland et al. 2015).

2.6 Fish Residence Time in Ponds Before Harvest

The length of the production phase changed over time corresponding with the fertilization regimen. During 1975 through 1986, fingerlings produced by natural spawning and organic fertilization were cultured in the ponds until fingerlings were harvested in the fall, either September or October (>90 days duration). During 1996 through 2017, after adopting the inorganic fertilization regimen, fingerlings were harvested in June (<90 days duration) (Table 1).

2.7 Fingerling Harvest Methods

For all these methods, fingerling harvest involved draining the ponds slowly through 1 m × 1 m outlet screens equipped with 1 mm mesh screening. Because pond depth decreases away from the outlet, draining forces the fish to move slowly toward the outlet which contains a rectangular, 0.75 m deep × 2 m × 2 m concrete "kettle" from which fingerlings can be dip netted and transferred to a tub resting on a mechanical scale. The fingerlings can then be transferred to another tub and the tare weight of the original tub and remaining water on the scale determined, allowing measurement of the net mass of fingerlings harvested from the pond. Counting the number of fish in a 500 g subsample enabled estimate of the total number and mean individual mass of the fingerlings harvested from that pond.

2.8 Statistical Analyses of Production Metrics

Use of different egg production methods, incubation systems, pond fertilization regimens, and pond durations resulted in four distinct production periods at St. Marys SFH (Table 1). During 1977 through 1984, plus 1986, Yellow Perch were produced by natural spawning in ponds, organic fertilization, and fish reared for >90 days in ponds. During 1996 through 2005, plus some ponds in 2012, fish were produced by natural spawning, inorganic fertilization, and <90 days in ponds. During 2006 through 2011, fish were produced by manual spawning with incubation to hatching using Tank Racks, inorganic pond fertilization, and <90 days in ponds. During 2012 through 2017, fish were produced by manual spawning and incubation to hatching using Heath Trays, inorganic pond fertilization, and <90 days in ponds.

We statistically compared all available production metrics across the four distinct production periods. Production metrics were averaged across ponds within years.

Thus, each production year was treated as a single, independent sample for each production metric. The number of fingerlings harvested from ponds, total harvested fish weight (kg), fish size at harvest (g fish^{-1} and total length, TL), and harvest yield (fish m^{-2} and kg ha^{-1}) were compared across all production periods using one-way ANOVAs or non-parametric Kruskal-Wallis (K-W) tests. K-W tests were used when the production metrics violated the assumption of homoscedasticity, which was determined from Levene's tests. For ANOVAs, Tukey's HSD was used to test for pairwise differences among production periods. For K-W tests, Dunn's post hoc tests were used to determine pairwise differences among production methods after Bonferroni correction for multiple comparisons. Given the large variation among ponds and years, pairwise differences were reported at $P < 0.1$ rather than the more traditional $P < 0.05$. The number of adults stocked into ponds, number of fry stocked into ponds, and the percent survival of fingerlings were each only available for two of four production periods (Table 2). Each of these metrics was compared between periods using independent-samples t-tests. We used SPSS Statistics version 26.0 (IBM, Armonk, NY) for all ANOVAs, K-W tests, post hoc tests, and t-tests.

The size and number of fish produced in the ponds displayed an inverse relationship, whereby a few large fish or many small fish were harvested. We plotted each pond according to the type of spawning, fertilization regimen, and fish residence time before harvest. We plotted the mean individual mass at harvest from each pond against the number of fish harvested from the same pond and performed a breakpoint analysis to identify threshold relationships between individual mass and number harvested using Kolmogorov-Smirnov (2DKS) tests (Garvey et al. 1998).

3 Results

Yellow Perch production from 1977 through 2017 was variable, but generally increased through time (Fig. 1). The most dramatic increase in fish production coincided with the use of Heath Trays to incubate fertilized eggs. Comparisons of production metrics for Yellow Perch across the four production periods are presented in Table 2. First, we present overall differences in production across the four distinct production periods. Second, we explore competition-based relationships between fish harvest densities and mean individual body sizes. Finally, we examine relationships between stocking and harvest rates for the periods that used manual spawning.

3.1 Differences in Yellow Perch Production Among the Four Production Periods

Yellow Perch production metrics changed dramatically through time as a result of changing the combinations of production methods (Table 1). Harvest density (fish

Table 2 Comparisons of Yellow Perch production metrics among fish production methods. Values represent mean ± SE. Nonparametric ANOVAs (i.e., Kruskal-Wallis tests) were used for production metrics that violated the assumption of homoscedasticity, which was determined from Levene's tests. Significant differences ($P < 0.1$, adjusted for multiple pairwise comparisons) among methods are indicated by different superscripts

Production metric	Natural spawning in ponds, >90 days in culture, organic fertilization	Natural spawning in ponds, <90 days in culture, inorganic fertilization	Manual spawning using tank racks, <90 days in culture, inorganic fertilization	Manual spawning using heath trays, <90 days in culture, inorganic fertilization	Statistical test	*P*-value
Number of production years	9	11	6	6	–	–
Years applied	1977–1984, 1986	1996–2005, 2012	2006–2011	2012–2017	–	–
Number of stocked adults	83 ± 17	36 ± 2	–	–	Indep. samp. t-test	0.028
Number of stocked fry	–	–	273,000 ± 20,000	615,000 ± 86,000	Indep. samp. *t*-test	0.003
Fry stocking density (fry m^{-2})	–	–	78 ± 6	177 ± 25	Indep. samp. *t*-test	0.003
Number of fingerlings harvested	134,000 ± 65,000^{a}	103,000 ± 17,000^{ab}	76,000 ± 12,000^{a}	183,000 ± 22,000^{b}	Kruskal-Wallis Test	0.048
Harvest survival (%)	–	–	28 ± 5	31 ± 3	Indep. samp. *t*-test	0.535
Fish mass at harvest (g $fish^{-1}$)	4.89 ± 1.93^{a}	0.35 ± 0.05^{b}	0.57 ± 0.07^{ab}	0.38 ± 0.08^{b}	Kruskal-Wallis Test	<0.001

Fish total length at harvest (mm)	59 ± 8^{a}	31 ± 2^{b}	38 ± 2^{ab}	31 ± 2^{b}	Kruskal-Wallis Test	0.001
Harvest density (fish m^{-2})	13 ± 4^{a}	29 ± 5^{b}	22 ± 4^{ab}	53 ± 6^{c}	One-way ANOVA	<0.001
Harvest weight (kg)	126 ± 32^{a}	28 ± 4^{b}	36 ± 5^{b}	52 ± 5^{ab}	Kruskal-Wallis Test	0.001
Harvest yield (kg ha^{-1})	168 ± 33^{a}	81 ± 13^{b}	103 ± 13^{ab}	150 ± 15^{ab}	One-way ANOVA	0.014

m^{-2}) was the highest and most consistent when hatchery staff employed Heath Trays to incubate fertilized eggs (Table 2). Using the combination of Heath Trays, a < 90 days culture period of Yellow Perch fingerlings in ponds, and inorganic fertilization (2012–2017), hatchery staff harvested on average 53 ± 6 fish m^2 from the ponds. By comparison, the other periods produced about one-quarter to one-half of this density. For example, the earliest period (1977–1986), which employed natural spawning in ponds, >90 days in ponds, and organic fertilization, was "boom and bust" in nature. During that time, production among ponds within a single year and production among years were extremely variable. Indeed, the coefficient of variation (CV = standard deviation/mean) for fish harvest density (fish m^{-2}) among years during 1977–1986 was 0.91 as compared to 0.30 during 2012–2017. The two intervening periods had intermediate harvest densities and CVs. These patterns are also reflected in the fish harvest yields (kg ha^{-1}) (Table 2).

Fish body sizes in length and mass at harvest reflect the patterns in fish harvest densities. We observed these patterns throughout our Yellow Perch production time series. During 2012–2017, when hatchery staff employed Heath Trays to hatch eggs, they had a large, steady supply of fry to stock ponds at high densities, resulting in a large number of consistently small (<0.05 g mass ind^{-1}) Yellow Perch fingerlings at harvest (Table 2; Fig. 3a, b, c). By contrast, the largest average fish sizes, but with the most variation among ponds, occurred during 1977–1986 (Fig. 3a, b). Yellow Perch fingerlings sizes at harvest during the two intervening production periods were most similar to the Heath Trays period (i.e., 2012–2017), having consistently small body sizes in length and mass (Fig. 3a, b).

Fish survival could only be compared between the Tank Racks period (2006–2011) and the Heath Trays period (2012–2017), when known numbers of Yellow Perch fry were stocked into ponds. There were no differences in fish survival between these periods, with average survival around 30% (Table 2). It is noteworthy that increasing fry stocking densities during 2012–2017, which was possible because of higher fry yields hatched from Heath Trays, did not reduce fish survival, nor did it appreciably reduce size at harvest in length or mass. In other words, switching from Tank Racks to Heath Trays led to a linear, proportionate increase in fish harvest densities without sacrificing fish sizes at harvest.

3.2 *Relationships Between Yellow Perch Harvest Densities and Mean Individual Body Sizes*

We sought to further explore relationships between fish harvest densities and fish sizes at harvest to identify production thresholds for managers. Identifying statistical breakpoints in our dataset provides managers a working model of how different ranges of harvest densities can result in fish harvests of different body sizes. For example, managers who desire the largest numbers of fish at harvest should use adequately high fry stocking densities to produce smaller fish sizes with limited

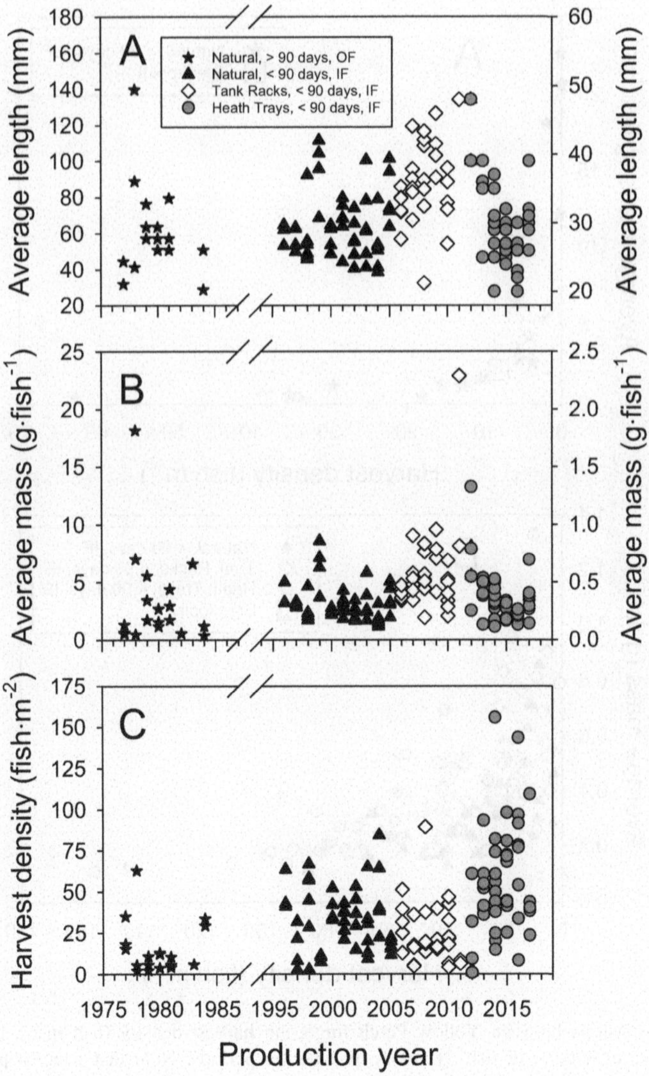

Fig. 3 Comparisons of Yellow Perch fingerling (**a**) total lengths (mm), (**b**) average individual mass (g), and (**c**) densities (fish m^{-2}) at pond harvest. Symbols represent values for individual ponds in each year. Note that fish lengths and masses are scaled differently before 1985 (left *y*-axis) as compared to after 1995 (right *y*-axis), because fish sizes were much larger during the earlier period. Production methods compared are natural spawning versus manual spawning (Tank Racks and Heath Trays), culture duration (<90 days and >90 days), and fertilization regimens (*OF* organic fertilization and *IF* inorganic fertilization)

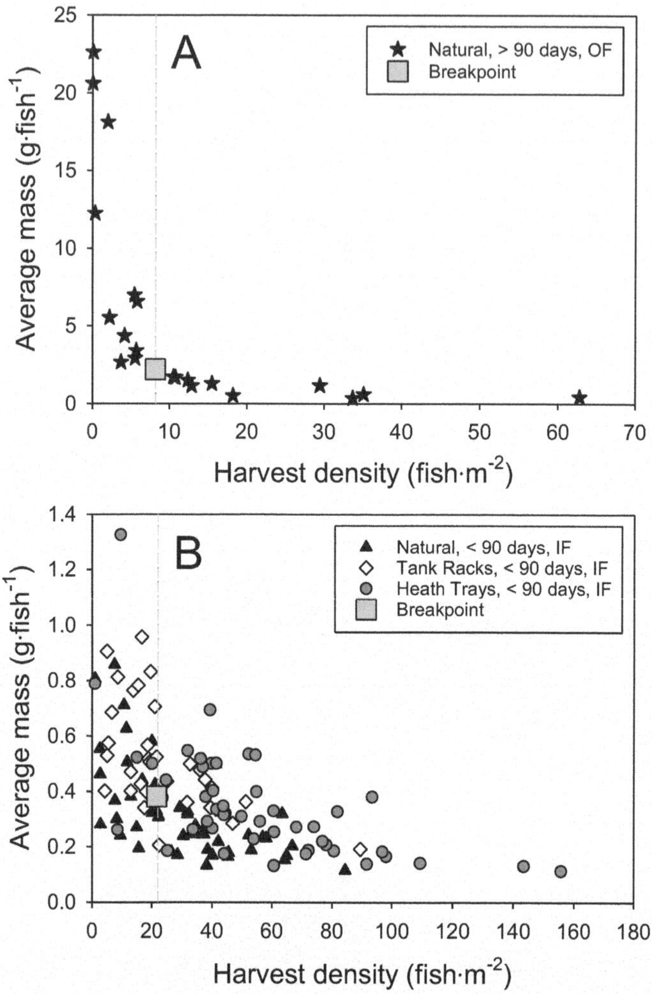

Fig. 4 Relationships between Yellow Perch fingerling harvest density (fish m^{-2}) and average individual size at harvest (g $fish^{-1}$) among production periods. Note that panel **a** presents the relationship for the early period of "natural spawning, >90 days in ponds, organic fertilization (OF)" separate from the other three periods (in panel **b**), because harvest densities were much lower and fish sizes at harvest were an order of magnitude larger during that early period. Each symbol represents results from an individual pond. Breakpoint (BP) symbols are shown to illustrate the two-dimensional (2D) breakpoints as determined using 2DKS tests. Note that panel B presents the 2D breakpoint for all three <90-day production periods combined and all using the inorganic fertilization (IF) regimen

individual growth imposed by intraspecific competition within ponds. We used 2DKS tests to identify significant breakpoints in these production metrics.

There were significant breakpoints between Yellow Perch harvest density (fish m^{-2}) and average mass of individual fish at harvest during all production periods

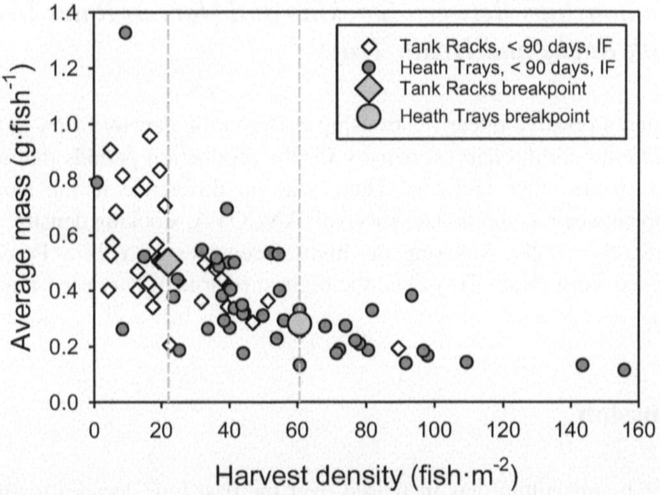

Fig. 5 Comparison of 2D breakpoint relationships between Yellow Perch harvest density and size for fish produced using Tank Racks versus Heath Trays. Each symbol represents results from an individual pond. Breakpoint (BP) symbols are shown to illustrate the two-dimensional (2D) breakpoints for each production method as determined using 2DKS tests (*IF* inorganic fertilization)

(Fig. 4a, b). During 1977–1986, ponds with harvest densities >8 fish·m^{-2} produced fish with individual mass <2.2 g (2DKS test; $D_{BKS} = 0.25$, $P < 0.001$; Fig 4a). Beginning in 1996, liquid inorganic N and P fertilization was adopted, and fish were harvested after <90 days in ponds. During the three production periods starting in 1996, ponds with harvest densities >22 fish·m^{-2} produced fish with individual mass <0.4 g ($D_{BKS} = 0.13$, $P < 0.001$; Fig. 4b). We also examined differences in these breakpoint relationships between the manual spawning methods of Tank Racks and Heath Trays (Fig. 5). During the Tank Racks period (2006–2011), ponds with harvest densities >22 fish·m^{-2} produced fish with individual mass <0.5 g ($D_{BKS} = 0.16$, $P = 0.001$). By comparison, during the Heath Trays period (2012–2017), ponds with harvest densities >61 fish·m^{-2} produced fish with mass < 0.3 g ($D_{BKS} = 0.16$, $P < 0.001$). Note that during the Heath Tray period, average individual fish mass at harvest was consistent all the way up to harvest density of nearly 160 fish m^{-2} still producing fish with mass of about 0.1 g (Fig. 5).

Managers can use our breakpoint results to achieve the desired individual size of harvested fingerlings and harvest densities by adjusting fry stocking densities. The analyses showed that individual mass at harvest declines more rapidly before the breakpoint, whereas individual mass declines relatively little after the breakpoint, allowing for many more somewhat smaller fish to be harvested.

3.3 Relationships Between Stocking and Harvest Rates Using Tank Racks and Heath Trays

We identified a positive linear relationship between the density of fry stocked into ponds and the fingerling harvest density for the production periods that used Tank Racks and Heath Trays (Fig. 6). There was no difference in the slope of the relationship between methods (i.e., survival; ANCOVA, stocking density × production method, $P = 0.28$). Stocking the highest densities of Yellow Perch fry into ponds hatched from Heath Trays had the highest returns of harvested fingerlings.

4 Discussion

Yellow Perch aquaculture has increased over the past four decades to supplement declines in wild populations and meet increased demands by anglers. As hatchery managers have been asked to keep up with the numbers of fingerlings requested by state agencies and grow-out facilities, new methods have developed over the years to help managers achieve these demands. We had the advantage of a 41-year, detailed, and unique longitudinal dataset of Yellow Perch production records from St. Marys SFH to analyze different production methods used over time and to help determine the best methods to produce Yellow Perch fingerlings to stock into reservoirs.

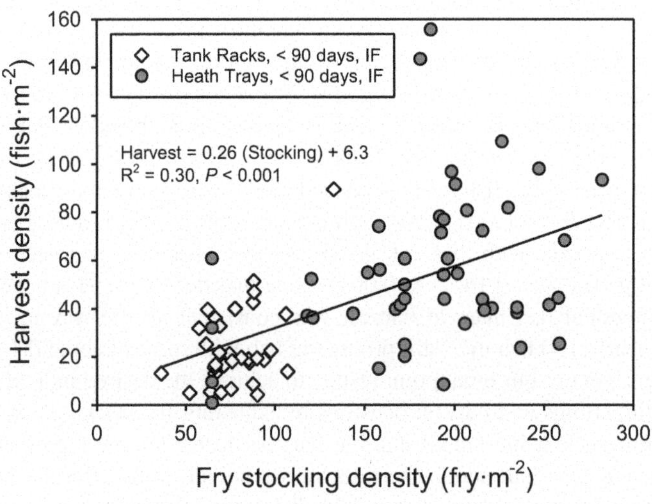

Fig. 6 Relationship between the number of Yellow Perch fry stocked into ponds and the number of fingerlings harvested from ponds using both the Tank Rack and Heath Tray production methods. Note the linear relationship between the numbers of fish stocked and harvested across the entire range of stocking densities. The slope of the best fit line is shown, with the slope representing overall survival for these methods of about 25% survival to harvest

Variation in Yellow Perch production among years resulted from differences in numbers of fish for stocking requested by the managers of the many state reservoirs, methods of fry production, pond fertilization techniques, and occasional excessively low pond temperatures occurring after fry were stocked in the ponds (Mort Pugh, St. Marys SFH manager, personal communication). Although our dataset does not have a typical statistical design and does not represent causal results from controlled experiments, we can draw conclusions from the comparisons between fry production methods, fertilization regimens, and culture duration in the ponds that may be used as guidelines to hatchery managers to increase fingerling production relative to their own facilities.

We observed a shift in fry production methods from natural spawning to a more controlled system of manual spawning, which decreased the variability in size of fingerlings at harvest, as all fry stocked into culture ponds occurred on the same day and with cohorts that hatched at the same time. Producing Yellow Perch eggs naturally is the least desirable method of egg production due to the variability of hatching, numbers produced versus harvested, and the possibility of cannibalism by early hatching fry on late hatching fry (Hart et al. 2006). The staff at St. Marys SFH used a modified method to manually spawn and incubate fertilized eggs. Instead of physically stripping eggs from females, St. Marys SFH staff allowed females to spawn in tanks after hCG injections, making this method less labor-intensive as eggs were fertilized in the tanks instead of dry-mixed in a bowl as Hart et al. (2006) suggest. This also decreased the risk of harming the eggs and female fish. Further, it was possible to calculate the survival of fingerlings in ponds by comparing the number of fingerlings produced with the number of fry stocked in the ponds. We found St. Marys SFH produced more fry when using Heath Trays as compared to Tank Racks. We also found that an increase in fry stocked using Heath Trays produced many more fingerlings, albeit at a smaller size, without affecting percent survival. Heath Trays require less physical labor than Tank Racks throughout the incubation period and are thus preferred by managers at St. Marys SFH. Thus, the system can be leveraged to increase fry density in ponds and still have stockable-sized fingerlings to stock into reservoirs or for feed training.

Inorganic pond fertilization regimens reduced variability in production while increasing survival and yield (Hartleb et al. 2012). Inorganic fertilization has worked well at St. Marys SFH despite the very high nutrient content of its reservoir water. Upon filling, inorganic N concentrations in the ponds could be as high as 1100 $\mu g\ N\ L^{-1}$, so only an appropriate amount of H_3PO_4 was added that week. N concentrations typically decreased to below the 600 $\mu g\ N\ L^{-1}$ target by 2 weeks later. The three Ohio warmwater hatcheries differ greatly in the nutrient content of their ponds upon first filling (St. Marys SFH >> Hebron SFH > Senecaville SFH) due to the variation in the fertility of their source water reservoirs. These results emphasize the importance of measuring the N and P content variation from pond to pond and week to week prior to calculating the appropriate amount of liquid N and P fertilizers to add. As percid hatcheries occur across the Great Lakes region, with differing source waters, we realize that there is no set concentration to apply to every hatchery; thus each hatchery must identify its own N and P regimens that are most beneficial for avoiding poor water quality conditions (e.g., low DO concentrations, blooms of cyanobacteria and filamentous algae) while promoting zooplankton production.

Hartleb et al. (2012) summarized effects of organic nutrients on pond water quality, including low DO, yet also discussed problems of accumulation of organic matter on the bottom of ponds. Organic matter may promote invertebrate habitat, but this has not been experimentally tested. Wu and Culver (1992) found that Yellow Perch diets shifted from zooplankton to invertebrates (such as chironomid larvae and pupae) when zooplankton densities were less than 10 ind L^{-1} and fish were at least 50 mm TL. However, all fingerlings were harvested <50 mm TL during the Tank Rack and Heath Tray periods at St. Marys SFH. Although organic matter may be beneficial in plastic-lined ponds to promote invertebrate habitat, we caution that adding large additions of organic matter to ponds causes severe water quality issues. For example, Filbrun et al. (2013b) found that commercial feed added to earthen catfish ponds did not enhance invertebrate abundance but caused hypoxia formation and overgrowth of nuisance filamentous algae.

After adopting the inorganic fertilization regimen and decreased culture duration (<90 days) (1996–2017), fish length was more consistent across all years and spawning methods. When fingerlings were reared in ponds for >90 days, zooplankton decreased, and Yellow Perch switched their diets to benthos, yet there is an insufficient supply of prey available in ponds to support growth (Briland et al. 2015). This can contribute to low survival and/or cannibalism. Long culture duration (>90 days) resulted in larger Yellow Perch fingerlings, yet lower densities at harvest. Shorter culture duration (<90 days) resulted in significantly higher harvest densities, though smaller fish.

Aquaculture production datasets generally reveal an inverse relationship between fish harvest density and individual fish size at harvest, reflecting the intensity of intraspecific competition in ponds. Less-intensive production methods and those associated with high mortality rates generally produce few, large fish at harvest. Reduced fish densities increase individual growth rates as competition for limited food resources in ponds is relaxed.

Revisiting our questions, first, if smaller fish are acceptable to managers, then using manual spawning in tanks, with incubation of eggs in Heath Trays, and an inorganic fertilization regimen for <90 days culture duration should produce the most fingerlings to stock into reservoirs or begin to feed train as food fish. Second, we have shown that Yellow Perch do respond to the same fertilization regimen used for other percids (Walleye and saugeye) and variability in production is likewise reduced.

We recognize all hatcheries differ in their source water, water quality and pond sizes, and production goals. Our comparisons may help hatchery managers identify appropriate production methods relative to their own site-specific goals. Accordingly, we have provided an overview of how spawning methods, gamete incubation methods, pond fertilization regimens, and fish residence time in the ponds impact production metrics.

Acknowledgments We thank the staff at the St. Marys SFH (ODNR-DOW) for access to historical Yellow Perch production records and clarifying their standard operating procedures and production methods used for producing Yellow Perch fry and fingerlings. This research would not have been possible without their cooperation.

References

Baldwin NA, Saalfeld RW, Dochoda MR, Buettner HJ, Eshenroder RL, O'Gorman R (2018) Commercial fish production in the Great Lakes 1867–2015. Available at: http://www.glfc.org/great-lakes-databases.php (September 2018)

Briland RD, Doyle CM, Culver DA (2015) Large-scale production of yellow perch, walleye, and hybrid walleye in ponds. In: Kestemont P, Dabrowski K, Summerfelt RC (eds) Biology and culture of percid fishes. Springer, New York, pp 469–498

Craig JF (1987) The biology of perch and related fish. Timber Press, Portland, OR, p 333

Craig JF (2000) Percid fishes: systematics, ecology, and exploitation. Blackwell Science, Inc, Malden, MA, p 352

Culver DA, Madon SP, Qin J (1993) Percid pond production techniques: timing, enrichment, and stocking manipulation. In: Anderson RO, Tave D (eds) Strategies and tactics for management of fertilized hatchery ponds. Food Products Press, New York, pp 9–31

Dabrowski K, Ciereszko RE, Ciereszko A, Toth GP, Christ SA, El-Saidy D, Ottobre JS (1996) Reproductive physiology of yellow perch (*Perca flavescens*): environmental and endocrinological cues. J Appl Ichthyol 12:139–148

Ellison DG, Franzin WG (1992) Overview of the symposium of walleye stocks and stockings. N Am J Fish Manag 12:271–275

Espey JL (2003) Acute toxicity of ammonia and nitrite to yellow perch, *Perca flavescens*. MS thesis, Zoology, North Carolina State University, Raleigh, NC. http://www.lib.ncsu.edu/resolver/1840.16/2910

Farmer T, Marschall EA, Dabrowski K, Ludsin SA (2015) Short winters threaten temperate fish populations. Nat Commun 6:1–10

Filbrun JE, Conroy JD, Culver DA (2013a) Understanding seasonal phosphorus dynamics to guide effective management of shallow, hypereutrophic Grand Lake St. Marys, Ohio. Lake Reserv Manag 29:165–178

Filbrun JE, Reynolds CA, Culver DA (2013b) Effects of feeding rate on habitat quality in fish rearing ponds. J World Aquacult Soc 44:198–209

Fox MG (1989) Effect of prey density and size on growth and survival of juvenile walleye (*Stizostedion vitreum vitreum*). Can J Fish Aquat Sci 46:1323–1328

Garvey JE, Marschall EA, Wright RA (1998) From star charts to stoneflies: detecting relationships in continuous bivariate data. Ecology 79:442–447

Hart SD, Garling DL, Malison JA (eds) (2006) Yellow perch (*Perca flavescens*) culture guide. North Central Regional Aquaculture Center Culture Series #103, Iowa State University, Ames, IA

Hartleb CF, Johnson JA, Held JA (2012) Walleye and yellow perch pond fertilization. In: Mischke CC (ed) Aquaculture pond fertilization: impacts of nutrient input on production. Wiley, Ames, IA. https://doi.org/10.1002/9781118329443.ch11

Helal H, Culver DA (1991) N:P ratio and plankton production in fish ponds. Verh Int Verh Theor Angew Limnol 24:1508–1511

Hinshaw JM (2006) Species profile: yellow perch, *Perca flavescens*. Southern regional aquaculture center publication no. 7204. Mississippi State University, Starkville, MS, p 6

Hokanson KEF (1977) Temperature requirements of some percids and adaptations to the seasonal temperature cycle. J Fish Res Board Can 34:1524–1550

Jacob A, Culver DA (2010) Experimental evaluation of the impacts of reduced inorganic phosphorus fertilization rates on juvenile saugeye production. Aquaculture 304:22–33
Lesser W, Vilstrup R (1979) The supply and demand for yellow perch 1915–1990. Research Bulletin R3006, College of Agriculture and Life Sciences, University of Wisconsin, Madison, WI, pp 1–13
Malison JA (2000) A white paper on the status and needs of yellow perch aquaculture in the North Central Region. East Lansing, MI. https://www.ncrac.org/files/project/files/whitepaperyellowperch.pdf
Malison JA (2003) A white paper on the status and needs of yellow perch aquaculture in the North Central Region. North Central Region Aquaculture Center, Michigan State University, East Lansing, MI, pp 1–18
Malison JA, Held JA, Kaatz SE (2011) Sex determination of yellow perch by external morphology. N Am J Aqua 73:285–287
Mitzner L (2002) Effectiveness of walleye fry and fingerling stocking in Rathbun Lake, Iowa 1990–1999. N Am J Fish Manag 22:1007–1013
Nelson WR, Walburg CH (1977) Population dynamics of yellow perch (*Perca flavescens*), sauger (*Stizostedion canadense*), and walleye (*S. vitreum vitreum*) in four main stem Missouri River reservoirs. J Fish Res Board Can 34:1748–1763
Page LM, Burr BM (1991) A field guide to freshwater fishes: North America north of Mexico. The Peterson field guide series, volume 42. Houghton Mifflin Company, Boston, MA, p 432
Riepe JR (1998) Yellow perch markets in the North Central Region: results of a 1996/97 survey. Department of Agricultural Economics, Purdue University, Bulletin 756, West Lafayette, Indiana, p 48
Tew K, Conroy JD, Culver DA (2006) Effect of lowered phosphorus fertilization rates on pond production of percid fingerlings. Aqua 255:436–446
Thorpe JE (1977) Morphology, physiology, behavior, and ecology of *Perca fluviatilis* and *Perca flavescens* Mitchill. J Fish Res Board Can 34:1504–1514
Wu L, Culver DA (1992) Ontogenetic diet shift in Lake Erie Age-0 yellow perch (*Perca flavescens*): a size-related response to zooplankton density. Can J Fish Aquat Sci 49:1932–1937

Evaluation of a Statewide Yellow Perch Bag Limit for Michigan

David F. Clapp, Andrew S. Briggs, Randall M. Claramunt, David G. Fielder, and Troy G. Zorn

Abstract The Michigan statewide Yellow Perch bag limit has received increasing attention in recent years as a result of law enforcement concerns, biological concerns, interjurisdictional management issues, and ongoing comments from agency staff and the public. As a result, Michigan Department of Natural Resources Fisheries Division undertook an evaluation of a proposed new statewide daily bag limit for Yellow Perch. The objectives for this review were (1) to evaluate the potential effects—social as well as biological—of a specific proposed regulation change (reduction to a 25-fish per day bag limit for Yellow Perch) and (2) to provide a blueprint for future regulation evaluations in Michigan. Based on social survey results and feedback from advisory committees, a majority of anglers preferred a reduced statewide bag limit. Analyses of creel survey data from inland and Great Lakes fisheries indicated that a reduced statewide bag limit would help achieve a balance between conservation and opportunity. Based on data from the 2017 fishery,

D. F. Clapp (✉)
Michigan Department of Natural Resources, Charlevoix Fisheries Research Station, Charlevoix, MI, USA
e-mail: clappd@michigan.gov

A. S. Briggs
Michigan Department of Natural Resources, Lake St. Clair Fisheries Research Station, Harrison Twp, MI, USA
e-mail: BriggsA4@michigan.gov

R. M. Claramunt
Michigan Department of Natural Resources, Oden State Fish Hatchery, Alanson, MI, USA
e-mail: claramuntr@michigan.gov

D. G. Fielder
Michigan Department of Natural Resources, Alpena Fisheries Research Station, Alpena, MI, USA
e-mail: fielderd@michigan.gov

T. G. Zorn
Michigan Department of Natural Resources, Marquette Fisheries Research Station, Marquette, MI, USA
e-mail: zornt@michigan.gov

J. C. Bruner, R. L. DeBruyne (eds.), *Yellow Perch, Walleye, and Sauger: Aspects of Ecology, Management, and Culture*, Fish & Fisheries Series 41,
https://doi.org/10.1007/978-3-030-80678-1_3

15% of harvested Yellow Perch would have been protected under a 25-fish bag limit, and only 1% of angling parties would have been affected (i.e., experienced a reduction in harvest). Previous analyses and reviews conducted in Michigan, in other midwestern states and provinces, and documented in literature also supported the proposal for reducing the statewide Yellow Perch bag limit. Based on this review, a statewide 25-fish Yellow Perch bag limit was recommended for its effectiveness in terms of optimizing angler satisfaction across a range of fisheries and balancing conservation with opportunity for Michigan resource users. Combining public opinion surveys with fishery-dependent and fishery-independent analyses provided valuable insights into the likely appropriateness of a proposed statewide regulation in meeting management objectives.

Keywords *Perca* · Fisheries · Regulation · Management

1 Introduction

Regulations are possibly the most visible and controversial tools used by resource agencies to manage fish populations (see AFS n.d.; Goeman et al. 1995; Mather et al. 1995). They are typically enacted to reduce mortality, leading to improvements in the abundance, size structure, or reproductive potential of a population. Regulations are ideally determined by considering appropriate factors and tailoring a specific regulation to ensure a sustainable fishery or facilitate recovery of a fish population that has declined significantly. However, due to limitations in knowledge of specific fisheries, lack of staffing levels needed to effectively implement and enforce system-specific regulations, or as a way of streamlining and simplifying regulations for the angling public, resource agencies often implement more generic, statewide regulations to provide some minimum level of protection across a wide range of fisheries. Successful application of a statewide regulation is difficult in Michigan due to the range of fish populations affected: Michigan biologists manage 4 Great Lakes, 11,000 inland lakes, and more than 70,000 miles of rivers and streams. A statewide regulation has the potential to influence the fishing behavior of over one million license holders. As a result, changes to these regulations require careful consideration.

The statewide Yellow Perch (*Perca flavescens)* bag limit has received increasing attention in recent years as a result of law enforcement concerns about enforcing special regulation exceptions on connected waters (e.g., tributaries, drowned river mouths), biological concerns in certain systems, interjurisdictional management issues, and ongoing comments from agency staff and the public. Michigan Department of Natural Resources (MDNR) Fisheries Division recently undertook an evaluation of a proposed new statewide daily bag limit for Yellow Perch (25 fish per day), providing an opportunity to describe the process involved in evaluating and recommending regulation changes, review potential social and biological effects of the proposed regulation change, and develop a blueprint for future regulation evaluations in Michigan for all species.

2 Why Regulations (Goals and Expected Effects)?

Regulations should be considered when angling harvest or other factors prevent the attainment of specific management goals (AFS n.d.; Goeman et al. 1995). They are typically enacted to manage social issues, prevent overfishing, or manipulate aquatic communities (Radomski et al. 2001). Social issues can include equitable distribution (harvest) of fish, distribution of anglers, managing expectations, ease of interpretation for anglers, law enforcement, or providing a quality or challenging experience (Noble and Jones 1993). For example, biologists would often like to enact very specific regulations for each waterbody, but the confusion or burden this would cause anglers and law enforcement officers is often not worth the marginal biological benefit that would be gained. Angler compliance is critical to the success of regulations (Gigliotti and Taylor 1990; Johnson and Martinez 1995), and wherever possible, the goal should be to make regulations consistent and easy to understand. In the case of bag limits, they often serve as targets for anglers or are used as a measure of angler success and satisfaction. By setting bag limits, biologists are in some sense setting expectations for anglers in that fishery (Radomski et al. 2001).

As Radomski et al. (2001) point out, overfishing is often difficult to define and in recreational fisheries is not always analogous to more traditional definitions applied to commercial fisheries where recruitment, growth, quality, economic, and community overfishing are factors that are considered (Colby et al. 1994; Schneider et al. 2007). Examples of recruitment and growth overfishing in recreational fisheries are rare or undocumented, but quality overfishing may occur (but, again, is not well documented; Radomski et al. 2001). This is especially the case for Yellow Perch (Schneider et al. 2007).

Yellow Perch are not typically thought of as a species that plays a significant role in directly manipulating aquatic communities, although there is some evidence that they serve as significant predators on small Bluegill (*Lepomis macrochirus*) (Schneider et al. 2007). More often, however, the converse is true, and fish communities are often manipulated by managers in hopes of improving Yellow Perch populations. The most common example of this type of manipulation is predator management in systems with sub-optimal perch populations, for example, stocking Walleye (*Stizostedion vitreum*) to control stunted Yellow Perch (Schneider 1983). A desire to decrease predation on Yellow Perch in Saginaw Bay (Michigan; Lake Huron) was part of the rationale behind lowering the Saginaw Bay Walleye length limit from 15 inches to 13 inches and raising the Walleye bag limit from 5 to 8 fish in 2015 (Fielder and Baker 2019). To further address depressed Yellow Perch populations, the Yellow Perch daily bag limit was reduced to 25 fish in conjunction with this change in Walleye regulations (Fielder and Baker 2019).

3 Yellow Perch Regulations (Local, Regional, National)

Creel bag limits tend to be lower for large predators and higher for smaller-bodied insectivorous or planktivorous fish (Radomski et al. 2001). Consistent with this general principle, Michigan Yellow Perch regulations (bag and size limits) have tended to be liberal. From Schneider et al. (2007):

> "...Daily possession limits were changed from unlimited to 25 (1903, in combination with many other species), retained at 25 (1915, in combination with panfish), increased to no limit (1962), then changed (1979) to the current limit of 50 per day. The MSL [minimum size limit] was increased from none to 5 inches (1903), to 6 inches (1915), to 7 inches (1929), and then back to no MSL (1949). There have been no statewide MSLs since 1949."

Michigan's approach to Yellow Perch statewide regulation has generally been similar to most other jurisdictions. A review of freshwater fishing regulations in 54 US states and Canadian provinces (Radomski et al. 2001) documented a median bag limit of 25 for Yellow Perch. At the time of the review described in this chapter, Michigan's statewide bag limit (50 fish) was somewhat more liberal than the median documented in Radomski et al. (2001).

Michigan does have some waterbody-specific exceptions to the current statewide regulation, and there are differing approaches to management across some systems—in particular for Great Lakes populations. Yellow Perch can and do move across state borders (see Glover et al. 2008), so in a sense the fishery is shared among anglers and commercial fishers in the states/provinces surrounding each of the Great Lakes. Great Lakes Yellow Perch regulations are at minimum discussed and coordinated among all jurisdictions on a lake (e.g., Makauskas and Clapp 2018). In some cases (e.g., Lake Erie), specific regulations arise directly from more intensive evaluations of Yellow Perch population dynamics, including modeling and quota setting (LE YPTG 2011; Belore et al. 2018).

4 Michigan's Regulation Implementation Process

As is the case for resource management decisions made by most state agencies, changes to Michigan fishing regulations involve a deliberative process. While the final step in enacting most new fishing regulations is issuance of a Fisheries Order by the MDNR Director, development and approval of these orders include multiple steps:

1. In March of each year, Fisheries Division initiates recommendations for new Fisheries Orders, or for renewal, rescissions, or amendments to old orders. Recommendations for new or changed regulations can originate from the general public, user groups, policy makers, legislators, or Fisheries Division staff. Recommendations, with supporting information, are prepared on a "Fisheries Orders Change" form.

2. Following (or prior to) recommendations, proposed regulation changes are discussed internally (among agency staff), with public advisory groups, and with the public at large. Public input can be gathered informally, through public meetings and hearings, or through online survey tools.
3. Once input is gathered from all interested parties, formal Fisheries Order memos are prepared for submission to the state Natural Resources Commission (NRC; www.michigan.gov/nrc). Fisheries Order memos are signed by all Resource Management Bureau chiefs (Fisheries, Wildlife, Forest Resources, Law Enforcement, Parks and Recreation) prior to submission to the NRC.
4. Proposed Fisheries Orders are discussed and revised following presentation to the NRC. Typically, this process will occur over the course of several monthly NRC meetings.
5. Following final revisions and approval, Fisheries Orders are signed by the MDNR Director. Signature is given under authority of 1994 PA 451 (Natural Resources and Environmental Protection Act); Part 411 and Part 487. Orders signed under Part 411 are in effect for a maximum of 5 years and must be reviewed at that time or any time prior to the 5-year limitation. These orders may go into effect immediately or, more commonly, at the beginning of the next fishing year (April 1). Fisheries Orders signed under the authority of Part 487 go into effect the April 1 following the date of the Director's signature. While there is no limitation on the duration of these orders, typically they follow the same 5-year cycle as the Fisheries Orders signed under the authority of Part 411.
6. Following final approval and signature, Fisheries Division staff prepare regulation changes for inclusion in the next edition of the Michigan Fishing Guide (i.e., public regulation guidebook). Emergency Fisheries Orders are published in newspapers of affected counties at least 21 days but not more than 60 days prior to the order taking effect.

5 Objectives

Regulations are not always proposed, deliberated, adopted, or evaluated in a consistent and unbiased manner (Graff 1977; Johnson and Martinez 1995). The objectives for the current review were (1) to evaluate the potential effects—social as well as biological—of a specific proposed regulation change (reduction to a 25-fish per day bag limit for Yellow Perch) and (2) to provide a blueprint for future regulation evaluations in Michigan. Management objectives for the proposed regulation change were to address angler concerns (the MDNR received significant comment over several years that Yellow Perch regulations were too liberal), to address law enforcement concerns (specifically in areas where regulations differed in adjacent bodies of water), to better distribute the Yellow Perch resource among anglers, and to build a stable, high-quality Yellow Perch fishery statewide. While it is difficult (if not impossible) for a statewide regulation to meet all of these objectives for all waterbodies, to the extent possible the review addressed these multiple desired outcomes for a Yellow Perch regulation—while also recognizing the need for exceptions to the statewide regulation for certain fisheries where managers are dealing with specific overriding circumstances.

6 Methods

In evaluating the proposed regulation change, results were analyzed from a recent public opinion survey, as well as from previous public and biologist surveys. Public input was also gathered informally, through public meetings and advisory workshops. Harvest data from inland and Great Lakes Yellow Perch fisheries were examined, results from previous fishery independent surveys and modeling efforts were reviewed, and comparison were made to Yellow Perch recreational harvest management in other Great Lakes states and Ontario. A brief description of each of these data collection and analysis efforts is provided here, along with a description of the criteria used in evaluating that information, relative to management objectives. Additional, more detailed, information on any of these data sets or analyses is available from the authors.

6.1 2018 Social Survey

The primary objective of the 2018 social survey was to gather social input on a proposed reduction in the statewide Yellow Perch daily bag limit for Michigan. The survey (copy available from authors, on request) consisted of 23 questions and one space for comments. Of the 23 questions, only 2 asked about the daily bag limit. The remaining questions were designed to understand the angler, with 5 questions about their demographics (i.e., age, gender, education, income level, and zip code), 5 questions on their motivation and satisfaction (e.g., reasons for fishing and satisfaction, desired management goals, and participation and/or membership in fishing organizations), and 11 questions asking about their experience in angling (e.g., experience fishing in Michigan or other states, species targeted or caught most often, type of waterbody typically fished, and the extent to which they fished for Yellow Perch). Because of the diversity in Yellow Perch populations (e.g., small ponds up to the Great Lakes) and in the fishery (e.g., beginning young anglers up to charter fisheries), the survey was designed to fully understand the angler and their relationship to fishing. The survey was also designed to evaluate differences in demographics between anglers who responded to the electronic survey and anglers represented in the 2017 license holder database (includes all 850,000 anglers that bought a fishing license; see below).

The survey was designed and published using SoGoSurvey software (https://www.sogosurvey.com/) and implemented by using the MDNR angler database that included 376,204 e-mail addresses from anglers that had purchased a MDNR fishing license during the 2014–2017 fishing seasons. The survey was made available via the Internet, and anglers were notified by e-mail of its availability. The survey was distributed to the 376,204 anglers on April 17, 2018, and responses were closed on May 1, 2018. The link to the survey was advertised by some angler groups on social media and on Internet bulletin boards, so some respondents may have been from outside the initial invited target survey audience. Responses were downloaded from SoGoSurvey for analyses, and spatial distributions of respondents were plotted via a map viewer in ArcGIS Online (Environmental Systems Research Institute; https://www.esri.com). Descriptions of spatial distributions in the Results section of this chapter are based on simple visual inspection and interpretation of the plots for broad-scale patterns.

Additional public social surveys concerning Yellow Perch regulations had been conducted in 2013 and 2017. Because these surveys were more local or regional in coverage, the results are not presented in detail in this chapter. However, the surveys and results are available from the authors, on request. In addition, a survey of MDNR Fisheries Division biologists was conducted in 2010; details concerning that survey are also available. The primary criterion in evaluating responses to all these surveys was whether there was significant acceptance or rejection of a specific proposed regulation change. Public opinion was not considered to the exclusion of other criteria (see below), but public acceptance of a regulation is important, especially with respect to later compliance with and enforceability of the regulation (Johnson and Martinez 1995).

6.2 *Other Public Input*

In Michigan, the MDNR Fisheries Division has several formal venues for sharing and discussing proposed management changes with the public. The Michigan Warmwater Resources Steering Committee (WRSC) is charged with providing input, advice, and recommendations to MDNR on issues related to Michigan's warmwater fisheries resources, as well as strategies and goals for managing those resources (including regulations). They are the primary advisory group to Fisheries Division concerning statewide warmwater regulation changes (including those targeting Yellow Perch). In addition to the WRSC, Fisheries Division staff meet regularly with Great Lakes basin-specific citizen advisory groups to discuss regional management issues. The proposed Yellow Perch statewide bag limit change was brought to the WRSC and to the Lake Erie/Lake St. Clair, Lake Huron, Lake Michigan, and Lake Superior Citizens Fishery Advisory Committees (CFAC) for discussion and input prior to formal presentation of the change to the Natural Resource Commission and Director. As with responses to social surveys described above, the primary criterion in evaluating non-survey comments was whether significant public input was received in favor of or opposed to the proposed regulation change. These meetings also provided a forum for exchange of information on other aspects of the proposed regulation change, including angler recruitment and retention, potential ecosystem-level effects, and future evaluation of success.

6.3 *Analysis of Fishery and Population Survey Data*

6.3.1 Impacts of Varying Daily Bag Limits on the Recreational Fishery

Michigan DNR conducts creel surveys on some Great Lakes and inland waterbodies as part of the Statewide Angler Survey Program, with specific annual coverage based on management need and available personnel resources. Great Lakes and inland surveys employ somewhat different methods (see Lockwood et al. 1999; Su and Clapp 2013), but both survey types result in a database of interviews of angling parties that include the number of anglers in the party and the number of fish harvested by species. Impacts of a reduced Yellow Perch daily bag limit on the recreational fishing experience were assessed by deriving the proportion of Yellow Perch harvest under a 50-fish daily limit that would have been foregone or "saved" had a lower bag limit been in place. In addition, the proportion of angler parties that would have been affected by a reduced bag limit (i.e., had some portion of their harvest curtailed) was calculated. Yellow Perch harvest statistics were derived by plotting the frequency harvested by collective parties under the 50-fish per individual bag limit. For example, a two-person fishing party would have a collective party limit of 100 Yellow Perch for that interview. Interview data (completed trip) was analyzed on a party basis because that is how interviews are conducted by creel clerks. Analysis was stratified first by all fishing parties interviewed and then by including only those angling parties that reported they were specifically targeting Yellow Perch; graphical results were limited to analyzing parties targeting Yellow Perch. The analysis of Yellow Perch harvest by party was performed for parties of one to five anglers. Rarely were there more than five anglers in a single fishing party. For each hypothetical limit of 0 to 50, the number of parties that had harvest exceeding that number was totaled to exhibit the proportion of parties that would have been affected by the lower hypothetical limit, where 50 is the level at which no parties are affected (i.e., existing bag limit) and zero is the level at which 100% of parties are affected (i.e., no harvest permitted). This analysis offered insights into inflection points on regulation impacts beyond just the proposed 25-fish daily limit, but results are summarized with special emphasis on the 25 Yellow Perch bag limit value. Analysis was limited to recreational fishing parties and did not reflect the charter boat fishery.

Analyses of creel survey interviews were not specific to any 1 year, but included all collections (1987–2017) for which data needed to complete the analysis were available, as the intent was to assess how lower daily bag limits could affect anglers across a range of fisheries. Recognizing they would vary by fishery type, results were organized by those from the Great Lakes and those from inland waterbodies

Table 1 Locations of recreational creel survey interviews used for analysis of effects of bag limit reductions by fishing party

Category	Lake or waterbody	County	Year(s)	Months	Area(ha)
Great Lakes	Saginaw Bay	Various	1987–1994	Jan–Dec	296,036
Great Lakes	Saginaw Bay	Various	1995–2004	Jan–Dec	296,036
Great Lakes	Saginaw Bay	Various	2005–2013	Jan–Dec	296,036
Great Lakes/channel	Lake St. Clair	Various	2016	Apr–Oct	111,370
Great Lakes	Little Bay de Noc	Delta	2017	Apr–Oct	12,141
Great Lakes	Les Cheneaux Islands	Mackinac	2017	Jan–Dec	11,650
Great Lakes/channel	St. Marys River	Chippewa	2017	May–Oct	75,038
Great Lakes	S. Lake Michigan	Various	2017	Apr–Oct	~580,000
Great Lakes	MI waters of Lake Erie	Various	2006–2016	Apr–Oct	approx. 160,000
Small inland	Thumb	Charlevoix	2007	Apr–Sep	484
Small inland	Cass & Union	Oakland	2016	Mar–Oct	708
Small inland	Corey & Pleasant	St. Joseph	2008	Apr–Oct	750
Small inland	Deer	Marquette	2016	May–Oct	950
Small inland	Croton Pond	Newaygo	2007	Apr–Oct	1209
Medium inland	East & West Gun	Barry	2014	Jan–May	2680
Medium inland	Brevort	Mackinac	2014	Jun–Sep	4233
Medium inland	Glen	Leelanau	2009	Jan–Mar	4871
Medium inland	Hamlin	Mason	2009	Jan–Mar	4990
Medium inland	Grand	Presque Isle	2004	May–Oct	5660
Large inland	Hubbard	Alcona	2007	Jan–Oct	8850
Large inland	Higgins	Roscommon	2001	Apr–Sep	9900
Statewide	Multiple[a]	Multiple	2017	Apr–Oct	

[a]Includes data from all waterbodies surveyed; Great Lakes and inland combined

(Table 1). Analyses of Great Lakes Yellow Perch fisheries included years of high (i.e., some years in Saginaw Bay), moderate, and low abundance using creel survey estimates of recreational harvest to categorize relative abundance. Analyses did include Lake Erie (for comparative purposes), even though a regulation exception is in place on this system due to the intensive, multiagency management conducted there (see above, Introduction). Analysis of southern Lake Michigan reflected data from a fishery with a 35-fish daily limit. Twelve inland lakes were assessed and organized by area in three categories: small (<2000 acres), medium (2000–6000 acres), and large (>6000 acres). Most analyses were limited to open water months (typically April–October), but winter ice fisheries were included when those data were available. Winter ice fishery data were typically available for Great Lakes fisheries such as those in Saginaw Bay and the Les Cheneaux Islands (Table 1), but less available for inland fisheries. As a result the analysis presented is somewhat conservative, in that the degree of protection from harvest in a (presumably) more intensively exploited winter ice fishery would be at least as great (if not greater) as the level of protection determined by implementing a more restrictive regulation for less exploited open-water fisheries.

Some have noted daily bag limits can serve as the basis for anglers to form expectations and determine the fishing experience satisfaction. Cook et al. (2001) suggest that the optimal daily bag limit for panfish be set where 10% of angling parties achieve that catch over time. To explore this principle of daily bag setting, the 10% bag limit value was also identified for each of the types of fisheries and waterbodies examined. For all recreational fishery data analyses, the evaluation assessed how the present (or recent past) fishery would have been affected by the proposed regulation change: Are significant numbers of anglers affected, and Would the regulation significantly influence harvest of Yellow Perch?

6.3.2 Inland Lake Fishing Effort vs. Yellow Perch Population Characteristics

Using data from MDNR's Statewide Angler Survey Program, Fish Collection System (FCS) database (statewide fishery-independent survey data), and published studies, relationships were graphed between fishery-independent indices of abundance of harvestable-sized Yellow Perch and how heavily inland lakes in Michigan were fished. The focus of these analyses was a statewide set of lakes previously studied by Schneider et al. (2007) and 22 lakes recently surveyed under MDNR's Large Lakes Program (Hanchin 2017). The goal was to ascertain any potential individual or population-level effects of the proposed regulation change for Yellow Perch.

Yellow Perch abundance was indexed by fishery-independent survey catch per unit effort (CPE) of Yellow Perch and fishery-independent survey CPE of "legal-sized" Yellow Perch (7 inches or more in length) from surveys in the FCS or Yellow Perch harvest rate from creel surveys. Limited data were typically available in the FCS database, but when more than one survey occurred for a lake, catch data were used from fish community surveys whose purpose was identified as "Status and Trends" or "General Surveys" (see Wehrly et al. 2015 for additional details). The preference for data summaries in decreasing order (reflecting their coverage and likelihood of providing good information on abundance of Yellow Perch) was all gear types combined; netting-based summaries; and, lastly, electrofishing-based summaries. Fishing intensity was quantified as angler effort per acre and Yellow Perch fishery rank (1 = heavily fished to 4 = rarely fished) from Schneider et al. (2007). The number of lakes used varied by comparison since the availability of fishery and survey data varied among lakes. A limitation of these data was that creel surveys and lake fish community (fishery-independent) surveys rarely were conducted in the same year.

Survey and creel data were also assembled for 22 lakes surveyed via MDNR Fisheries Division's Large Lakes Survey Program (Hanchin 2017), analyzing relationships by lake size, angler effort, angler harvest, and size structure of Yellow Perch caught in the fishery-independent survey. The lakes were notably larger in this set, with a median surface area of 7360 acres, compared to 657 acres for lakes in the Schneider et al. (2007) set. Angling effort was often more than 20 angler hours per acre in the Schneider et al. (2007) set of smaller lakes, but usually less than this value in the large lakes (median angler hours per acre was 21 and 8 for the sets of smaller and large lakes).

6.4 Other Information Sources

Several additional historical and ongoing studies of Yellow Perch management in Michigan and other Great Lakes states were also summarized, with the objective of examining whether the proposed regulation change was consistent with successful management actions in other jurisdictions, as well as to gather additional insights into potential effects of the proposed regulation change. Some of these additional information sources were specific evaluations of recreational regulations, whereas others (e.g., catch-at-age models) involved development of tools that biologists could use to better understand Yellow Perch population dynamics. Employment of these tools may have in some cases led to recommendations for changing recreational fishing regulations but could also have led to recommendations for other management actions or recognition that more restrictive management was not needed.

- ***Lake Erie Yellow Perch Task Group process:*** Since 1980, the Lake Erie Yellow Perch Task Group (LE YPTG) has served Lake Erie managers (Lake Erie Committee [LEC]) by conducting assessments and analyses to describe the status of Yellow Perch stocks in Lake Erie and to provide information and advice to the LEC for development of recommended allowable harvests (RAH) of Yellow Perch from the lake. This process is documented in the LE YPTG terms of reference (LE YPTG 2011) and in annual reports to the LEC (see Belore et al. 2018).

- ***Lake Michigan statistical catch-at-age and decision analysis projects:*** Similar to efforts to address Yellow Perch management in Lake Erie, Lake Michigan Yellow Perch researchers and managers have collaborated since the mid-1990s to improve understanding and management of Yellow Perch populations in Lake Michigan. These efforts were documented in Wilberg et al. (2005), Irwin et al. (2008, 2011), and Lake Michigan Yellow Perch Task Group/Inshore Working Group annual reports (see Makauskas and Clapp 2018).
- ***Saginaw Bay statistical catch-at-age model:*** The Saginaw Bay Yellow Perch population is subject to both recreational and state-licensed commercial fisheries. Historically the collective annual yields have exceeded 450,000 kg but have declined in recent decades due to effects of invasive species and food web changes (Fielder et al. 2014; Baldwin et al. 2009). In an effort to supply fishery managers and decision makers with information about recruitment, mortality rates, and other metrics of population dynamics, the MDNR initiated the development of a statistical-catch-at-age (SCAA) model, similar to the one developed for Lake Michigan by Wilberg et al. (2005).
- ***Previous Michigan DNR regulation analyses:*** There is a long history of research and analysis of fishing regulations in Michigan, including Yellow Perch. Historical (Schneider and Lockwood 1979) and contemporary (Lake St. Clair, Saginaw Bay) Yellow Perch regulation analyses were reviewed to gain additional insights into the potential advantages and disadvantages to implementing a statewide Yellow Perch bag limit.
- ***Other state resource agency processes:*** In recent years, several midwestern states and provinces have undertaken evaluations of Yellow Perch and/or panfish regulations. In completing the review, current information was compiled on Yellow Perch regulations in other jurisdictions. Similar regulation reviews or studies conducted in Minnesota (Cook et al. 2001), Ontario (McShane and Bowman 2003), South Dakota (Isermann et al. 2007), and Wisconsin (Mosel et al. 2015; Rypel 2015; Rypel et al. 2016) were also evaluated.

7 Results

7.1 2018 Social Survey

A total of 18,743 survey responses were submitted, which was 5.0% of the number of anglers initially notified via email. Based on the survey responses, there was a higher proportion of anglers from the 55 and older age classes responding to the survey compared to the age distribution of anglers in the 2017 license database (Fig. 1). Approximately 27.7% of the survey responders were between ages 55 and 64, but only 19.0% of the 2017 anglers were ages 55 to 64. In contrast, only 2.6% of the survey responses compared to 11.8% of the 2017 anglers were between the ages

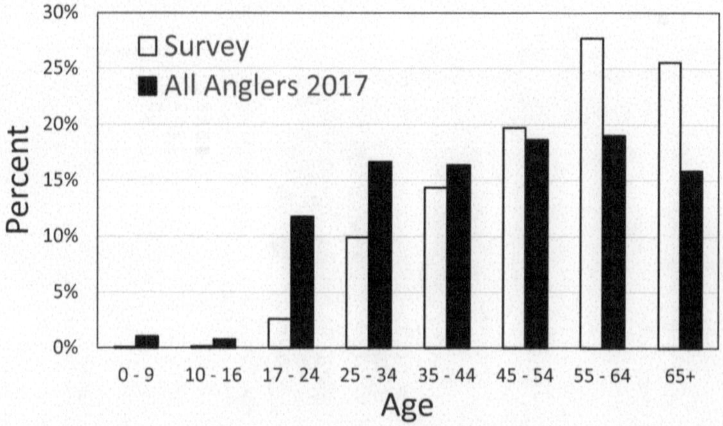

Fig. 1 Age distribution of social survey respondents and 2017 anglers

of 17 and 24. In addition to the bias toward responses from older age groups, female anglers were substantially underrepresented in the survey, as approximately 21% of the 2017 anglers were female compared to only 6% of the respondents in the electronic survey.

While there did appear to be some age and gender bias in survey responses, surveyed anglers were spatially well distributed (Fig. 2), and there wasn't strong evidence of a spatial bias in the responses from anglers expressing a preference for either the existing regulation (50-fish daily bag limit Fig. 2b) or for the proposed regulation (25 fish; Fig. 2c). Based on visual inspection of response distributions, it did appear that a slightly higher percentage of respondents in the western Upper Peninsula preferred a reduced daily bag limit (Fig. 2c); this is consistent with informal public comments received from the Upper Peninsula Sportsmen's Alliance and Lake Superior CFAC (see Sect. 7.2). For the current analysis, responses were combined spatially to evaluate overall angler preference for Yellow Perch regulations.

Overall, the survey respondents favored a reduced limit, with approximately 55% of the anglers preferring a bag limit of 25 and 9% preferring a limit of 10 perch per day (Fig. 3). In combination, 63.8% of the anglers preferred a reduced limit of 25 perch or fewer per day. In contrast, approximately 27% of the anglers preferred the 50-fish limit, while the remaining 9% were not sure or did not respond. These results are consistent with data from previous social surveys (2013, 2017), indicating a high level of acceptance for a 25-Yellow Perch bag limit.

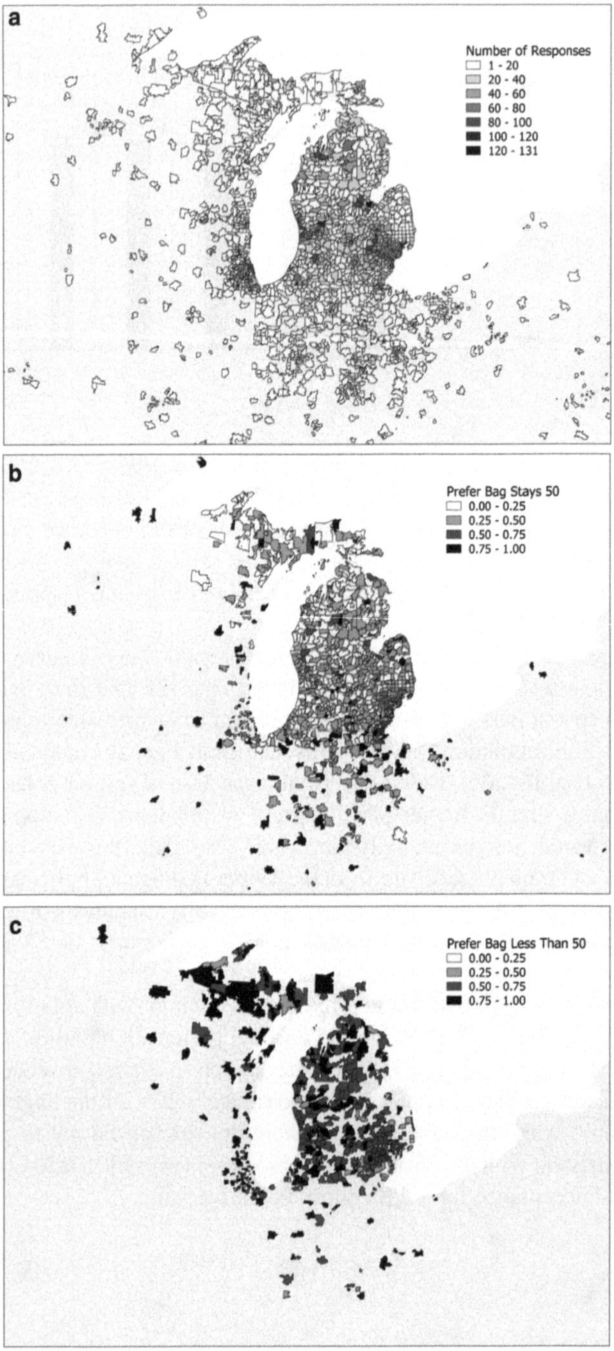

Fig. 2 (**a**) Geographic distribution of social survey respondents, based on reported zip code. All zip code regions outlined in black had at least one survey response. (**b**) Geographic distribution of respondents (percent of total, with total including all respondents for a given zip code; see Fig. 3) expressing a preference for the existing Yellow Perch bag limit (50 fish). (**c**) Geographic

7.2 *Other Public Input*

The Michigan WRSC reviewed the issue of potentially reducing the daily bag limit for Yellow Perch multiple times; most notable were two discussions in 2013 and another discussion in 2018. When a bag limit reduction was discussed initially in 2013, it did not receive majority support, although the committee was interested in additional information on the biological and social aspects of a regulation change for Yellow Perch. In the second 2013 discussion, a minority of committee members supported a lower bag limit or potentially combining Yellow Perch and panfish into a 25-fish limit. At that time, the Upper Peninsula (Michigan) Sportsmen's Alliance also reported that they were pursuing the issue for inland lakes, as they believed that a reduction was warranted. It was noted that MDNR agreed to utilize the Statewide Angler Survey Program and social surveys to investigate angler success and opinions on the Yellow Perch regulations. In 2018, the WRSC was provided with the results from these surveys and was informed of a large-scale review of the statewide regulation (this chapter) that included a statewide online survey of Michigan's fishing license holders. At the 2018 meeting, the WRSC was generally supportive of the survey and of using the responses to help inform a potential regulation change for Yellow Perch.

Input from the Lake Erie/Lake St. Clair, Lake Huron, Lake Michigan, and Lake Superior Great Lakes Citizens Fishery Advisory Committees contributed significantly to the development and final recommendation for a change to the statewide Yellow Perch bag limit regulation. At the Lake Michigan CFAC, advisors seemed comfortable with the existing (50-fish) statewide limit being applied in Lake Michigan, but they also expressed significant interest in reducing the bag limit to 25 Yellow Perch on inland lakes. Members of the Lake Huron CFAC were generally in favor of the proposed 25-fish statewide bag limit. In the minutes from their April 2018 meeting, Lake Huron advisors recognized the importance of Yellow Perch fisheries in the state—specifically emphasizing the importance of Yellow Perch as a great entry fishery for young anglers.

In the Lake Erie/Lake St. Clair CFAC, 65% of meeting participants were in favor of retaining the current statewide limit, while 35% were in favor of a 25-fish statewide bag limit. Similar to Lake Huron advisors, Lake Erie/Lake St. Clair participants recognized the importance of the Yellow Perch fishery in southeast Michigan, the need to be proactive in management, and that some important biological factors (including Yellow Perch density and forage reductions) could be influencing Yellow Perch populations outside of angler harvest. Some concerns expressed with respect to the proposed reduction included potential effects of release mortality and the thought that reducing the bag limit may be viewed as reflecting

Fig. 2 (continued) distribution of respondents (percent of total, with total including all respondents for a given zip code; see Fig. 3) expressing a preference for a reduced Yellow Perch bag limit (25 fish or less)

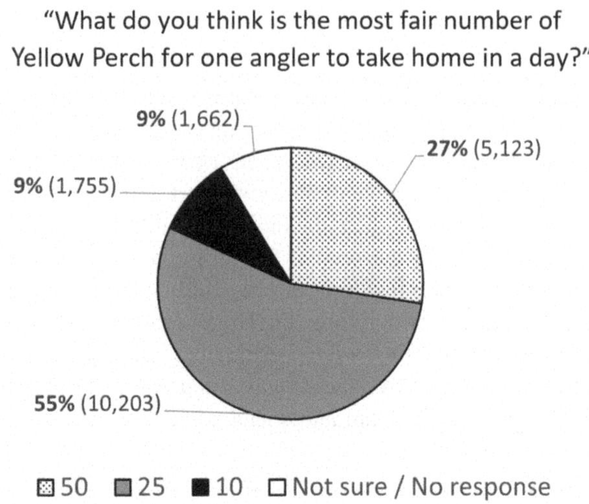

Fig. 3 Social survey response (percent, with number of respondents in parentheses) to "fair bag limit" question

Yellow Perch densities and might therefore lead to reduced participation in the fishery.

At the April 2018 meeting of the Lake Superior CFAC, a majority of those in attendance were in favor of reducing the statewide Yellow Perch bag limit to 25 fish, although a few members who did not actively pursue Yellow Perch had no opinion. Angling groups represented on the Lake Superior CFAC (e.g., Upper Peninsula Sportsmen's Alliance) also expressed support for the proposed 25-fish limit, and some Lake Superior CFAC members advocated for combining a size limit (similar to the existing Yellow Perch regulation on Lake Gogebic, a Michigan inland lake with restrictions on harvest of fish 12 inches or greater in length) with the proposed statewide bag limit.

7.3 Analysis of Fishery and Population Survey Data

7.3.1 Impacts of Varying Daily Bag Limits on the Recreational Fishery

Analysis of impacts of a reduced daily Yellow Perch bag limit (25 fish) by party size and waterbody type varied from a low of 0% Yellow Perch saved (small inland lakes) to a high of 16.5% Yellow Perch saved (Lake Erie; Table 2, All Parties). Statewide in 2017, had the daily bag limit been set at 25 fish, 14.3% of the harvested Yellow Perch would have been protected. Similarly, the proportion of parties affected had the proposed 25 Yellow Perch daily bag limit been in place varied from 0% (small inland lakes) to a high of 13.0% in Lake Erie. Statewide in 2017, only 1.3% of angling parties would have been affected (Table 2, All Parties).

When various fisheries were examined not just by a hypothetical 25-fish limit but across all possible reduced limits, thresholds of effect were indicated for both percent

Table 2 Proportion of Yellow Perch harvested that would have been protected (saved from harvest) and proportion of angling parties affected if a 25 Yellow Perch daily bag limit had been in effect instead of the 50-fish limit (35-fish limit for Southern Lake Michigan)

		All parties		Parties targeting Yellow Perch	
Location	Description	% parties affected[a]	% Perch "saved"[a]	% Parties affected[a]	%Perch "saved"[a]
Saginaw Bay	High-density years	5.0	11.6	9.3	11.8
Saginaw Bay	Mod-density years	4.1	13.5	9.1	12.9
Saginaw Bay	Low-density years	1.5	8.4	4.9	8.5
Lake St. Clair	2016	2.3	2.3	3.9	3.3
Little Bay de Noc	2017	0.2	0.2	0.6	0.2
Les Cheneaux Islands	2017	1.8	6.2	3.4	6.2
St. Marys River	2017	0.3	2.1	2.1	2.4
S. Lake Michigan	2017	0.6	3.2	6.2	5.7
Lake Erie	2006–2016	13.0	16.5	17.1	17.7
Small inland lakes	Various	0.0	0.0	0.0	0.0
Medium inland lakes	Various	0.6	3.2	1.7	4.7
Large inland lakes	Various	0.5	6.6	–[b]	–[b]
Statewide	2017	1.3	14.3	8.4	14.9

[a]Estimates are from creel survey interview analysis and are based on available data for the location/systems indicated. For example, estimates for inland lakes combined all available data across years for lakes in that system category. Statewide estimates utilized all data available from surveys conducted in 2017, combined across system type

[b]Insufficient data were available to conduct analyses for parties targeting Yellow Perch in large inland lakes

of Yellow Perch saved and percent of parties affected. Generally, waters with higher density Yellow Perch fisheries had higher thresholds of effect (e.g., greater than 40-fish limit in Saginaw Bay during high-density Yellow Perch years; Figure 4a), while waters with lower Yellow Perch densities had a lower threshold of effect (e.g., approximately 15-fish limit in small inland lakes, Figure 4b), indicating that to reduce harvest via a lower bag limit in those fisheries, the limit would have to be set much lower. A complete set of outcomes across lakes and waterbodies examined is available from authors, on request.

Another way Yellow Perch fisheries were characterized using creel data was by calculating an optimal or target bag limit, where 10% of parties are likely to "limit out" (Cook et al. 2001). Calculated optimal bag limits ranged from 7 Yellow Perch

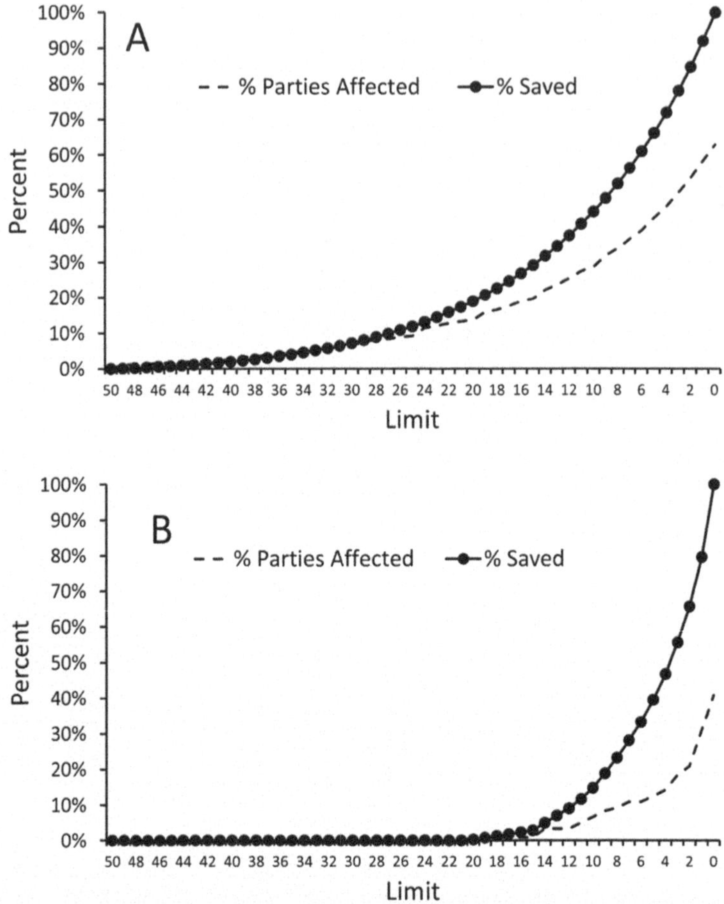

Fig. 4 Proportion of Yellow Perch protected (saved) by hypothetical lower daily bag limits and proportion of angling parties targeting Yellow Perch affected by the same daily limit for (**a**) Saginaw Bay during years of high Yellow Perch density (1987–1994) and (**b**) small inland lakes (<2000 acres), as derived by analysis of creel survey interviews

for small inland lakes up to 37 for Lake Erie (Table 3). The value calculated using data collected statewide in 2017 was 25 Yellow Perch. In this type of analysis, the better the Yellow Perch fishing, the higher the catch achieved by the 10% of targeted fishing parties. This approach is based on optimizing angler satisfaction and suggests a 25-fish limit is the most appropriate across a range of fishery types.

Table 3 Yellow Perch daily bag limit by lake or waterbody type calculated to optimize angler party satisfaction using the 10% attainment guideline from Cook et al. (2001). Analysis is based on data from angler parties targeting Yellow Perch

Location	10% limit targeted
Great Lakes	
Lake St. Clair	17
Lake Erie	37
Les Cheneaux Islands	17
Little Bay de Noc	7
Saginaw Bay (high-density years)	25
Saginaw Bay (med-density years)	24
Saginaw Bay (low-density years)	18
St. Marys River	16
Southern Lake Michigan	20
Inland	
Small inland lakes	7
Medium inland lakes	11
Large inland lakes	–[a]
Statewide **(excluding Lake Erie)**	25

[a]Insufficient data were available to conduct analyses for parties targeting Yellow Perch in large inland lakes

7.3.2 Inland Lake Fishing Effort vs. Yellow Perch Population Characteristics

Analyses of inland lake survey data suggested intense levels of angling have the potential to reduce Yellow Perch populations in inland lakes (Fig. 5). The wedge-shaped distribution of data points suggested as angling pressure increases, the potential for anglers to experience high harvest rates of Yellow Perch (shown along the upper edge of the distribution) becomes increasingly limited. Note these plots do not identify other factors that could keep anglers from having maximum catch rates at any particular level of angling pressure (Terrell et al. 1996; Thompson et al. 1996). Angling pressure on a per acre basis reached higher levels (e.g., over 20 angler hours per acre annually) in smaller lakes, potentially making them more vulnerable to declines in Yellow Perch harvest rates (Fig. 6). While larger lakes tended to support larger Yellow Perch populations based on creel harvest data (Fig. 7), they also appeared to return lower harvests to perch anglers on a per acre basis (Fig. 8). Collectively, these results suggested population-level effects, positive or negative, of Yellow Perch bag limits may be most noticeable in smaller lakes where perch populations are less abundant, more likely to be subjected to intense angling pressure, and more vulnerable to overharvest. Bag limits generally have less pronounced effects in large inland lakes or the Great Lakes, although there are exceptions (e.g., see Saginaw Bay in some years). These conclusions were consistent with limited data from Schneider et al. (2007) which suggested lakes with higher fishery-independent survey catch rates of Yellow Perch had moderate (rather than heavy) angler use (Fig. 9).

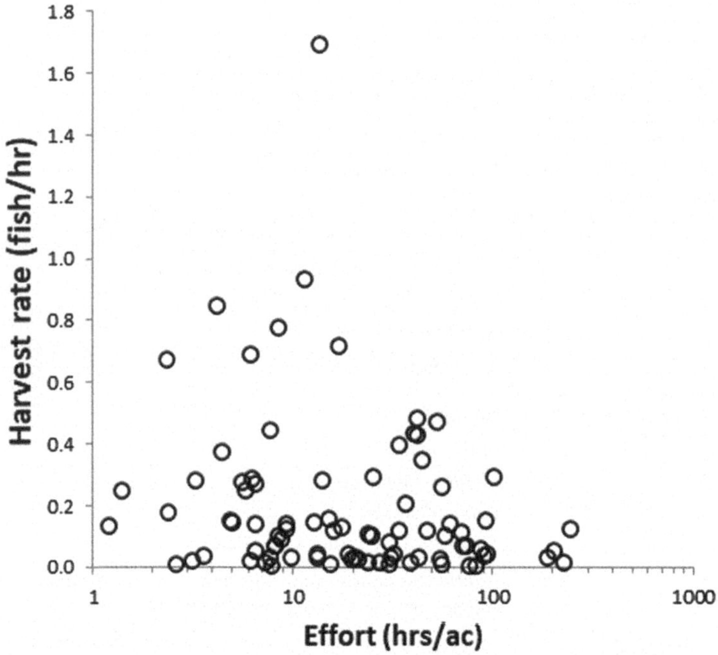

Fig. 5 Relationship between angling effort and harvest rate for Michigan inland lake Yellow Perch fisheries. Note $\log_{10}$ transformed *x*-axis

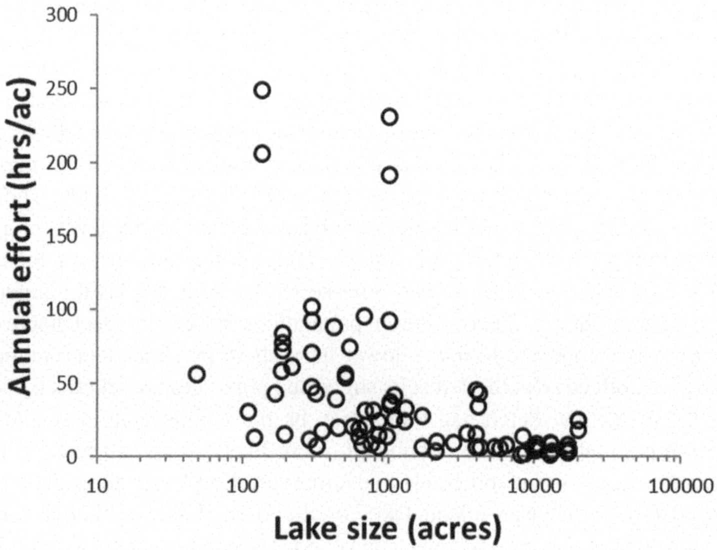

Fig. 6 Relationship between lake size and annual angling effort per acre for inland lakes in Michigan. Note $\log_{10}$ transformed *x*-axis

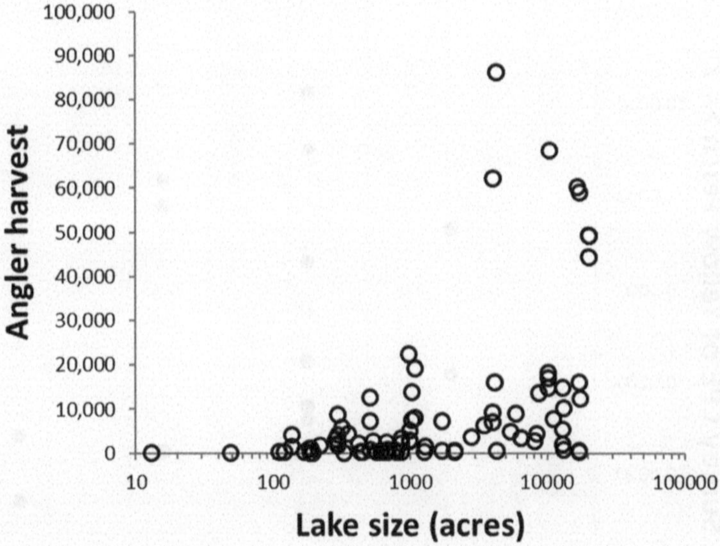

Fig. 7 Relationship between lake size and annual (or open water) angler harvest of Yellow Perch for inland lakes in Michigan. Note $\log_{10}$ transformed *x*-axis

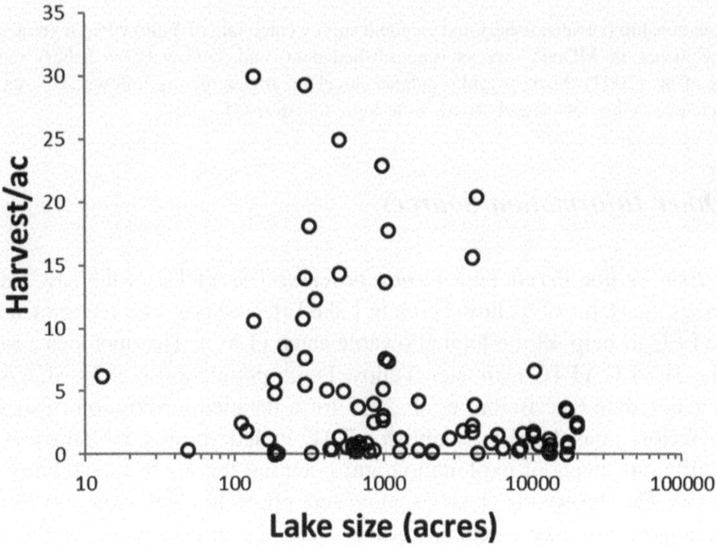

Fig. 8 Yellow Perch harvest per acre for inland lakes in Michigan. Note $\log_{10}$ transformed *x*-axis

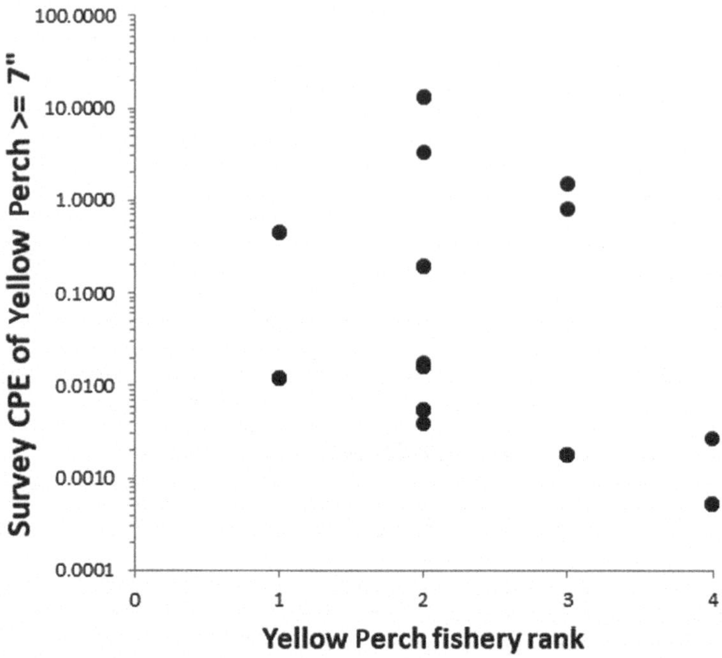

Fig. 9 Relationship between fishery-independent survey catch rate of Yellow Perch greater than or equal to 7 inches in MDNR surveys (unpublished data) and Yellow Perch fishery rank from Schneider et al. (2007). Fishery ranks indicate level of angler use as follows: 1 = extensive; 2 = moderate; 3 = low; 4 = rarely used. Note $\log_{10}$ transformed *y*-axis

7.4 Other Information Sources

- ***Lake Erie Yellow Perch Task Group process:*** The YPTG of the LEC annually describes the status of Yellow Perch in Lake Erie and provides recommendations to the LEC to help set the total allowable catch (TAC). This includes a range of RAHs. The LE YPTG estimates Yellow Perch population size by SCAA using interagency data (see Belore et al. 2018 for a detailed description of methods). Other factors considered to establish a TAC include biological risks associated with different levels of exploitation and potential threats to the stability of the resource. The lake-wide TAC is allocated proportionately among individual states and the province of Ontario using a sharing formula based on the relative lake surface area within each jurisdiction. Individual jurisdictions are then responsible for setting their own fishing regulations to ensure that they do not exceed the established TAC for their waters.

  The LE YPTG process provides an example of a situation or system in which an exemption to the statewide 25-fish bag limit might be acceptable or even preferred. Lake Erie managers undertake an intensive assessment process to develop a TAC for that system which is then shared among jurisdictions. In this case,

Michigan managers may elect to retain a 50-fish bag limit or choose some other regulation as a means of fully utilizing the state's portion of the annual allowable catch.

- ***Lake Michigan statistical catch-at-age and decision analysis projects:*** In the early 2000s, members of the Lake Michigan Yellow Perch Task Group worked with researchers from the Michigan State University Quantitative Fisheries Center to develop SCAA models for the southern Lake Michigan Yellow Perch population (Wilberg et al. 2005). These efforts provided detailed information on the status of southern Lake Michigan Yellow Perch populations during the 1980s and early 1990s and importantly demonstrated that unsustainably high fishing mortality rates were a substantial contributing cause of the rapid population decline observed on Lake Michigan in the mid-1990s (Clapp and Dettmers 2004; Marsden and Robillard 2004). The work of Wilberg et al. (2005) also served as the basis for a subsequent, more detailed analysis of harvest policy options for Lake Michigan Yellow Perch. Irwin et al. (2008) conducted stakeholder workshops and used decision analysis and simulation modeling to investigate the relative costs and benefits of implementing state-dependent (i.e., varying with some measure of fish population status) versus constant fishing mortality harvest policies. The results of this work indicated that state-dependent policies, while somewhat more difficult to implement in terms of agency effort, produced higher average harvests and lower frequency of years with low Yellow Perch spawning stock biomass than constant fishing mortality policies. However, these state-dependent policies also resulted in more frequent years with low harvests. As Irwin et al. (2008) pointed out, choice of a harvest policy is strongly dependent on managers' (and the public's) risk aversion and the relative concerns about low stock biomass versus low harvest levels.

  The Lake Michigan analyses provided evidence that recreational fisheries can exert significant harvest pressure on Yellow Perch populations, even in systems as large as the Great Lakes. They also provided additional tools needed to develop more system-specific regulations if conditions (e.g., Yellow Perch population status, changes to interagency management strategies) warrant.
- ***Saginaw Bay statistical catch-at-age model:*** The Saginaw Bay Yellow Perch SCAA model is in development. However, lack of up-to-date data on the age structure of the commercial portion of the harvest prevented the model from generating reliable estimates. The MDNR embarked on an annual commercial catch sampling program in 2014 to build up this requisite age-structure data. The model will be refitted and tested in 2021 with the new data, at which point it will be used to evaluate the appropriateness of the new statewide regulation for Saginaw Bay and potentially increase specificity in management of the Saginaw Bay Yellow Perch population and fisheries.
- ***Previous MDNR regulation analyses:***

  Lake St. Clair:

  Researchers at the MDNR Lake St. Clair Fisheries Research Station modeled the effects of reducing the bag limit from 50 to 25, 15, and 10 fish on Lake St. Clair Yellow Perch abundance, catch, and harvest (T. Wills and M. Thomas,

Table 4 Simulated percent change in Yellow Perch abundance, catch, and harvest (overall and for large Yellow Perch >25.4 cm) after reducing the daily bag limit from 50 to 25, 15, and 10 fish in Michigan waters of Lake St. Clair

	Bag limit		
	25 fish	15 fish	10 fish
Overall			
Abundance	0.3%	0.6%	1.0%
Catch	0.5%	0.9%	1.4%
Harvest	−23.1%	−46.5%	−70.1%
Yellow Perch >25.4 cm			
Abundance	2.2%	4.5%	6.9%
Catch	2.2%	4.4%	6.6%
Harvest	−21.7%	−44.5%	−68.5%

MDNR unpublished data). The analysis was conducted using an Excel tool developed by Dr. Daniel Hayes at Michigan State University; model inputs were hooking mortality, catchability, fishing effort, annual recruitment, voluntary release rate (the proportion of fish caught that are voluntarily released if they are of legal size), minimum size for harvest, annual survival rate, mean length-at-age, and initial population size.

Data from Thomas and Towns (2011) illustrated that 20%, 40%, and 60% of the Lake St. Clair Yellow Perch harvest would have been protected under a 25-fish, 15-fish, and 10-fish daily bag limit, respectively. Accordingly, changes to daily bag limits were examined by modifying the voluntary release rate input to simulate a reduced bag limit. In other words, fish that would previously have been harvested under more liberal bag limits were "put back" under more restrictive bag limits by adjusting voluntary release. The starting voluntary release value was set at 15% (see below) for a 50-fish daily bag limit and increased by 20% for each successive bag limit reduction (i.e., 35% voluntary release for 25-fish limit [simulating a 20% reduction in harvest], 55% voluntary release for 15-fish limit [simulating a 40% reduction in harvest], and 75% voluntary release for 10-fish limit [simulating a 60% reduction in harvest]). Hooking mortality estimates came from a study at the Les Cheneaux Islands, Lake Huron (Lucchesi 1988). Yellow Perch catchability and fishing effort were estimated from Thomas and Towns (2011). Since the number of Yellow Perch released was not recorded by Thomas and Towns (2011), the initial voluntary release rate input of 15% was estimated from their data for Walleye and assumed to be similarly low for Yellow Perch, as release of harvestable-size percids is thought to be uncommon. Annual recruitment, survival, mean length-at-age, and initial population size were based on spring trawl data in Lake St. Clair from 1993 to 2011 (Michigan DNR, unpublished data). Simulations were made using the assumptions that natural mortality and growth were not density dependent, hooking mortality was equal for all fish sizes, and fish <4 inches (10.2 cm) long were not harvested.

Simulations suggested reducing the daily bag limit for Yellow Perch on Lake St. Clair would result in minimal changes to overall Yellow Perch abundance (total population) and catch even though angler harvest would decrease substantially, particularly at a 10-fish daily bag limit (Table 4). Increases in abundance and catch of large >10 inches (>25.4 cm) Yellow Perch were also low, with harvest decreasing at a similar rate for all sizes of Yellow Perch (Table 4). Estimates for changes in Yellow Perch population size and catch were low due to few (<10%) anglers harvesting daily bag limits, while "saved" fish were still subject to natural and hooking mortality.

Saginaw Bay:

In 2014, the Lake Huron Basin Team (internal Fisheries Division committee) initiated a review of Yellow Perch management in Saginaw Bay out of concern over steep declines in both the commercial and recreational harvests as well as in the population. Although researchers lacked a working Yellow Perch population model at the time, enough was known to identify that mortality issues were limiting the population and it was not a reproduction problem. A suite of management actions were settled upon that were intended to provide better survival of Yellow Perch assuming a mortality bottleneck (attributed mainly to predation) was occurring between age 0 (young-of-the-year) and age 1 (yearling). These actions included liberalizing recreational Walleye regulations, implementing Double-crested Cormorant (*Phalacrocorax auritus*) management, and cooperating with US Fish and Wildlife Service and other agencies on a proposal to reintroduce Cisco (*Coregonus artedi*) into Saginaw Bay, as alternate forage for Walleye. Additional strategies were also implemented to reduce fishing mortality on Yellow Perch and help better position the adult population to rebuild once (if and when) the survival of juvenile Yellow Perch improved. Actions to reduce fishing mortality included relocating (through license/permit provisions) some commercial Yellow Perch fishing effort on Saginaw Bay to Lake Whitefish (*Coregonus clupeaformis*) effort outside of the bay and application of a reduced (25 fish) daily bag limit on Yellow Perch (this occurred ahead of the statewide reduction and was part of the impetus for examining Yellow Perch regulations statewide). The reduction of the daily bag limit was based on the same analysis summarized in Table 2; this analysis indicated that the level of reduced harvest could amount to as much as 11.6%. It was the conclusion of the Lake Huron Basin Team that this regulation change would help position the Yellow Perch population to rebuild once other bottlenecks were resolved.

- ***Other State Resource Agency Processes***: Review of Yellow Perch regulations in midwestern states and provinces indicated that management approaches have stayed somewhat consistent over the past 15+ years. In 8 states and the Province of Ontario, statewide bag limits ranged from 20 fish to no limit, and none of the jurisdictions had a statewide minimum size limit (Table 5). All jurisdictions provided for system-specific exemptions to the statewide regulation (typically exemptions were more restrictive regulations, compared to statewide).

Several midwestern states and the Province of Ontario have recently undertaken evaluations of their Yellow Perch and/or panfish regulations. Minnesota and Wisconsin Yellow Perch fisheries are probably the ones most similar to those occurring in Michigan, and review focused on analyses conducted in those states. The review

Table 5 Regional summary of Yellow Perch bag-limit and minimum size limit (MSL) regulations in 2018

State/province	Total regulations (includes statewide and site-specific regulations)	Statewide or majority regulation
Illinois	4	No bag; No MSL
Indiana	1	No bag; No MSL
Michigan	5	Bag = 50; No MSL
Minnesota	10	Bag = 20; No MSL
New York	6	Bag = 50; No MSL
Ohio	2	Bag = 30; No MSL
Ontario	6	Bag = 50 (Sport license), 25 (Conservation license); No MSL
Pennsylvania	4	Bag = 50[a]; No MSL
Wisconsin	6	Bag = 25[a]; No MSL

[a]In combination with other "panfish" species

in Minnesota led managers to reduce the statewide bag limit from 100 to 20 Yellow Perch. While managers there acknowledged that this regulation change would lead to only a modest (<5%) reduction in harvest (Cook et al. 2001), they also recognized important social aspects related to fisheries regulation. They pointed out that setting a high bag limit may cause anglers to have an unrealistic expectation for potential harvest (Cook et al. 2001), and therefore chose to revise bag limits to more closely match the level of harvest that more anglers could attain. In Wisconsin, a 70-year retrospective analysis of size-structure data collected from 19 gamefish species indicated that panfish species (including Yellow Perch) have experienced "substantial and sustained" erosions in size structure over the period of data collection (1944–2012; Rypel et al. 2016). The Wisconsin review also indicated that excessive harvest by anglers is a likely factor in the size structure decline and that regulations, in combination with ecosystem-based management approaches, may help to improve panfish size structure. Isermann et al. (2007) reached a similar conclusion from their study of South Dakota Yellow Perch fisheries, indicating restrictive harvest limits can lead to improvements in size structure (but only when angler harvest is the dominant source of mortality).

8 Discussion

The Michigan regulation review critically evaluated four measures of the appropriateness of a Yellow Perch regulation change (public opinion, recreational fishery effects, population level effects, related management in other jurisdictions); it indicated a majority of the public were in favor of the proposed change, the new regulation would have a measurable biological effect (at least in some systems) while negatively influencing the harvest of a small percentage of anglers, and the regulation was consistent with successful management in other jurisdictions and with previous peer-reviewed evaluations. The comprehensive nature of this evaluation provided decision makers with the tools and background needed to make an informed decision on management of an important statewide fishery.

Analyses of creel survey data from inland and Great Lakes fisheries indicated a reduced statewide bag limit would help to achieve a balance between conservation of Yellow Perch and angler opportunity. Based on the 2017 fishery statewide, 15% of harvested Yellow Perch would have been protected under a 25-fish bag limit, and only 1% of angling parties would have been affected (i.e., experienced a reduction in harvest). Analyses from inland systems also suggested intense angling pressure does have the potential to reduce Yellow Perch populations, especially in smaller lakes (150 acres or less). As a result, positive population-level effects of reduced Yellow Perch bag limits (or negative effects of higher bag limits) may be more likely to occur in these systems. In analyzing a proposed statewide bag limit, harvest regulations were evaluated to determine which among a range of possible regulations would be most applicable to the largest number of perch fisheries in the state. The creel analysis indicated clearly a 25-fish Yellow Perch bag limit would be effective in terms of optimizing angler satisfaction across a range of fishery types (see Table 3).

Previous analyses and reviews conducted in Michigan, in other midwestern states, and in Ontario and documented in literature lend support to Michigan's proposal for reducing the statewide Yellow Perch bag limit. Even in large systems (e.g., Great Lakes), Yellow Perch population declines attributable to high fishing mortality have been observed (see Wilberg et al. 2005; Irwin et al. 2008). Excessive harvest has also been shown to negatively influence Yellow Perch size structure (Rypel et al. 2016). Given these biological considerations, as well as consideration of social factors surrounding these popular fisheries, most midwestern states have chosen to lower panfish bag limits over historically more liberal regulations or no regulations at all (see Cook et al. 2001; McShane and Bowman 2003; Rypel et al. 2016).

8.1 *Evaluating Social and Biological Effects of a New Statewide Yellow Perch Regulation*

The objective of this review was to evaluate the potential effects—social as well as biological—of moving to a statewide 25-fish per day bag limit for Yellow Perch. Social issues were a primary impetus for review of the statewide regulation, and the proposed revised regulation at least partially addresses all these issues (Table 6). The revised statewide limit (in combination with other proposed changes) certainly decreases complexity of regulations and also addresses specific enforcement concerns that were raised by MDNR Law Enforcement Division and the public. A 25-fish per day bag limit is consistent with Yellow Perch regulations in neighboring jurisdictions, as well as with regulations for common panfish species. The bag limit

Table 6 Evaluation of the adequacy of a statewide 25-fish per day bag limit in addressing social and biological concerns for Michigan Yellow Perch fisheries. Social and biological factors are discussed in more detail earlier in this report, as well as in Noble and Jones (1993) and Radomski et al. (2001)

Concern	Addressed	Partially addressed	Not addressed
Social			
Regulation complexity, enforcement	X		
Consistency across species and jurisdictions	X		
Distribution of harvest among anglers		X	
Providing opportunities, options		X	
Public concern over populations, angler satisfaction		X	
Biological			
Prevent overexploitation/improved abundance		X	
Improved size structure		X	
Protected/improved reproduction			X
Increased harvest			X
Positive effects on fish community			X

partially addresses factors like distribution of harvest and angler satisfaction, but the degree to which the proposed regulation is successful in influencing these factors can only be fully determined through post-implementation evaluation. A requirement for ensuring that social considerations are properly addressed is to involve interested parties, keep them informed on management of the fishery, and provide them with opportunities for input (Cochrane 2002); this requirement was fulfilled to the extent possible (see Methods).

While social issues had a significant influence in determining the need for a new statewide Yellow Perch bag limit, the revised statewide regulation also does a reasonable job of addressing at least some biological concerns (Table 6). Moving from a 50-fish to a 25-fish bag limit will reduce harvest for most fisheries and may lead to modest improvements in size structure (Coble 1988; Rypel 2015; Rypel et al. 2016). While effects on other parts of the fish community were not specifically addressed in this review, these effects can certainly be evaluated in post-implementation studies. Finally, effects of the proposed regulation on reproduction were not evaluated, as this was not raised as a concern by any of the constituencies providing input.

Whether or not a regulation has a biological effect is a matter of some debate. Any fish harvested (or saved from harvest) has some biological impact; the important questions are (1) is that impact measurable, and (2) what is the importance of that impact relative to environmental and population-level factors that might also be exerting an influence. A reduction in the Michigan statewide Yellow Perch bag limit to 25 fish will certainly have a measurable impact on harvest (14.3% based on combined statewide creel data); the importance of that impact to Yellow Perch populations remains to be seen. Mosel et al. (2015) set their threshold for biological

Table 7 Hypothetical effects of bag limit regulations at varying densities of Yellow Perch and varying levels of fishing intensity

	Fishing intensity		
Perch density	Low	Med	High
Low	Little limiting effect	Little limiting effect	Little limiting effect
Moderate	Some limiting effect	Considerable limiting effect	Some limiting effect
High	Little limiting effect	Little limiting effect	Some limiting effect

effect (with respect to harvest reduction) at 25%. Michigan's 25-fish statewide bag limit will almost certainly not achieve this 25% threshold in most fisheries; to achieve that level of harvest reduction would likely require a bag limit of less than 10 Yellow Perch (Radomski et al. 2001; Cook et al. 2001; Jacobson 2005), which would be unacceptable to many Michigan constituents (and managers). However, moving from a 50-fish statewide bag limit to a 25-fish bag limit, while having a smaller effect, still moves populations in a direction desirable to managers; this more reasonable bag limit also serves as a prerequisite to overall management effectiveness.

To further refine ideas around potential regulation effects on Yellow Perch populations, theoretical models were explored describing potential effects given a range of Yellow Perch density and fishing intensity scenarios. One model suggests a nonlinear function between fishing intensity and biological effect, dependent on relative Yellow Perch density (Table 7). At high Yellow Perch density, one might assume slow growth (stunting) so retained catch (true harvest) would probably be low. At low density, growth rates might be faster, but anglers would be unlikely to harvest 25 Yellow Perch under these conditions (because there are fewer perch). At moderate densities, the fishery might experience medium to high fishing intensity (i.e., effort), fishing will be very good, and more anglers harvesting limits will be likely (at least at times). In other words, at either end (low or high density), perch populations will tend to "self-regulate" the fishery, but it is when the population is well balanced in terms of density and growth rate that one might see a population more responsive to harvest effects. The proposed statewide bag limit will be somewhat conservative in these situations, but the goal is to sustain the state's Yellow Perch fishery for a longer period and avoid boom-and-bust cycles. Additional work in developing this and related models is certainly warranted.

8.2 Implementing and Evaluating a Regulation Change

Discussions with several advisory groups centered around the need to evaluate regulations to ensure that they have the desired effect; Fisheries Division is strongly supportive of and committed to evaluating management actions (see MDNR 2018). Compliance is often an issue when implementing new regulations (Gigliotti and Taylor 1990; Caroffino 2013), but MDNR Fisheries Division already has excellent systems (e.g., creel clerks, charter reporting, law enforcement) in place to allow us to evaluate the extent and effect of non-compliance, especially on Great Lakes waters. Pre-implementation review relied heavily on historical creel survey data, and these data will likewise be used in evaluating effectiveness of the new statewide bag limit, post-implementation. A limitation for future evaluations involving creel data is that the largest number of Yellow Perch fisheries occurs on smaller inland lakes that are not surveyed annually like the Great Lakes. Managers have the ability to request and implement "exception" regulations if they feel that a statewide regulation is not meeting local management needs and objectives (see "Michigan's Regulation Implementation Process" in Introduction); one way to evaluate the proposed statewide regulation will be to track the number of additional exceptions that are requested in years following implementation. Finally, the 2018 social survey was useful in evaluating public opinion concerning the proposed regulation change; a follow-up survey, along with regular discussions with advisory groups, will be used to measure public satisfaction going forward.

One concern with any regulation change (in this case, reduction of bag limits) is that the regulation may negatively influence Michigan resource-based economies. In several areas of the state (e.g., Lake Erie, the Les Cheneaux Islands, specific inland lakes), local economies rely heavily on Yellow Perch-based recreational fisheries. In the Les Cheneaux Islands area, for example, Yellow Perch recreational fisheries account for from $2 to $6 million in local economic activity (Diana et al. 1987; D. Fielder, MDNR unpublished data). On Lake Erie, more than 37,000 charter trips were reported from 1990 to 2009—many of these trips targeted Yellow Perch. Charter trips contributed an economic impact of more than $47.5 million to Lake Erie coastal communities during this period (O'Keefe and Miller 2011). It is important in development and evaluation of both statewide and system-specific regulations to take potential economic factors into account (Noble and Jones 1993); review of a new statewide regulation will evaluate local economic effects to the extent possible.

While the current proposed regulation will primarily affect recreational anglers, Yellow Perch are an important commercial species in some parts of the Great Lakes, and changes to Yellow Perch populations can have consequences for those commercial fisheries and associated businesses. In Saginaw Bay, the Yellow Perch commercial fishery has declined in parallel with the recreational fishery. Both fisheries suffer from the same problem (i.e., high juvenile mortality rates limiting recruitment to the fishery). In competing fisheries (both recreational and commercial targeting the same population, as in Saginaw Bay), there is sometimes reluctance to embrace more restrictive harvest regulations believing that the benefits will all be gained by one side or the other. In Saginaw Bay, the suite of management actions implemented around 2015 included changes for both the recreational and commercial fisheries. The commercial fishery in Saginaw Bay is over-capitalized and currently self-regulating in that many licenses (operations) are idle (Fielder et al. 2014). It is hoped that the Saginaw Bay Yellow Perch SCAA model under development will aid in understanding the individualized effects of both fisheries and lead to better management of each.

9 Conclusions

Developing a statewide regulation that is effective across a variety of waterbodies and fisheries is a significant challenge for fisheries managers. In Michigan, there are vast differences in productive capacity among aquatic systems—from the Great Lakes to a 75-acre inland lake—and no single regulation will be perfect for every situation. Based on this review, a statewide 25-fish Yellow Perch bag limit was recommended for its effectiveness in terms of balancing angler satisfaction and opportunity (across a range of fishery types) with conservation of this important species. A review combining public opinion surveys with fishery-dependent and fishery-independent analyses provided valuable insights into the likely appropriateness of this regulation in meeting management objectives. A similar comprehensive review (including use of these tools and analyses) is recommended in other cases where changes to fishing regulations are being considered.

Acknowledgments The review of Michigan's statewide Yellow Perch regulation benefitted from the efforts of many individuals. We thank members of the public and citizen advisory groups who provided thoughtful comments and input throughout the process of developing the regulation proposal. Special thanks to former Fisheries Division Chief John Scott for the excellent historical perspective he provided on Yellow Perch management in Michigan. Tracy Claramunt (Fisheries Division Statewide Angler Survey Program biologist) supplied all the creel data we analyzed; her timely response to requests and significant input are greatly appreciated. Neal Godby and Dave Borgeson (Fisheries Division Northern Lake Huron Management Unit) provided input and oversight related to collection of 2013 and 2017 social survey questions, and all of the Northern Lake Huron creel clerks contributed significantly through their conscientious data collection efforts and positive daily interactions with Lake Huron anglers. George Madison contributed by summarizing a regulation review conducted by Fisheries Division biologists in 2010. For the inland fishery data collections used in our analysis, data were collected and entered by Fisheries Division management unit staff and creel clerks throughout the state. Grace Zimmerman (Fisheries Division State Worker, Marquette) provided assistance with data compilation. Patrick Hanchin (Fisheries Division Tribal Coordination Unit supervisor/Lake Superior Basin Coordinator) oversaw the Large Lakes Survey Program, coordinating survey activities and data analyses. Sara Thomas reviewed information related to the Lake Erie Yellow Perch Task Group, and Todd Wills and Mike Thomas contributed significantly through their work on the Lake St. Clair daily bag limit analysis. We thank Dustin Isenhoff, social science specialist with MDNR Marketing and Outreach Division, for assistance in the design and implementation of the 2018 email survey. Also, we thank Michael Rubley, James Tchorzynski, Brian Maki, and the Resource Assessment Section of the Forest Resource Division for their assistance with the online viewer applications for the 2018 survey results and Ben Turschak (MDNR fisheries research biologist, Charlevoix Fisheries Research Station) for assistance in construction of maps displaying survey results. Dr. Daniel Hayes, Philip Schneeberger, and David Lucchesi reviewed drafts of this report and provided helpful comments and advice. Finally, we acknowledge the many contributions to our understanding of Yellow Perch biology and management made by Institute for Fisheries Research (Ann Arbor) research biologist James C. Schneider, who retired in 2000 after 38 years of service to the anglers of Michigan. Jim was widely recognized as one of the premier percid biologists in North America, and his outstanding work is chronicled in close to 70 research publications, many of which were consulted in developing the recommendations contained in this report.

References

American Fisheries Society (n.d.) Policy Statement #28, "Special fishing regulations for managing freshwater sport fisheries" [online]. Available from http://fisheries.org/wp-content/uploads/2015/05/policy_28f.pdf (accessed September 2020)

Baldwin NA, Saalfeld RW, Dochoda MR, Buettner HJ, Eshenroder RL (2009) Commercial fish production in the Great Lakes 1867–2006 [online]. Available from http://www.glfc.org/great-lakes-databases.php. Accessed September 2020

Belore M, Cook A, Faust M, Hosack M, Knight C, May C, Robinson J, Thomas S (2018) Report of the Lake Erie Yellow Perch Task Group, March 2018. Presented to the Standing Technical Committee, Lake Erie Committee of the Great Lakes Fishery Commission. Ann Arbor, MI

Caroffino DC (2013) Angler compliance with Lake Trout length limit regulations in Great Lakes waters. N Am J Fish Manag 33:1203–1209

Clapp DF, Dettmers JM (2004) Yellow perch research and management in Lake Michigan: evaluating progress in a cooperative effort, 1997-2001. Fisheries 29(11):11–19

Coble DW (1988) Effects of angling on Bluegill populations: management implications. N Am J Fish Manag 8:277–283

Cochrane KL (ed) (2002) A fishery manager's guidebook. Management measures and their application. FAO Fisheries Technical Paper. No. 424. FAO, Rome, 231 p

Colby PJ, Lewis CA, Eschenroder RL, Haas RC, Hushak LJ (1994) Walleye rehabilitation guidelines for the Great Lakes area. Great Lakes Fisheries Commission, Ann Arbor, MI

Cook MF, Goeman TJ, Radomski PJ, Younk JA, Jacobson PC (2001) Creel limits in Minnesota: a proposal for change. Fisheries 26:19–26

Diana JS, Jones CA, Lucchesi DO, Schneider JC (1987) Evaluation of the Yellow Perch fishery and its importance to the local economy of the Les Cheneaux Islands area. Grant LRP-8C-7, Final Report to the Coastal Management Program, Michigan Department of Natural Resources, 59 p

Fielder DG, Baker JP (2019) Recovery of Saginaw Bay Walleye, Lake Huron. In: Krueger CC, Taylor WW, Youn S (eds) From catastrophe to recovery: stories of fishery management success. American Fisheries Society, Bethesda, MD, pp 411–430

Fielder DG, Kolb TL, Goniea TM, Wesander DL, Schrouder KS (2014) Fisheries of Saginaw Bay, Lake Huron 1986-2010. Michigan Department of Natural Resources, Fisheries Division, Fisheries Report 02. Ann Arbor, MI

Gigliotti LM, Taylor WW (1990) The effect of illegal harvest on recreational fisheries. N Am J Fish Manag 10:106–110

Glover DC, Dettmers JM, Wahl DH, Clapp DF (2008) Yellow perch stock structure in Lake Michigan: an analysis using mark-recapture data. Can J Fish Aquat Sci 65:1919–1930

Goeman TJ, Bonneau DL, Baccante D, Clark RD, Willis DW, Novinger GD (1995) Special fishing regulations for managing freshwater sport fisheries. Fisheries 20(1):6–8

Graff DR (1977) Special regulations - time for evaluation. Pennsylvania Angler June, pp 11–15

Hanchin PA (2017) A summary and analysis of the Large Lakes Survey Program in Michigan 2001-2010. Michigan Department of Natural Resources, Fisheries Division, Fisheries Report 25. Ann Arbor, MI

Irwin BJ, Wilberg MJ, Bence JR, Jones ML (2008) Evaluating alternative harvest policies for Yellow Perch in southern Lake Michigan. Fish Res 94:267–281

Irwin BJ, Wilberg MJ, Jones ML, Bence JR (2011) Applying structured decision making to recreational fisheries management. Fisheries 36:113–122

Isermann DA, Willis DW, Blackwell BG, Lucchesi DO (2007) Yellow Perch in South Dakota: population variability and predicted effects of creel limit reductions and minimum length limits. N Am J Fish Manag 27:918–931

Jacobson PC (2005) Experimental analysis of a reduced daily Bluegill limit in Minnesota. N Am J Fish Manag 25:203–210

Johnson BM, Martinez PJ (1995) Selecting harvest regulations for recreational fisheries: opportunities for research/management cooperation. Fisheries 20(10):22–29

LE YPTG (Lake Erie Yellow Perch Task Group) (2011) Terms of reference for the Yellow Perch Task Group (YPTG) of the Lake Erie Committee [online]. Available from http://www.glfc.org/pubs/lake_committees/Lake%20Erie%20Committee%20Yellow%20Perch%20Task%20Group%20TOR.pdf. Accessed September 2020

Lockwood RN, Benjamin DM, Bence JR (1999) Estimating angling effort and catch from Michigan roving and access site angler survey data. Michigan Department of Natural Resources, Fisheries Division, Fisheries Research Report 2044. Ann Arbor, MI

Lucchesi DO (1988) A biological analysis of the Yellow Perch population in the Les Cheneaux Islands, Lake Huron. Michigan Department of Natural Resources, Fisheries Division, Fisheries Research Report 1958. Ann Arbor, MI

Makauskas D, Clapp D (2018) Status of Yellow Perch in Lake Michigan, 2016–2017. Report to the Lake Michigan Committee, Sault Ste Marie, ON

Marsden JE, Robillard SR (2004) Decline of Yellow Perch in southwestern Lake Michigan, 1987-1997. N Am J Fish Manag 24:952–966

Mather ME, Parrish DL, Stein RA, Muth RM (1995) Management issues and their relative priority within state fisheries agencies. Fisheries 20(10):14–21

McShane L, Bowman J (2003) A literature review of Yellow Perch regulations in North America. Ontario Ministry of Natural Resources, Peterborough, ON, 8 p

Michigan Department of Natural Resources (2018) Charting the course: Fisheries Division's framework for managing aquatic resources (2018–2022 Fisheries Division Strategic Plan), Lansing, MI, 15 p

Mosel KJ, Isermann DA, Hansen JF (2015) Evaluation of daily creel and minimum length limits for Black Crappie and Yellow Perch in Wisconsin. N Am J Fish Manag 35:1–13

Noble RL, Jones TW (1993) Managing fisheries with regulations. In: Kohler CC, Hubert WA (eds) Inland fisheries management in North America. American Fisheries Society, Bethesda, MD, pp 383–402

O'Keefe DM, Miller SR (2011) 2009 Michigan charter fishing study. Michigan Sea Grant Report MICHU-11-200. 23 p

Radomski PJ, Grant GC, Jacobson PC, Cook MF (2001) Visions for recreational fishing regulations. Fisheries 26(5):7–18

Rypel AL (2015) Effects of a reduced daily bag limit on Bluegill size structure in Wisconsin lakes. N Am J Fish Manag 35:388–397

Rypel AL, Lyons J, Tober Griffin JD, Simonson TD (2016) Seventy-year retrospective on size-structure changes in the recreational fisheries of Wisconsin. Fisheries 41:230–243

Schneider JC (1983) Experimental walleye-perch management in a small lake. Michigan Department of Natural Resources, Fisheries Division, Fisheries Research Report 1905. Ann Arbor, MI

Schneider JC, Lockwood RN (1979) Effects of regulations on the fisheries of Michigan lakes. Michigan Department of Natural Resources, Fisheries Division, Fisheries Research Report 1872. Ann Arbor, MI

Schneider JC, O'Neal RP, Clark, Jr. RD (2007) Ecology, management, and status of Walleye, Sauger, and Yellow Perch in Michigan. Michigan Department of Natural Resources, Fisheries Division, Special Report 41. Ann Arbor, MI

Su Z, Clapp D (2013) Evaluation of sample design and estimation methods for Great Lakes angler surveys. Trans Am Fish Soc 142:234–246

Terrell JW, Cade BS, Carpenter J, Thompson JM (1996) Modeling stream fish habitat limitations from wedge-shaped patterns of variation in standing stock. Trans Am Fish Soc 125:104–117

Thomas MV, Towns GL (2011) Summary of creel survey results for the St. Clair River, Lake St. Clair, and the Detroit River, 2002-05. Michigan Department of Natural Resources, Fisheries Division, Fisheries Technical Report 2011-1. Ann Arbor, MI

Thompson JD, Weiblen G, Thomson BA, Alfaro S, Legendre P (1996) Untangling multiple factors in spatial distributions: lilies, gophers, and rocks. Ecology 77:1698–1715

Wehrly KE, Hayes DB, Wills TC (2015) Status and trends of Michigan inland lake resources, 2002-2007. Michigan Department of Natural Resources, Fisheries Division, Fisheries Report 08. Ann Arbor, MI

Wilberg MJ, Bence JR, Eggold BT, Makauskas D, Clapp DF (2005) Yellow Perch dynamics in southwestern Lake Michigan during 1986-2002. N Am J Fish Manag 25:1130–1152

Distribution and Abundance of Pelagic Larval Yellow Perch in Lake St. Clair (USA/Canada) and Adjoining Waters

Robin L. DeBruyne, Taaja R. Tucker, Clara Lloyd, Andrew S. Briggs, Megan Belore, and Edward F. Roseman

Abstract Yellow Perch is one of the most sought-after species in the recreational fisheries of lakes St. Clair and Erie and is commercially important in lakes Huron and Erie. Long-term ichthyoplankton surveys revealed high densities of larval Yellow Perch originating from Lake St. Clair drifting through the Detroit River to Lake Erie that were found to contribute to the western Lake Erie stock. We examined the density and distribution of larval Yellow Perch in Lake St. Clair and the Detroit River to identify spawning and nursery areas and ecological factors influencing their early life history. Lake-wide pelagic larval sampling occurred weekly during 2018 and 2019 using paired bongo nets at 35 Lake St. Clair and 11 Detroit River sites beginning in mid-March through mid-August. Yellow Perch first appeared in samples on May 8, 2018, and April 22, 2019, when mean lake temperatures reached 7–11 °C and quickly peaked in density (May 14–21; 9–13 °C). Significant density hot spots were present in Mitchell's Bay (northeast Lake St. Clair) and in Anchor Bay (northwest Lake St. Clair) in 2018 and 2019 and corresponded with shallower areas containing high levels of submerged aquatic vegetation. Weekly density trends in the Detroit River sites followed the trends in Lake St. Clair, but were generally lower in density, indicating most pelagic Yellow Perch likely originated upstream in Lake St. Clair. Larval densities were higher in 2018, which were reflected in higher

R. L. DeBruyne (✉)
Department of Environmental Sciences, Lake Erie Center, University of Toledo, Oregon, OH, USA
e-mail: robin.debruyne@utoledo.edu

T. R. Tucker · E. F. Roseman
U.S. Geological Survey Great Lakes Science Center, Ann Arbor, MI, USA

C. Lloyd
Paul Smith's College, Paul Smiths, NY, USA

A. S. Briggs
Michigan Department of Natural Resources, Lake St. Clair Fisheries Research Station, Harrison Twp, MI, USA

M. Belore
Ontario Ministry of Natural Resources and Forestry, Wheatley, ON, Canada

J. C. Bruner, R. L. DeBruyne (eds.), *Yellow Perch, Walleye, and Sauger: Aspects of Ecology, Management, and Culture*, Fish & Fisheries Series 41,
https://doi.org/10.1007/978-3-030-80678-1_4

young-of-year fall trawl catches near Anchor Bay. By combining measurements of larval Yellow Perch density, water temperature, aquatic vegetation, and water clarity, we identified two areas providing suitable habitat for larval Yellow Perch survival and growth through mid-summer, when larvae metamorphosed to the demersal juvenile stage. We provide a contemporary assessment of early life stage of Yellow Perch and the factors associated with their distribution and abundance to aid future management of this species and its habitats in Lake St. Clair.

Keywords Yellow Perch · Lake St. Clair · Larvae · Distribution · Temperature · Aquatic vegetation · Great Lakes connecting channel

1 Introduction

Assessing the spatial and temporal distribution of early life history stages of freshwater fish is important for understanding spawning timing and locations (McKenna et al. 2008; Sesterhenn et al. 2014; Tucker et al. 2018), habitat selection (Paradis et al. 2014; Massicotte et al. 2015), and interactions with their environment (Bacheler et al. 2011; Janssen et al. 2014; Carreon-Martinez et al. 2015). Systems with optimal habitat and available prey have lower starvation rates and promote growth and survival of larval fish, resulting in successful fish recruitment (Houde 1987; Graeb et al. 2004; Zhao et al. 2009). Nearshore and offshore areas in both lentic and lotic systems can be important in the life history and production of fishes because of the variation of habitat and food resources (Robinson et al. 1998; Roseman and O'Brien 2013; Fraker et al. 2015). Variability in ecosystem conditions can significantly affect larval fish survival due to their high sensitivity to environmental disturbances during this life stage (Houde 1989).

The St. Clair-Detroit River System (SCDRS) is a Great Lakes connecting channel bordering the United States and Canada encompassing the St. Clair River, Lake St. Clair, and the Detroit River that connects Lake Huron to Lake Erie (Fig. 1). The SCDRS contains important fish spawning and nursery habitats and serves as a major migration route for fishes (Goodyear et al. 1982; Hondorp et al. 2014; Kessel et al. 2018; Faust et al. 2019; Colborne et al. 2019). Important for spawning and nursery habitat, the aquatic macrophyte community in Lake St. Clair has transitioned in the past century from minimal bottom coverage to extensive bottom coverage not observed since the 1800s (Thomas and Haas 2012). Even so, many spawning and nursery habitats, especially in the rivers, have been lost or altered as a result of accommodations for commercial navigation, industry, and urban development (Bennion and Manny 2011). The recreational fishery for all species in Lake St. Clair was valued at $26.3 million in 2017, with Yellow Perch (*Perca flavescens*) being the third (United States) or second to fourth (Canada) most popular species targeted by anglers (Castle 2018). Yellow Perch catch rates during the 2017 open water season were the highest for any species (Castle 2018). The Yellow Perch fishery is typically concentrated in Lake St. Clair, emphasizing the need to

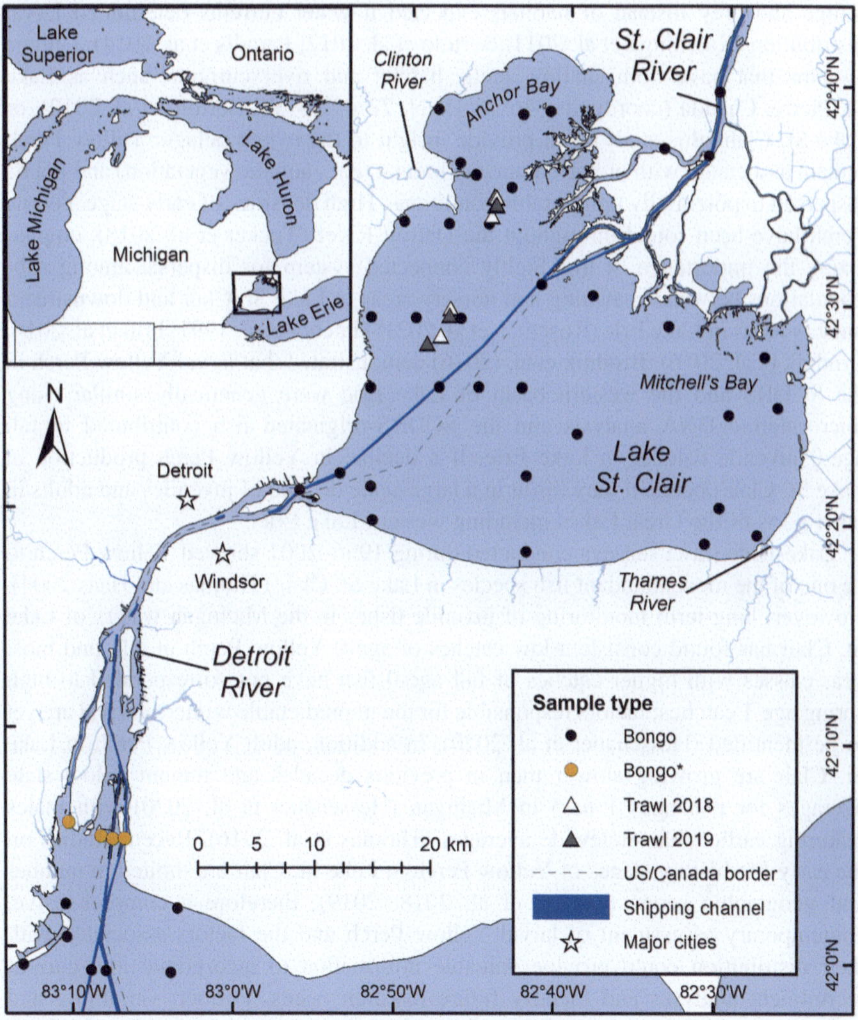

Fig. 1 Map of the lower St. Clair River, Lake St. Clair, and Detroit River showing surface larval (circles) and young-of-year trawl (triangle) sampling sites for 2018 and 2019. "Bongo*" samples indicate four Detroit River bongo sites sampled throughout the seasons in 2018 and 2019

understand the population and recruitment dynamics in this system to effectively manage this valuable fishery across the SCDRS.

Pelagic larval fish distribution is often determined by water currents transporting larvae until larvae develop the ability to control their position in the water column (Hjort 1914). This relationship has been documented for percids in rivers and lakes (Mion et al. 1998; Zhao et al. 2009; Fraker et al. 2015), including Yellow Perch (Roseman et al. 2021; Höök et al. 2006). Larval Yellow Perch distribution has been associated with specific habitats, such as submerged aquatic vegetation, that provide refuge and prey instead of habitats expected if water currents determined larval distribution (Kaemingk et al. 2011; Bertolo et al. 2012; Paradis et al. 2014). Unique systems that have both shallow lentic habitat and river currents, such as Lake St. Pierre, Canada (coordinates 46°12′15′ N, 72°49′56′ W; Bertolo et al. 2012), or Lake St. Clair (this study), can provide insight to the extent pelagic Yellow Perch remain associated with preferred nursery habitat (e.g., aquatic vegetation) and not be dispersed to potentially unfavorable conditions. High densities of early stage Yellow Perch have been found throughout the Detroit River (Tucker et al. 2018), emphasizing the importance of this highly connected system for dispersal among subpopulations between spawning and nursery areas in Lake St. Clair and downstream nursery areas in Lake Erie (Roseman et al. 2021; Hatcher et al. 1991; Pritt et al. 2014; Brodnik et al. 2016). Brodnik et al. (2016) demonstrated that larval Yellow Perch in the SCDRS and the western basin of Lake Erie were genetically similar using microsatellite DNA analysis and the SCDRS-originated fish contributed to fall age-0 juvenile catches in Lake Erie. If a decline in Yellow Perch production in Lake St. Clair occurs, it may result in a large-scale decline of juveniles and adults in other parts of the Great Lakes including western Lake Erie.

Lake-wide trawl surveys conducted during 1996–2001 showed Yellow Perch to be one of the most abundant fish species in Lake St. Clair (Thomas and Haas 2004). However, long-term monitoring of juvenile fishes in the Michigan waters of Lake St. Clair has found consistent low catches of age-0 Yellow Perch in fall, and most year classes with higher catches of fall age-0 fish have not corresponded to high spring age-1 catches; factors responsible for the unpredictable winter survival are yet to be identified (Hessenauer et al. 2020). In addition, adult Yellow Perch in Lake St. Clair are growing slower than in previous decades and remain below state averages for fish ages 1 to 5 in Michigan (Hessenauer et al. 2020) with males maturing earlier than statewide averages (Thomas et al. 2016). Recent studies on the early-life history stages of Yellow Perch in Lake St. Clair are limited in number and geographic scope (Tucker et al. 2018, 2019); therefore a comprehensive, contemporary assessment of larval Yellow Perch and the factors associated with their distribution could provide valuable information to incorporate into current recruitment analyses and identify future research needs. Further, variability and uncertainty of climate change effects on physical processes, ecosystems, habitats, populations, and individuals (Whitney et al. 2016; Myers et al. 2017; Kao et al. 2020) emphasize the value of contemporary population dynamics information for valuable commercial and recreational fish populations (Ludsin et al. 2014; DeVanna Fussell et al. 2016; Collingsworth et al. 2017).

The goal of our research was to assess the temporal and spatial distribution and densities of pelagic larval stages and young-of-year Yellow Perch in Lake St. Clair, including the lower St. Clair River (immediately upstream) and the upper Detroit River (immediately downstream). Using these larval data, we examine relationships between density of larval Yellow Perch and environmental factors such as temperature, submerged aquatic vegetation, and water depth. The temporal distribution and densities of pelagic Yellow Perch were also compared with those in the lower Detroit River and river mouth to assess connectivity between Lake St. Clair and western

Lake Erie. Understanding the movement and drift patterns of the early life stages of Yellow Perch in these connected areas can help elucidate Yellow Perch spawning activity and larval habitat conditions in the SCDRS, providing new information to aide future management of this species and its habitats throughout the SCDRS.

2 Methods

2.1 *Study Area*

The St. Clair River, Lake St. Clair, and Detroit River form part of the border between Michigan, the United States, and Ontario, Canada (Fig. 1). The St. Clair River extends 62 km from the outlet of Lake Huron to Lake St. Clair where the channel splits to form a large delta. Lake St. Clair is relatively shallow (mean depth 3.7 m) with a surface area of 1114 km^2 and drains into the Detroit River (51 km long), which provides 80% of water for Lake Erie (Bolsenga and Herdendorf 1993). Common fish species native to these bodies of water include Yellow Perch, Walleye (*Stizostedion vitreum*), Lake Sturgeon (*Acipenser fulvescens*), Lake Whitefish (*Coregonus clupeaformis*), Muskellunge (*Esox masquinongy*), and various suckers (family Catostomidae) and minnows (family Cyprinidae) (Roth et al. 2013). The SCDRS supports a valuable recreational fishery and is heavily utilized by commercial shipping traffic which contributed to the introduction of several non-native plant and animal species such as Round Goby (*Neogobius melanostomus*), Common Reed (*Phragmites australis* subsp. *australis*), and dreissenid mussels (Zebra Mussel *Dreissena polymorpha* and Quagga Mussel *D. bugensis*), with the first detections of Zebra Mussels in the United States occurring in 1988 in Lake St. Clair (Hebert et al. 1989; Jude et al. 1992; Arzandeh and Wang 2003).

2.2 *Field Collection*

Pelagic larval fishes were sampled weekly in two areas: (1) Lake St. Clair (30 sites within Lake St. Clair, 3 sites upstream in the St. Clair River, and 2 sites in the upper Detroit River and directly upstream in the navigational channel) and (2) lower Detroit River (11 sites in the lower Detroit River; Fig. 1). Sampling occurred during March 22–July 18, 2018, and March 26–August 14, 2019. After the week of May 28, the remaining lower Detroit River samples were collected from only four sites in 2018 (Table 1; Fig. 1). Weekly sampling occurred until larval Yellow Perch were no longer present in the sampling gear. Collections were made during daylight hours using paired bongo nets (two 60-cm diameter conical nets, 3.3 m long, 500-μm mesh) fitted with flow meters to estimate volume of water sampled (Pritt et al. 2015; Tucker et al. 2018). All samples were collected in the upper two meters of the water

Table 1 Number of sample tows, mean weekly surface water temperature (°C), and mean pelagic Yellow Perch (YEP) density per 1000 m^3/tow/site/week by life stage captured in the lower St. Clair River, Lake St. Clair, and upper Detroit River (collectively LSC) and the lower Detroit River (DTR) during 2018 and 2019. Week dates used for display were kept consistent between the 2 years (2018 = Monday date, 2019 = Tuesday date). YS Yolk sac, "–" weeks sampling did not occur

	Sample tows		Temp. (°C)		LSC density			DTR density		
Week	LSC	DTR	LSC	DTR	YS	Larvae	All YEP	YS	Larvae	All YEP
2018										
Mar 19	–	11	–	1.6	–	–	–	0.0	0.0	0.0
Mar 26	30	11	2.0	2.2	0.0	0.0	0.0	0.0	0.0	0.0
Apr 2	35	11	2.0	2.3	0.0	0.0	0.0	0.0	0.0	0.0
Apr 9	35	11	2.2	2.6	0.0	0.0	0.0	0.0	0.0	0.0
Apr 16	35	11	2.4	3.3	0.0	0.0	0.0	0.0	0.0	0.0
Apr 23	35	11	5.5	7.5	0.0	0.0	0.0	0.0	0.0	0.0
Apr 30	35	11	7.3	8.0	0.0	0.0	0.0	0.0	0.0	0.0
May 7	35	11	11.2	12.6	95.0	52.8	147.8	43.2	1.9	45.1
May 14	35	11	10.8	12.9	451.3	1755.0	2206.4	109.0	194.9	303.9
May 21	35	11	13.5	15.1	461.3	1116.5	1577.9	47.3	525.4	572.6
May 28	35	10	18.7	18.9	51.4	56.0	107.3	10.1	127.6	137.7
Jun 4	35	4	14.7	15.8	9.7	82.4	92.2	1.4	39.6	41.0
Jun 11	35	4	16.8	18.4	4.3	22.4	26.7	0.0	41.1	41.1
Jun 18	35	4	18.5	18.8	0.8	17.1	17.9	0.0	13.1	13.1
Jun 25	35	4	18.1	19.3	0.8	15.7	16.5	0.0	1.6	1.6
Jul 2	35	4	22.5	21.8	1.2	0.6	1.8	0.0	0.0	0.0
Jul 9	35	4	22.0	22.8	0.3	0.0	0.3	0.0	0.0	0.0
Jul 16	35	4	22.8	23.7	0.0	0.0	0.0	0.0	0.0	0.0
Jul 30	–	–	–	–	–	–	–	–	–	–
Aug 13	–	–	–	–	–	–	–	–	–	–
2019										
Mar 19	–	–	–	–	–	–	–	–	–	–
Mar 26	35	11	2.2	2.2	0.0	0.0	0.0	0.0	0.0	0.0
Apr 2	35	11	2.6	3.1	0.0	0.0	0.0	0.0	0.0	0.0
Apr 9	35	11	4.9	6.0	0.0	0.0	0.0	0.0	0.0	0.0
Apr 16	35	11	5.2	6.2	0.0	0.0	0.0	0.0	0.0	0.0
Apr 23	35	11	7.0	7.4	0.0	0.0	0.0	1.0	0.0	1.0
Apr 30	35	11	7.1	7.6	19.2	0.0	19.2	2.1	0.6	2.7
May 7	35	11	9.9	9.5	76.8	11.2	88.4[a]	69.6	7.1	76.7
May 14	35	11	8.6	9.1	121.5	273.7	395.4[b]	14.5	6.4	20.9
May 21	35	11	11.7	12.3	142.6	747.0	889.6	24.8	118.2	143.1
May 28	35	11	13.7	15.0	102.9	339.0	441.9	19.5	161.7	181.3
Jun 4	35	11	14.0	15.0	26.8	122.1	148.9	13.1	436.4	449.5
Jun 11	35	11	15.6	17.5	0.1	205.4	205.5	5.0	248.6	253.6
Jun 18	35	11	16.9	17.3	3.0	131.8	134.9	0.0	66.5	66.5
Jun 25	35	12	18.1	20.3	0.0	13.9	13.9	0.0	0.0	0.0
Jul 2	35	11	21.1	22.8	0.0	1.4	1.4	0.0	2.0	2.0
Jul 9	–	–	–	–	–	–	–	–	–	–

(continued)

Table 1 (continued)

	Sample tows		Temp. (°C)		LSC density			DTR density		
Week	LSC	DTR	LSC	DTR	YS	Larvae	All YEP	YS	Larvae	All YEP
Jul 16	35	11	22.7	24.5	0.0	0.0	0.0	0.0	0.0	0.0
Jul 30	35	11	22.9	24.0	0.0	0.0	0.0	0.0	0.0	0.0
Aug 13	35	11	22.8	23.4	0.0	0.0	0.0	0.0	0.0	0.0

[a]Includes two unknown life stage YEP
[b]Includes one unknown life stage YEP

column by towing the bongo nets in a circular pattern into the current (where applicable) at approximately 3.0 km/h for 5 min (Pritt et al. 2015; Tucker et al. 2018). Surface water temperature (°C), water depth (m), and Secchi disc depth (m) were collected at all sample sites. All samples were preserved in the field using 95% ethanol. Samples were brought to the laboratory at the US Geological Survey Great Lakes Science Center in Ann Arbor, Michigan, and larval fish were removed and identified to lowest possible taxon (Auer 1982). Pelagic Yellow Perch were classified into developmental stage (i.e., yolk sac or larvae), and a random subset of individuals ($n = 30$) of each larval stage from a sample were measured to the nearest 0.01 mm. Pelagic larval counts and water volume sampled were used to estimate larval density per tow (number of fish per cubic meter per tow [No. fish/m^3/tow]). In rare cases, individual samples were lost ($n = 8$), with only one sample representing a tow.

Bottom trawl surveys for fall forage fish have been conducted in Lake St. Clair by the Michigan Department of Natural Resources since 1996 to provide indices of abundance, sizes, and distribution of demersal fishes (Thomas and Haas 2012). Trawl sampling was limited to two areas of the Michigan waters due to extensive macrophyte coverage, and the number of sites sampled varied between the years (six sites in August 2018, five sites in August 2019). The two-seam otter trawl used for all tows had a 10.66-m headrope with 4.6-m wings and was 18.9 m in overall length. The net was constructed of 76-, 38-, and 32-mm graded stretched-measure mesh from gape to cod end, with a 9-mm stretched-mesh liner in the cod end. The net was towed along the bottom for 10 min by a single warp and 45.7-m bridle at approximately 3.7 km/h. After the trawl was pulled onto the boat, all fish were identified to species, enumerated, a subset of each species were measured for total length, and a group weight by species and age class was taken.

2.3 Data Analysis

All statistical analyses were performed in R v.4.0.2 (R Core Team 2020). Temporal trends in pelagic larval fish density were examined by calculating the weekly mean larval density (No. fish/m^3/tow) at each sampling site. A hot spot analysis was used to assess pelagic Yellow Perch spatial distributions in Lake St. Clair and the lower Detroit River by identifying areas that had significant clustering of Yellow Perch larvae. The hot spot analysis considers pelagic larval fish density at each sampling site within the context of the weight of surrounding sites to examine if densities of Yellow Perch are significantly clustered (Nelson and Boots 2008; Tucker et al. 2018, 2019). The Getis-Ord-Gi statistic was calculated to identify significant hot spots (most dense) sites ($Z > 1.96$) or cold spots (less dense) sites ($Z < -1.96$) (Bivand and Piras 2015). Hot spot analysis was performed using the *spdep* R package (Bivand et al. 2019) using inverse distance weighting and a distance band such that each site had at least one neighbor. Hot spot analysis was performed for each week when sampling occurred using the mean density of pelagic Yellow Perch per site (No. fish/m^3/tow/site).

Boosted regression tree (BRT) analysis was used to determine which environmental variables influenced mean pelagic larval Yellow Perch densities (No. fish/1000 m^3/tow) at the Lake St. Clair sites. A machine learning process, BRTs combine many regression trees to increase predictive power (De'Ath 2007). BRTs are useful for performing exploratory analyses when a variety of potential independent variables are collected but their relative impact on the dependent variable is not yet known and when nonlinear effects or missing values are present (Elith et al. 2008). Independent variables included year, water temperature, water depth, Secchi disc depth, latitude, longitude, percent submerged aquatic vegetation (SAV) cover, and percent SAV volume in the water column. SAV metrics obtained for each site were extracted from a spatial layer of average SAV values collected across Lake St. Clair during 2008–2011 by the Michigan Department of Natural Resources (Thomas and Haas 2012) (www.glahf.org), while all other values were collected during larval sampling. Data extraction was performed in ArcMap v10.7.1. Of the variables included in analysis, only SAV volume and SAV cover were moderately correlated ($r = 0.61$). BRTs were generated with the *dismo* R package v1.1–4 (Hijmans et al. 2017), using the Gaussian distribution, learning rate = 0.005, bag fraction = 0.5, and tree complexity = 2 (Elith et al. 2008). Data from 1212 larval tows (2422 samples) were used in analyses, and mean pelagic larval Yellow Perch densities were $\log_{10}$-transformed for the Gaussian model. Model performance was assessed using percent deviance explained and the "pseudo-R-square" or the CV correlation squared (Riley et al. 2015). Importance of each independent variable is determined by its percent relative influence on the model or the percent contribution to the reduction to the model's loss function. Partial dependence plots were created to show each variable's marginal effect on modeled larval density. BRTs were created for all pelagic Yellow Perch collected, and separately for yolk sac and larval stage fish, for a total of three BRT models.

3 Results

Larval Yellow Perch were detected from May 8, 2018, to July 11, 2018, and April 22, 2019, to July 3, 2019. At the Lake St. Clair sites, a total of 73,848 larval fish of all species were collected in 590 tows (1179 samples) from 35 sites in 2018, with 29,858 larvae identified as Yellow Perch (Table 1). In 2019, 69,064 total fish larvae were collected with 16,294 identified as Yellow Perch (622 tows, 1243 samples). At the 11 lower Detroit River sites, a total of 6056 larval fish were collected in 148 tows (294 samples) in 2018, and in 2019 15,190 larval fish were captured in 200 tows (396 samples). Yellow Perch represented 2086 of fish caught in 2018 and 2340 in 2019. Larval Yellow Perch in Lake St. Clair sites ranged from 4.5 to 23 mm TL. Lengths of yolk sac-staged fish remained consistent across time for both years; however, yolk sac-staged larvae were captured into July during 2018 compared to only mid-June in 2019 (Fig. 2). Larvae-stage Yellow Perch were captured from early May through the end of the bongo net sampling period in July.

In 2018, pelagic Yellow Perch were consistently concentrated in the east portion of Lake St. Clair, along the Canadian shoreline near the Thames River and in Mitchell's Bay, and in the Anchor Bay region in the northwest portion of Lake St. Clair, with few pelagic Yellow Perch captured in the lower St. Clair River or upper Detroit River (Fig. 3). The highest densities of larval Yellow Perch were found in areas having lower Secchi disc depths (<2.5 m) and shallower depths (Fig. 4). Lake-wide mean weekly densities were greatest during mid- to late May, when lake-wide mean water temperatures were 10.8–13.5 °C (Table 1). The lower Detroit River sites captured pelagic Yellow Perch concurrent with sites in Lake St. Clair beginning in early May, but at lower densities than those near Anchor Bay or Mitchell's Bay (Fig. 3). The hot spot analysis results revealed no significant cold spots ($Z < -1.96$) throughout the study period; however significant hot spots ($Z > 1.96$) of high larval Yellow Perch densities were found during May, June, and July in Mitchell's Bay and near the Clinton River delta in the Anchor Bay area (Fig. 5). Hot spots for Mitchell's Bay occurred primary during the month of May, while hot spots near Anchor Bay occurred primarily in June until early July. No hot spots were detected in the lower Detroit River sites in 2018.

Even though approximately half the number of larval Yellow Perch were captured in 2019 compared to 2018, the patterns in catch locations were similar. Larval Yellow Perch densities were highest in the east portion of Lake St. Clair near the Thames River and in Mitchell's Bay beginning a week earlier than in 2018 (Fig. 3), corresponding with the earlier warming of the lake (Table 1). The Anchor Bay region of Lake St. Clair also had high densities of larval Yellow Perch in 2019 (Fig. 3). The highest densities of larval Yellow Perch were again in areas with lower Secchi disc depths (<2.5 m) and shallow areas along the shorelines. The first pelagic yolk sac Yellow Perch was captured late April in the lower Detroit River and was captured in the lower through early July, similar to catches in Lake St. Clair. Significant hot spots ($Z > 1.96$) were found again in Mitchell's Bay and near the Thames River in the southeast (Fig. 5) where they were first detected at the end of

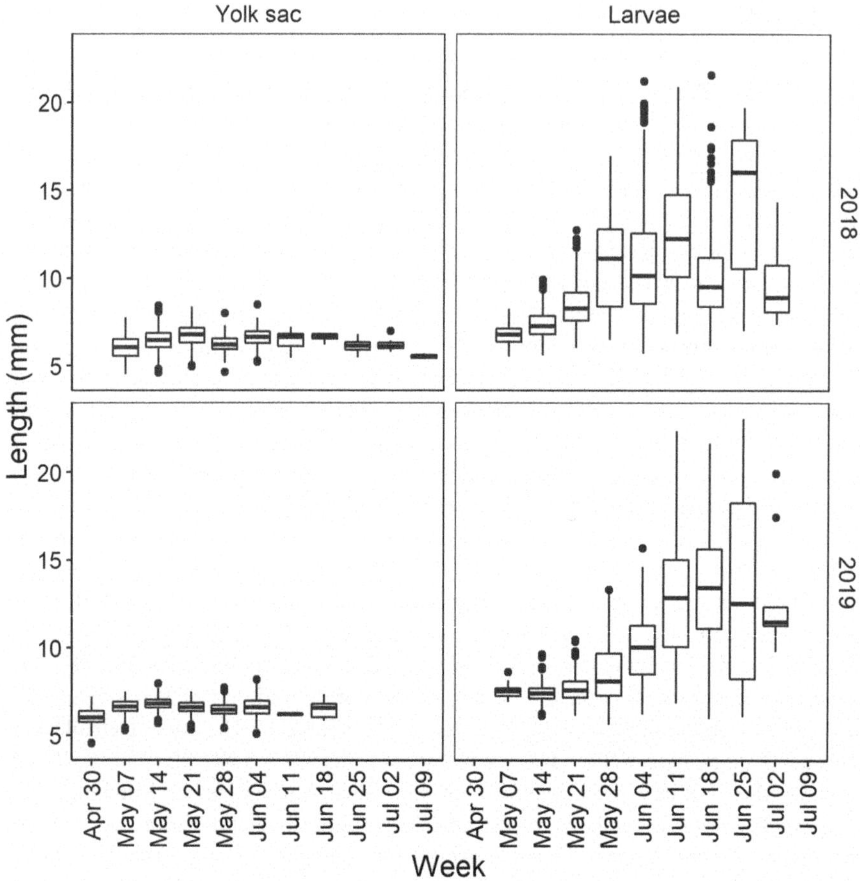

Fig. 2 Weekly Yellow Perch total lengths (mm) by developmental stage during 2018 (top panels) and 2019 (bottom panels) captured in the Lake St. Clair area. Note that not all larval fish captured were measured (see Methods for subsample description)

April and only until mid-May. Significant hot spots were also found near Anchor Bay and the Clinton River delta during late May through early July, but less frequently than in 2018. A single hot spot was detected in the lower Detroit River in late April from a single larval fish captured at one site. No significant cold spots ($Z < -1.96$) were found in 2019. Similar to 2018, lake-wide mean weekly densities were greatest during mid- to late May, when lake-wide mean water temperatures were 8.6–13.7 °C (Fig. 6; Table 1).

The mean trawl density of young-of-year Yellow Perch from the 6 sites sampled on August 25, 2018, was 407 fish/ha (total = 1277 fish), with a mean total length of 54.8 mm. The mean trawl density of young-of-year Yellow Perch from the 5 sites sampled on August 25, 2019, was 109 fish/ha (total = 211 fish), with a mean total length of 50.4 mm.

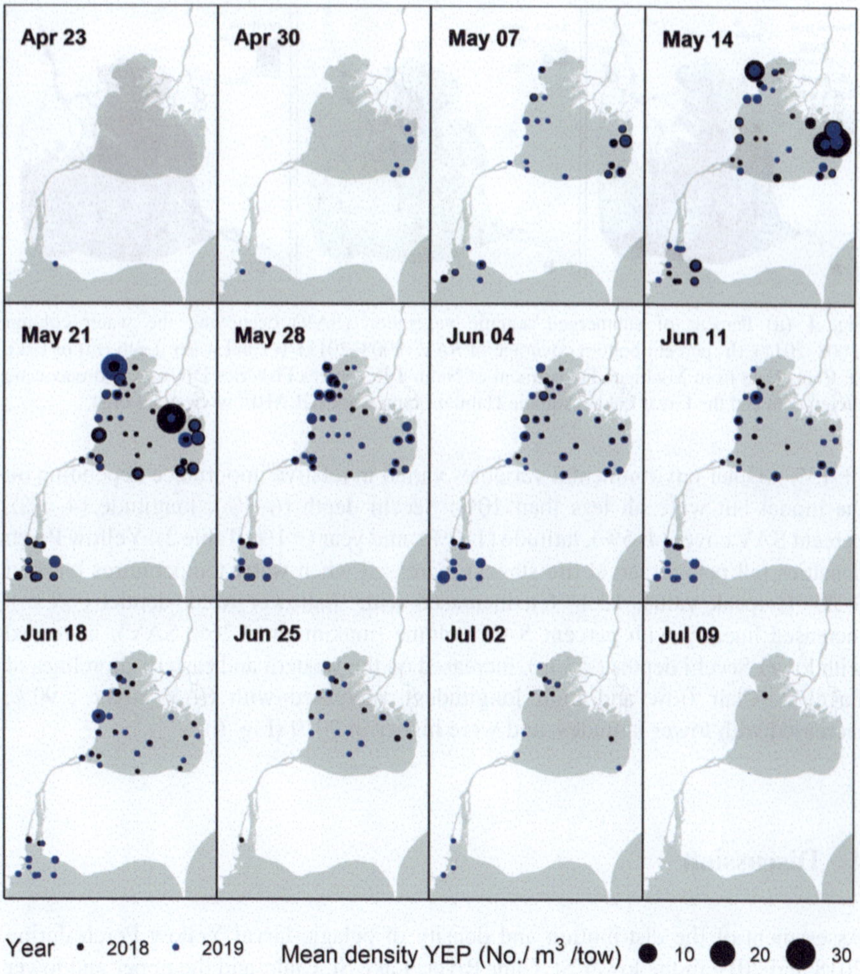

Fig. 3 Mean larval Yellow Perch (YEP) density (No./m^3/tow) at each sample site by week in the lower St. Clair River, Lake St. Clair, upper Detroit River, and lower Detroit River in 2018 (black circles) and 2019 (blue circles). Size of the circle indicates the density of Yellow Perch; the smallest circle represents 0–10 Yellow Perch/m^3/tow. Sites with zero Yellow Perch catches in a week are not shown

Boosted regression tree models identified relationships between mean density per tow of all pelagic Yellow Perch, larvae only, and yolk sac only fish and a variety of environmental variables. Model performance and relative influence values of explanatory variables were similar for all three models (Table 2), and we focus on results for all pelagic Yellow Perch in partial dependence plots (Fig. 6). Water temperature was the most important environmental variable explaining pelagic Yellow Perch density per tow in all three models (>53% relative influence), followed by percent SAV volume of the water column (11–14%) or water depth

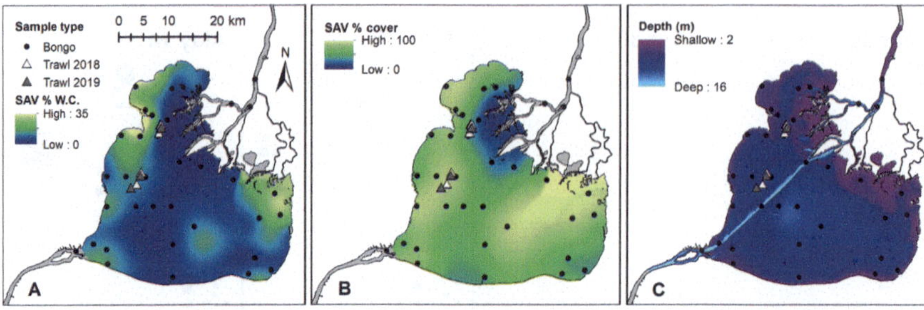

Fig. 4 (**a**) Percent of submerged aquatic vegetation (SAV) occupying the water column (2008–2011), (**b**) percent bottom coverage of SAV (2008–2011), (**c**) and water depth (m) of Lake St. Clair. Data from Michigan Department of Natural Resources Fisheries Division hydroacoustic survey data and the Great Lakes Aquatic Habitat Framework (GLAHF; www.glahf.org)

(9–13%). Other environmental variables varied in relative importance depending on the model but were all less than 10%: Secchi depth (6–7%), longitude (4–7%), percent SAV cover (4–5%), latitude (1–3%), and year (<1%; Table 2). Yellow Perch densities (all pelagic larval life stages) increased when water temperatures reached 9–20 °C (peak value: 13.4 °C), increased with shallower water depths (<9 m), increased linearly with percent SAV volume (maximum at 26% SAV), increased with lower Secchi depths (<3 m), increased on the western and eastern shorelines of Lake St. Clair (low and high longitudes), increased with SAV cover >90%, increased with lower latitudes, and were higher in 2019 (Fig. 6).

4 Discussion

Assessment of the distribution and density of pelagic larval Yellow Perch during 2018 and 2019 in the lower St. Clair River, Lake St. Clair, and the upper and lower Detroit River shows two regions of Lake St. Clair primarily utilized by larval Yellow Perch, even during years of contrasting larval densities. Larval Yellow Perch were concentrated in the warmer waters of the Anchor Bay and Mitchell's Bay regions in Lake St. Clair, with little presence in the lower St. Clair River or upper Detroit River. Additionally, swift-moving river waters likely reduced pelagic larval residence times compared to the more protected and vegetated bay areas near spawning areas or where pelagic larvae may have been retained by recirculating lake currents. Pelagic Yellow Perch densities in the lower Detroit River were lower than the Anchor Bay and Mitchell's Bay areas, but the temporal patterns and stages present coincided with those in Lake St. Clair. Temperature was found to have the highest percent influence on modeled pelagic larval densities in Lake St. Clair, which is consistent with other studies of larval Yellow Perch in the Great Lakes (Weber et al. 2011; Redman et al. 2011; Janetski et al. 2013). The fall young-of-year densities from the trawl surveys were higher in 2018 than 2019, possibly indicating that a significant proportion of

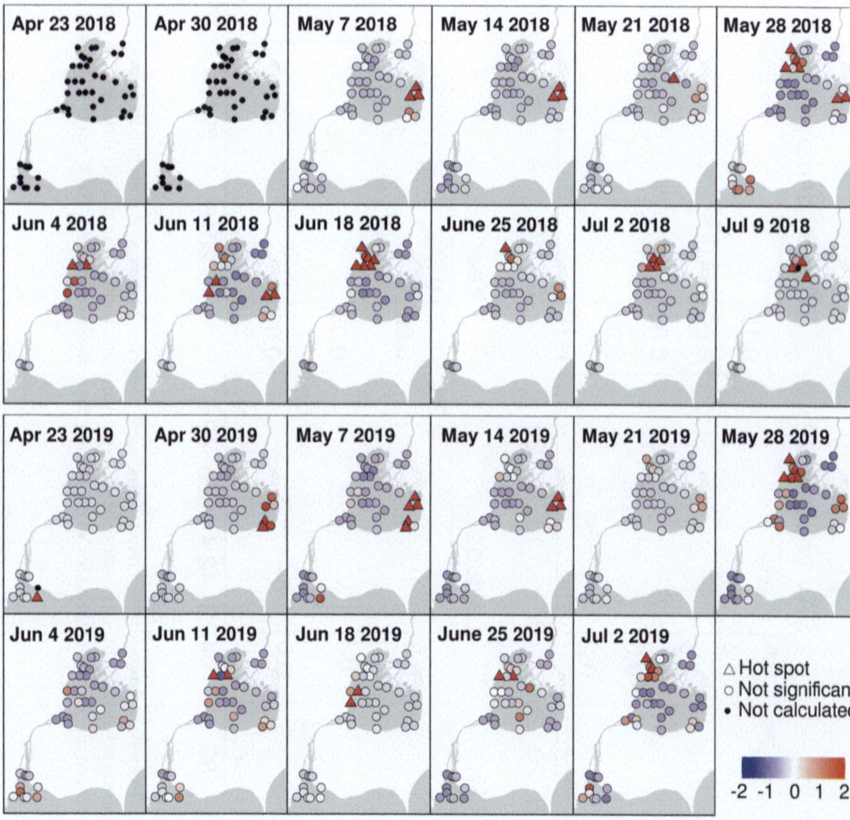

Fig. 5 Weekly hot spot analysis results for pelagic Yellow Perch in the lower St. Clair River, Lake St. Clair, and Detroit River during 2018 (top) and 2019 (bottom). Symbols indicate if a site was a significant hot spot ($Z > 1.96$) or not a significant hot spot ($Z < 1.96$). Significant cold spots ($Z < -1.96$) were not found, so it is not noted in legend. Black dots represent no Yellow Perch captured during that week at a sample location (2018) or no sampling performed at the location (2019)

larval Yellow Perch remained in Lake St. Clair through their first summer after hatching during 2018.

Emergence of larval Yellow Perch in Lake St. Clair initially occurred in Mitchell's Bay in both years where densities increased over time and had the highest recorded densities in 2018. Waters in Mitchell's Bay were the warmest in the lake and increased in temperature more quickly (mean surface water >12 °C week of May 7 in 2018 and 2019) than the rest of lake, especially when compared to the St. Clair River sites and sites near the larger delta channels (mean surface water >10 °C week of May 21 in 2018 and May 28 during 2019). In addition to shallow depths of Mitchell's Bay and Anchor Bay, tributaries may contribute to warming water temperatures because the landscape runoff is typically warmer than cooler waters of the St. Clair River. Based on our collection of yolk sac larvae, minimal egg incubation time (10 days; Whiteside et al. 1985), and weekly mean water

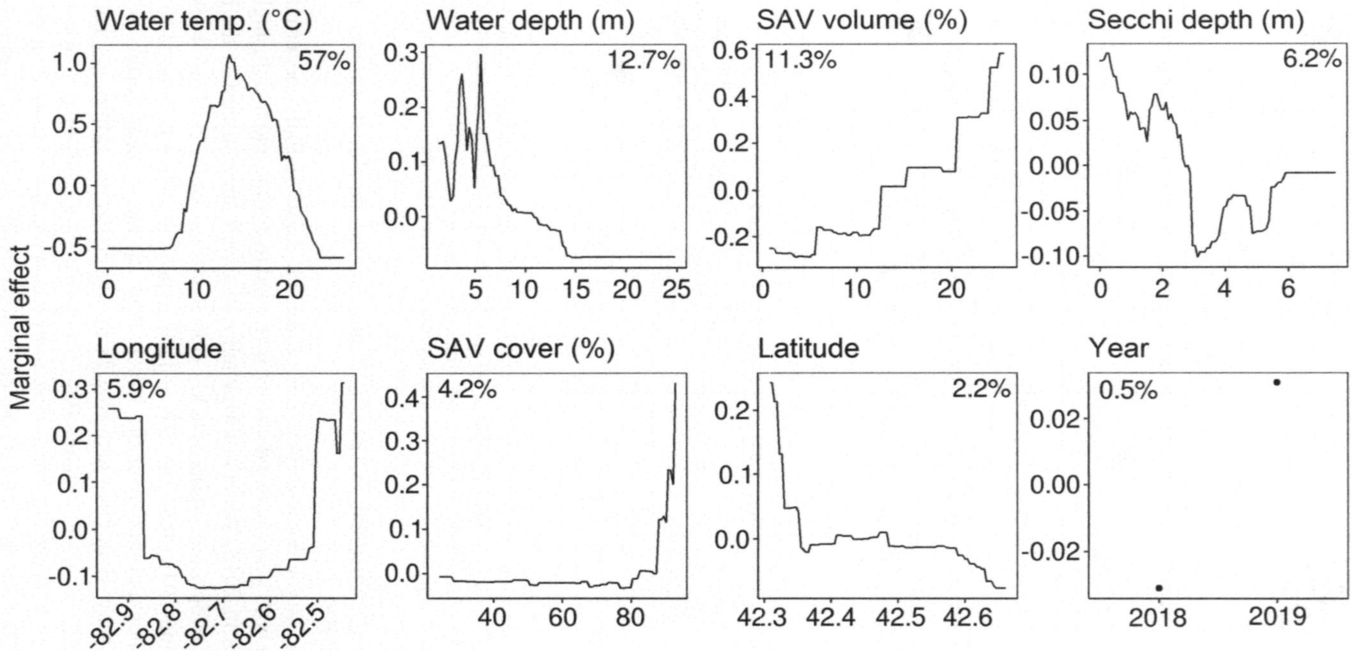

Fig. 6 Partial dependence plots showing boosted regression tree (BRT) model contributions (relative influence %, upper right corner of plot) and their modeled marginal effect on pelagic Yellow Perch density (yolk sac and larvae stage) at Lake St. Clair sites for the eight variables included in the BRT model. "Water temp." = water temperature. "SAV volume (%)" is the percent submerged aquatic vegetation (SAV) occupying the water column, and "SAV cover (%)" is the percent SAV on the lake bottom

Table 2 Boosted regression tree model statistics and relative influence values for models of Yellow Perch yolk sac, larvae, and all pelagic fish (all larval and yolk sac Yellow Perch) from the Lake St. Clair sites. "Water temp." = water temperature. "SAV volume" is the percent submerged aquatic vegetation (SAV) occupying the water column, and "SAV cover" is the percent SAV on the lake bottom

Model	Proportion deviance explained	CV proportion deviance explained	Pseudo-R-square	Number of trees	Relative influence (%)
Yolk sac	0.52	0.36	0.36	3750	Water temp. (61.1%), SAV volume (14.3%), water depth (8.7%), Secchi depth (6.2%), longitude (4.4%), SAV cover (3.6%), latitude (1.2%), year (0.5%)
Larvae	0.66	0.53	0.54	5600	Water temp. (53.3%), water depth (12.7%), SAV volume (11%), Secchi depth (7.2%), longitude (6.6%), SAV cover (5.1%), latitude (3.4%), year (0.7%)
All Yellow Perch	0.67	0.55	0.55	5600	Water temp. (57%), water depth (12.7%), SAV volume (11.3%), Secchi depth (6.2%), longitude (5.9%), SAV cover (4.2%), latitude (2.2%), year (0.5%)

temperatures at sites with earliest larval catches, we believe spawning occurred in parts of Lake St. Clair at <10 °C (between 5.7 and 8.2 °C in 2018 and 6.7–7.3 °C in 2019), which is similar to Matt et al. (2021) and is within the temperature range reported by Hokanson (1977). The warmer water temperatures during April 2019 (Table 1) may have reduced incubation time of eggs leading to a shorter hatch window compared to 2018, indicated by capture of yolk sac larvae into early July in 2018 (Fig. 2). The cooler summer temperatures in 2019 or changes in prey availability and abundance may be factors influencing the smaller-sized young-of-year at the end of August compared to 2018 (Power and van den Heuvel 1999; Bremigan et al. 2003), although neither zooplankton abundance nor larval diets were assessed as part of this study. Persistence of sac fry larvae into July during 2018 is unusual but not unique. Using back-calculated hatch dates from otolith daily increment analysis, Fitzgerald et al. (2001) inferred that some Yellow Perch juveniles collected in seines from nearshore areas in Lake St. Clair hatched in early to mid-July 1998, although the natal source of these fish was speculative.

During their early life stages, Yellow Perch have been shown to shift between nearshore and offshore habitats to obtain food, optimize growth conditions, or seek refuge from predators (Whiteside et al. 1985; Post and McQueen 1988; Bacheler et al. 2011; Manning et al. 2013; Paradis et al. 2014). In areas with water currents, larvae can be directed by physical factors away from spawning locations (Mion et al. 1998; Beletsky et al. 2004, 2007; Roseman et al. 2005, 2021; Weber et al. 2011). With established currents in Lake St. Clair influenced by the St. Clair and Detroit rivers (Schwab et al. 1989; Anderson et al. 2010), we anticipated larval fish would be moving through Lake St. Clair on these currents; however, this is not what we found in 2018 or 2019. This lack of movement, indicated by the lack of shifting higher densities or hot spots into the central area of Lake St. Clair, may have been influenced by nearshore vegetation retaining Yellow Perch and inhibiting their advection (Kaemingk et al. 2011) or older more developed larval fish actively maintaining position within the nearshore habitat (Paradis et al. 2014). Paradis et al. (2014) found larval Yellow Perch more strongly associated with specific habitat (i.e., aquatic macrophytes) than juvenile Yellow Perch.

The geographic scope of Tucker et al. (2018) only included sites in the St. Clair and Detroit rivers and river mouths and concluded that the significant hot spots found near the head of the Detroit River were likely caused by larvae transported from Lake St. Clair. By including Lake St. Clair in the analyses with the river sites, Anchor Bay and Mitchell's Bay were identified as the significant hot spots in the SCDRS because of the relatively higher density of pelagic larvae found in the lake. This study supports the conclusions of Tucker et al. (2018) and Roseman et al. (2021) that pelagic Yellow Perch are transported downstream based on the corresponding temporal and developmental stage overlap between pelagic Yellow Perch captured in the lower Detroit River with those in Lake St. Clair. There are small wetlands in the Detroit River previously found to have early-stage Yellow Perch larvae present (McDonald et al. 2014); however based on temperature in the Detroit River during 2018 and 2019, pelagic larval Yellow Perch would have been drifting weeks prior to those originating from Lake St. Clair.

There is little evidence that our sampling design missed sources of Yellow Perch larvae significant to Lake St. Clair. Upstream of Lake St. Clair are the much colder St. Clair River and Lake Huron. Except for the delta, the St. Clair River does not possess large areas of vegetation to serve as spawning areas or nursery areas (Goodyear et al. 1982), and only a few individual Yellow Perch larvae were captured in the St. Clair River delta during this study or by Tucker et al. (2018). Recruits originating from Lake Huron would emerge and drift later than in Lake St. Clair based on spawning temperature preferences. Therefore, the limited sampling locations in the St. Clair River were not considered a limitation to our sampling design. The Detroit River flows south into western Lake Erie, and export of Yellow Perch has been estimated at 319–690 million larvae annually between 2010 and 2015 (Roseman et al. 2021). Microsatellite analysis determined Yellow Perch from Lake St. Clair drift downstream and contribute to the juvenile abundance in Lake Erie (Brodnik et al. 2016). Contributions from other larger tributaries, such as the Clinton River, Michigan, and Thames River, Ontario, into Lake St. Clair are possible though not documented. The Clinton River flows into Anchor Bay, and those sites had high larval densities and were significant hot spots; therefore, we cannot rule out the Clinton River as a spawning location contributing to the larval Yellow Perch found in Lake St. Clair.

The limited bottom trawl sampling in Lake St. Clair due to heavy vegetation coverage (Fig. 4) prevented using trawl data for statistically meaningful hot spot analysis, estimating lake-wide fall age-0 densities, and analytical comparisons with the pelagic larval densities. The continued lack of suitable trawl sites supports our assumption that the vegetation coverage data from 2008 to 2011 (Fig. 4) could effectively represent vegetation coverage conditions during our sampling years in the BRT analysis. Future considerations to improve our analyses could include updating the aquatic vegetation maps with a seasonal component, incorporating additional sites throughout the lake suitable for trawling to complete a hot spot analysis for demersal young-of-year Yellow Perch or additional days of sampling to track age-0 Yellow Perch abundance and growth through their developmental stages (Forney 1971; Henderson and Nepszy 1988; Irwin et al. 2009; Manning et al. 2013).

Anchor Bay and Mitchell's Bay are regions of Lake St. Clair that experience the greatest aquatic macrophyte density, diversity, and growth (Fig. 4; Thomas and Haas 2012) and were the first areas where larval Yellow Perch were detected in late April or early May. While some vegetation persists over winter, new vegetation in these regions begins growing in April with peak aquatic vegetation densities occurring July through August (Schloesser et al. 1985; Thomas and Haas 2012) and is likely attracting adult Yellow Perch to use these areas for spawning (Parker et al. 2009). In our BRT analyses, the amount of SAV in the water column was the second or third most important predictor of larval and yolk sac Yellow Perch densities after water temperature. Therefore, the consistent high densities of pelagic Yellow Perch in regions with dense macrophytes could be a result of a combination of behavioral choices by the adult spawners to promote the retention of larvae once hatched and by the larvae to remain near these areas with warmer water temperatures and aquatic vegetation, presumably for increased prey availability or refuge.

The lower densities of larval Yellow Perch in 2019 could be from reduced spawning, lower hatching success, lower larval survival, or larval transport out of Lake St. Clair. Yellow Perch recruitment has been found to be controlled by Walleye predation in Lake Erie (Hartman and Margraf 1993) and Oneida Lake (Forney 1974; Nielsen 1980), but the comparatively low Walleye abundance in Lake St. Clair (Hessenauer et al. 2020) may not exert enough predation pressure to reduce the pelagic and young-of-year densities as seen between 2018 and 2019. It is possible that a larger proportion of larvae were advected out of the system in 2019, but sampling at the Detroit River river mouth sites stopped at the end of May in 2018, so comparisons cannot be made between years. Anchor and Mitchell's bays had lower water transparency in 2018, compared to other areas of Lake St. Clair, which could decrease larval predation and mortality when pelagic Yellow Perch have little mobility to actively evade predators (Houde 1969; Graeb et al. 2004; Carreon-Martinez et al. 2015), increasing week-to-week survival compared to 2019. Unfortunately, adult spawner abundance, egg density, and hatching success were not estimated during these study years; therefore we are not able to definitively identify the factors responsible for the reduced larval Yellow Perch density in 2019.

By comparing measurements of larval Yellow Perch density, water temperature, aquatic vegetation, and water clarity, we identified two high-density larval hot spot areas that may be significant spawning and nursery locations for Yellow Perch in Lake St. Clair. These areas are providing suitable habitat for larval Yellow Perch survival and growth through mid-summer, when larvae metamorphosed to the demersal juvenile stage. Fall bottom trawl sampling in these areas confirmed Yellow Perch remain in these areas through their first summer. Conducting spawning and egg assessments of Yellow Perch in April could confirm adult spawning site selection. Expanding bottom trawl sampling or other fish community sampling techniques, temporally and geographically, could aid in identifying when bottleneck(s) in recruitment occur and provide a year-class strength index for Yellow Perch in Lake St. Clair.

Acknowledgments The authors are grateful for the field and laboratory assistance provided by the numerous dedicated and proficient technicians and students who helped collect and process the thousands of larval fish samples that support this analysis. This work was supported by the Great Lakes Restoration Initiative, Michigan Department of Natural Resources (MDNR), and the US Geological Survey, Great Lakes Science Center. The MDNR bottom trawl survey is funded by the Sport Fish Restoration Fund (Study 230488, Project F-61-R, Michigan). Portions of this work were used as a senior thesis for C. Lloyd at Paul Smith's College under the guidance of Dr. Celia Evans. Handling of fishes followed the guidelines provided in the "Guidelines for the use of fishes in research" published by the American Fisheries Society, Bethesda, MD. Any use of trade, product, or firm names is for descriptive purposes only and does not imply endorsement by the US Government.

References

Anderson EJ, Schwab DJ, Lang GA (2010) Real-time hydraulic and hydrodynamic model of the St. Clair River, Lake St. Clair, Detroit River system. J Hydraul Eng 136:507–518. https://doi.org/10.1061/(ASCE)HY.1943-7900.0000203

Arzandeh S, Wang J (2003) Monitoring the change of *Phragmites* distribution using satellite data. Can J Remote Sens 29:24–35. https://doi.org/10.5589/m02-077

Auer NA (1982) Identification of larval fishes of the Great Lakes basin with emphasis on the Lake Michigan drainage. Great Lakes Fishery Commission, Ann Arbor, MI

Bacheler NM, Paoli TJ, Schacht GM (2011) Controls on abundance and distribution of yellow perch: predator, water quality, and density-dependent effects. Trans Am Fish Soc 140:989–1000. https://doi.org/10.1080/00028487.2011.603979

Beletsky D, Schwab D, Mason D, Rutherford E, McCormick M, Vanderploeg H, Janssen J (2004) Modeling the transport of larval yellow perch in Lake Michigan. Estuar Coast Model 2003:439–454

Beletsky D, Mason DM, Schwab DJ, Rutherford ES, Janssen J, Clapp DF, Detmars JM (2007) Biophysical model of larval yellow perch advection and settlement in Lake Michigan. J Great Lakes Res 33(4):842–866. https://doi.org/10.3394/0380-1330(2007)33[842:BMOLYP]2.0.CO;2

Bennion DH, Manny BA (2011) Construction of shipping channels in the Detroit River: history and environmental consequences. US Geological Survey, Ann Arbor, MI

Bertolo A, Blanchet FG, Magnan P, Brodeur P, Mingelbier M, Legendre P (2012) Inferring processes from spatial patterns: the role of directional and non–directional forces in shaping fish larvae distribution in a freshwater lake system. PLoS One 7:e50239. https://doi.org/10.1371/journal.pone.0050239

Bivand R, Piras G (2015) Comparing implementations of estimation methods for spatial econometrics. J Stat Softw 63:1–36. https://doi.org/10.18637/jss.v063.i18

Bivand R, Altman M, Anselin L, Assunção R, Berke O, Bernat A, Müller W (2019) Spdep: spatial dependence: weighting schemes, statistics. version 1.1-3

Bolsenga SJ, Herdendorf CE (eds) (1993) Lake Erie and Lake St. Clair handbook. Wayne State University Press, Detroit, MI

Bremigan MT, Dettmers JM, Mahan AL (2003) Zooplankton selectivity by larval yellow perch in Green Bay, Lake Michigan. J Great Lakes Res 29:501–510

Brodnik RM, Fraker ME, Anderson EJ, Carreon-Martinez L, DeVanna KM, Heath DD, Reichert JM, Roseman EF, Ludsin SA (2016) Larval dispersal underlies demographically important intersystem connectivity in a Great Lakes yellow perch (*Perca flavescens*) population. Can J Fish Aquat Sci 73:416–426. https://doi.org/10.1139/cjfas-2015-0161

Carreon-Martinez LB, Walter RP, Johnson TB, Ludsin SA, Heat DD (2015) Benefits of turbid river plume habitat for Lake Erie yellow perch (*Perca flavescens*) recruitment determined by Juvenile to larval genotype assignment. PLoS One 10:e0125234. https://doi.org/10.1371/journal.pone.0125234

Castle D (2018) Fishing for answers: restoration in the St. Clair-detroit river system improves angling opportunities. MSc Thesis, Central Michigan University

Colborne SF, Hondorp DW, Holbrook CM, Lowe MR, Boase JC, Chiotti JA, Wills TC, Roseman EF, Krueger CC (2019) Sequence analysis and acoustic tracking of individual lake sturgeon identify multiple patterns of river–lake habitat use. Ecosphere 10(12):e02983. https://doi.org/10.1002/ecs2.2983

Collingsworth PD, Bunnell DB, Murray MW, Kao Y-C, Feiner ZS, Claramunt RM, Lofgren BM, Höök TO, Ludsin SA (2017) Climate change as a lon-term stressor for the fisheries of the Laurentian Great Lakes of North America. Rev Fish Biol Fisheries 27:363–391. https://doi.org/10.1007/s11160-017-9480-3

De'Ath G (2007) Boosted trees for ecological modeling and prediction. Ecology 88:243–251

DeVanna Fussell KM, Smith REH, Fraker ME, Boegman L, Frank KT, Miller TJ, Tyson JT, Arend KK, Boisclair D, Guildford SJ, Hecky RE, Höök TO, Jensen OP, Llopiz JK, May CJ, Najjar RG, Rudstam LG, Taggart CT, Rao YR, Ludsin SA (2016) A perspective on needed research, modeling, and management approaches that can enhance Great Lakes fisheries management under changing ecosystem conditions. J Great Lakes Res 42(4):743–752. https://doi.org/10.1016/j.jglr.2016.04.007

Elith J, Leathwick JR, Hastie T (2008) A working guide to boosted regression trees. J Anim Ecol 77:802–813

Faust MD, Vandergoot CS, Brenden TO, Kraus RT, Hartman T, Krueger CC (2019) Acoustic telemetry as a potential tool for mixed-stock analysis of fishery harvest: a feasibility study using Lake Erie walleye. Can J Fish Aquat Sci 76:1019–1030. https://doi.org/10.1139/cjfas-2017-0522

Fitzgerald D, Dale A, Thomas M, Sale P (2001) Application of otolith analyses to investigate broad size distributions of young yellow perch in temperate lakes. J Fish Biol 58:248–263

Forney JL (1971) Development of dominant year classes in a yellow perch population. Trans Am Fish Soc 100(4):739–749. https://doi.org/10.1577/1548-8659(1971)100<739:DODYCI>2.0.CO;2

Forney JL (1974) Interactions between yellow perch abundance, walleye predation, and survival of alternate prey in Oneida Lake, New York. Trans Am Fish Soc 103:15–24. https://doi.org/10.1577/1548-8659(1974)103<15:IBYPAW>2.0.CO;2

Fraker ME, Anderson EJ, May CJ, Chen K-Y, Davis JJ, DeVanna KM, DuFour MR, Marschall EA, Mayer CM, Miner JG, Pangle KL, Pritt JJ, Roseman EF, Tyson JT, Zhao Y, Ludsin SA (2015) Stock-specific advection of larval walleye (*Sander vitreus*) in western Lake Erie: Implications for larval growth, mixing, and stock discrimination. J Great Lakes Res 41:830–845. https://doi.org/10.1016/j.jglr.2015.04.008

Goodyear CD, Edsall TA, Ormsby-Dempsey DM, Moss GD, Polanski PE (1982) Atlas of spawning and nursery areas of Great Lakes fishes. US Fish and Wildlife Service, Washington, DC Report FWS/OBS-82/52

Graeb BD, Dettmers JM, Wahl DH, Cáceres CE (2004) Fish size and prey availability affect growth, survival, prey selection, and foraging behavior of larval yellow perch. Trans Am Fish Soc 133:504–514

Hartman K, Margraf F (1993) Evidence of predatory control of yellow perch (*Perca flavescens*) recruitment in Lake Erie, USA. J Fish Biol 43:109–119. https://doi.org/10.1111/j.1095-8649.1993.tb00414.x

Hatcher CO, Nester RT, Muth KM (1991) Using larval fish abundance in the St. Clair and Detroit rivers to predict year-class strength of forage fish in Lakes Huron and Erie. J Great Lakes Res 17:74–84. https://doi.org/10.1016/S0380-1330(91)71343-9

Hebert PD, Muncaster BW, Mackie GL (1989) Ecological and genetic studies on *Dreissena polymorpha* (Pallas): a new mollusc in the Great Lakes. Can J Fish Aquat Sci 46:1587–1591. https://doi.org/10.1139/f89-202

Henderson BA, Nepszy SJ (1988) Recruitment of Yellow Perch (*Perca flavescens*) affected by stock size and water temperature in Lakes Erie and St. Clair, 1965–85. J Great Lakes Res 14: 205–215. https://doi.org/10.1016/S0380-1330(88)71549-X

Hessenauer J-M, Briggs A, Utrup B, Wills T (2020) Status of the fisheries in Michigan waters of Lake Erie and Lake St. Clair, 2019. Michigan Department of Natural Resources, Lansing, MI, pp 1–20

Hijmans R, Phillips S, Leathwick J, Elith J (2017) Package 'dismo'. Species Distribution Modeling. R package version 1:1–4

Hjort J (1914) Fluctuations in the great fisheries of northern Europe viewed in the light of biological research. ICES Conseil Permanent Internatlonal Pour 1, 'Explora Tion De La Mer Rapports Et Procrs-Verbaux 20:3–228

Hokanson KEF (1977) Temperature requirements of some percids and adaptations to the seasonal temperature cycle. J Fish Res Board Can 34:1524–1550

Hondorp DW, Roseman EF, Manny BA (2014) An ecological basis for future fish habitat restoration efforts in the Huron-Erie Corridor. J Great Lakes Res 40:23–30. https://doi.org/10.1016/j.jglr.2013.12.007

Höök TO, McCormick MJ, Rutherford ES, Mason DM, Carter GS (2006) Short-term water mass movements in Lake Michigan: implications for larval fish transport. J Great Lakes Res 32:728–737

Houde ED (1969) Sustained swimming ability of larvae of walleye (*Stizostedion vitreum vitreum*) and yellow perch (*Perca flavescens*). J Fish Board Can 26:1647–1659

Houde ED (1987) Fish early life dynamics and recruitment variability. Am Fish Soc Symp Ser 2: 17–29

Houde ED (1989) Subtleties and episodes in the early life of fishes. J Fish Biol 35:29–38

Irwin BJ, Rudstam LG, Jackson JR, VanDeValk AJ, Forney JL, Fitzgerald DG (2009) Depensatory mortality, density-dependent growth, and delayed compensation: disentangling the interplay of mortality, growth, and density during early life stages of Yellow Perch. Trans Am Fish Soc 138: 99–110. https://doi.org/10.1577/T07-256.1

Janetski DJ, Ruetz CR, Bhagat Y, Clapp DF (2013) Recruitment dynamics of age-0 yellow perch in a drowned river mouth lake: assessing synchrony with nearshore Lake Michigan. Trans Am Fish Soc 142:505–514. https://doi.org/10.1080/00028487.2012.756432

Janssen J, Marsden JE, Hrabik TR, Stockwell JD (2014) Are the Laurentian Great Lakes great enough for Hjort? ICES J Mar Sci 71(8):2242–2251. https://doi.org/10.1093/icesjms/fst220

Jude DJ, Reider RH, Smith GR (1992) Establishment of Gobiidae in the Great Lakes basin. Can J Fish Aquat Sci 49:416–421. https://doi.org/10.1139/f92-047

Kaemingk MA, Jolley JC, Willis DW, Graeb BDS (2011) Exploring spatial distributions of larval yellow perch Perca flavescens, bluegill Lepomis macrochirus and their prey in relation to wind. J Fish Biol 78:1132–1151. https://doi.org/10.1111/j.1095-8649.2011.02924.x

Kao Y-C, Rogers MW, Bunnell DB, Cowx IG, Qian SS, Anneville O, Beard TD Jr, Brinker A, Britton JR, Chura-Cruz R, Gownaris NJ, Jackson JR, Knagur K, Kolding J, Lukin AA, Lynch AJ, Mercado-Silva N, Moncayo-Estrada R, Njaya FJ, Ostrovsky I, Rudstam LG, Sandström

ALE, Sato Y, Siguayro-Mamani H, Thorpe A, van Zwieten PAM, Volta P, Wang Y, Weiperth A, Olaf LF, Weyl OLF, Young JD (2020) Effects of climate and land-use changes on fish catches across lakes at a global scale. Nat Commun 11:2526. https://doi.org/10.1038/s41467-020-14624-2

Kessel ST, Hondorp DW, Holbrook CM, Boase JC, Chiotti JA, Thomas MV, Wills TC, Roseman EF, Drouin R, Krueger CC (2018) Divergent migration within lake sturgeon (*Acipenser fulvescens*) populations: multiple distinct patterns exist across an unrestricted migration corridor. J Anim Ecol 87:259–273. https://doi.org/10.1111/1365-2656.12772

Ludsin SA, DeVanna KM, Smith REH (2014) Physical–biological coupling and the challenge of understanding fish recruitment in freshwater lakes. Can J Fish Aquat Sci 71:775–794. https://doi.org/10.1139/cjfas-2013-0512

Manning NF, Mayer CM, Bossenbroek JM, Tyson JT (2013) Effects of water clarity on the length and abundance of age-0 yellow perch in the Western Basin of Lake Erie. J Great Lakes Res 39: 295–302. https://doi.org/10.1016/j.jglr.2013.03.010

Massicotte P, Bertolo A, Brodeur P, Hudon C, Mingelbier M, Magnan P (2015) Influence of the aquatic vegetation landscape on larval fish abundance. J Great Lakes Res 41:873–880

Matt KJ, Welsh SA, Smith DM (2021) Spawning characteristics of yellow perch during periods of water level fluctuations in a hydropower reservoir. In: Bruner JC, DeBruyne RL (eds) Ecology, management, and culture of Yellow Perch, Walleye and Sauger. Springer, Heidelberg, pp 3–32

McDonald EA, McNaught AS, Roseman EF (2014) Use of main channel and two backwater habitats by larval fishes in the Detroit River. J Great Lakes Res Suppl 40:69–80. https://doi.org/10.1016/j.jglr.2013.10.001

McKenna JE, Davis BM, Fabrizio MC, Savino JF, Todd TN, Bur M (2008) Ichthyoplankton assemblages of coastal west-central Lake Erie and associated habitat characteristics. J Great Lakes Res 34:755–769. https://doi.org/10.1016/S0380-1330(08)71616-2

Mion JB, Stein RA, Marschall EA (1998) River discharge drives survival of larval walleye. Ecol Appl 8:88–103. https://doi.org/10.1890/1051-0761(1998)008[0088:RDDSOL]2.0.CO;2

Myers BJE, Lynch AJ, Bunnell DB, Chu C, Falke JA, Kovach RP, Krabbenhoft TJ, Kwak TJ, Paukert CP (2017) Global synthesis of the documented and projected effects of climate change on inland fishes. Rev Fish Biol Fisheries 27:339–361. https://doi.org/10.1007/s11160-017-9476-z

Nelson TA, Boots B (2008) Detecting spatial hot spots in landscape ecology. Ecography 31:556–566. https://doi.org/10.1111/j.0906-7590.2008.05548.x

Nielsen LA (1980) Effect of walleye (*Stizostedion vitreum vitreum*) predation on juvenile mortality and recruitment of yellow perch (*Perca flavescens*) in Oneida Lake, New York. Can J Fish Aquat Sci 37:11–19. https://doi.org/10.1139/f80-002

Paradis Y, Bertolo A, Mingelbier M, Brodeur P, Magnan P (2014) What controls distribution of larval and juvenile yellow perch? The role of habitat characteristics and spatial processes in a large, shallow lake. J Great Lakes Res 40:172–178. https://doi.org/10.1016/j.jglr.2013.12.001

Parker AD, Stepien CA, Sepulveda-Villet OJ, Ruehl CB, Uzarski DG (2009) The interplay of morphology, habitat, resource use, and genetic relationships in young yellow perch. Trans Am Fish Soc 138:899–914

Post JR, McQueen DJ (1988) Ontogenetic changes in the distribution of larval and juvenile yellow perch (*Perca flavescens*): a response to prey or predators? Can J Fish Aquat Sci 45:1820–1826

Power M, van den Heuvel MR (1999) Age-0 yellow perch growth and its relationship to temperature. Trans Am Fish Soc 128:687–700

Pritt JJ, DuFour MR, Mayer CM, Roseman EF, DeBruyne RL (2014) Sampling little fish in big rivers: larval fish detection probabilities in two Lake Erie tributaries and implications for sampling effort and abundance indices. Trans Am Fish Soc 143(4):1011–1027. https://doi.org/10.1080/00028487.2014.911204

Pritt JJ, Roseman EF, Ross JE, DeBruyne RL (2015) Using larval fish community structure to guide long-term monitoring of fish spawning activity. N Am J Fish Manag 35(2):241–252. https://doi.org/10.1080/02755947.2014.996687

R Core Team (2020) R: A language and environment for statistical computing. R Foundation for Statistical Computing. Vienna. https://www.R-project org/[Google Scholar]

Redman RA, Czesny SJ, Dettmers JM, Weber MJ, Makauskas D (2011) Old tales in recent context: current perspective on yellow perch recruitment in Lake Michigan. Trans Am Fish Soc 140 (5):1277–1289. https://doi.org/10.1080/00028487.2011.620480

Riley SC, Tucker TR, Adams JV, Fogarty LR, Lafrancois BM (2015) Factors associated with the deposition of *Cladophora* on Lake Michigan beaches in 2012. J Great Lakes Res 41 (4):1094–1105

Robinson AT, Clarkson RW, Forrest RE (1998) Dispersal of larval fishes in a regulated river tributary. Trans Am Fish Soc 127:772–786

Roseman EF, O'Brien TP (2013) Spatial distribution of pelagic fish larvae in the northern main basin of Lake Huron. Aquat Ecosyst Health Manage 16(3):311–321. https://doi.org/10.1080/14634988.2013.824348

Roseman EF, Taylor WW, Hayes DB, Tyson JT, Haas RC (2005) Spatial patterns emphasize the importance of coastal zones as nursery areas for larval walleye in western Lake Erie. J Great Lakes Res 31(Suppl 1):28–44. https://doi.org/10.1016/S0380-1330(05)70288-4

Roseman EF, DuFour M, Pritt J, Fischer JL, DeBruyne RL, Bennion DH (2021) Export of pelagic fish larvae from a large Great Lakes connecting channel. Adv Limnol 66:20. https://doi.org/10.1127/adv_limnol/2021/0060

Roth BM, Mandrak NE, Hrabik TR, Sass GG, Peters J (2013) Fishes and decapod crustaceans of the Great Lakes basin. Great Lakes fisheries policy and management, pp 105–135

Schloesser DW, Edsall TA, Manny BA (1985) Growth of submersed macrophyte communities in the St. Clair--Detroit River system between Lake Huron and Lake Erie. Can J Bot 63:1061–1065. https://doi.org/10.1139/b85-145

Schwab DJ, Clites AH, Murthy CR, Sandall JE, Meadows LA, Meadows GA (1989) The effect of wind on transport and circulation in Lake St. Clair. J Geophys Res Oceans 94(C4):4947–4958

Sesterhenn TM, Roswell CR, Stein SR, Klaver P, Verhamme E, Pothoven SA, Höök TO (2014) Modeling the implications of multiple hatching sites for larval dynamics in the resurgent Saginaw Bay walleye population. J Great Lakes Res Suppl 40:113–122

Thomas MV, Haas RC (2004) Status of the Lake St. Clair Fish Community and Sport Fishery, 1996-2001. Michigan Department of Natural Resources, Fisheries Division Research Fisheries Research Report 2067, Lansing, MI, 1–26p

Thomas MV, Haas RC (2012) Status of Lake St. Clair submerged plants, fish community, and sport fishery. Michigan Department of Natural Resources Fisheries Research Report 2099, Lansing, MI, 1–90p

Thomas M, Wills T, Hessenauer J-M (2016) Status of the Fisheries in Michigan Waters of Lake Erie and Lake St. Clair, 2015. Michigan Department of Natural Resources, Lansing, MI, 1–27p

Tucker TR, Roseman EF, DeBruyne RL, Pritt JJ, Bennion DH, Hondorp DW, Boase JC (2018) Long-term assessment of ichthyoplankton in a large North American river system reveals changes in fish community dynamics. Can J Fish Aquat Sci 75:2255–2270. https://doi.org/10.1139/cjfas-2017-0511

Tucker TR, DeBruyne RL, Roseman EF, Larson D, McNaught AS (2019) Assessment of larval fish assemblages and nursery habitat in the St. Clair River delta. J Great Lakes Res 45(4):762–776

Weber MJ, Dettmers JM, Wahl DH (2011) Growth and survival of age-0 yellow perch across habitats in southwestern Lake Michigan: early life history in a large freshwater environment. Trans Am Fish Soc 140:1172–1185

Whiteside MC, Swindoll CM, Doolittle WL (1985) Factors affecting the early life history of yellow perch, *Perca flavescens*. Environ Biol Fish 12:47–56

Whitney JE, Al-Chokhachy R, Bunnell DB, Coldwell CA, Cooke SJ, Eliason EJ, Rogers M, Lynch AJ, Paukert CP (2016) physiological basis of climate change impacts on North American Inland Fishes. Fisheries 41:332–345. https://doi.org/10.1080/03632415.2016.1186656

Zhao Y, Jones ML, Shuter BJ, Roseman EF (2009) A biophysical model of Lake Erie walleye (*Sander vitreus*) explains interannual variations in recruitment. Can J Fish Aquat Sci 66:114–125. https://doi.org/10.1139/F08-188

Part II
Walleye (*Stizostedion vitreum*)

Using Genomic Data to Guide Walleye Management in the Great Lakes

Peter T. Euclide, Jason Robinson, Matthew Faust, Stuart A. Ludsin, Thomas M. MacDougall, Elizabeth A. Marschall, Kuan-Yu Chen, Chris Wilson, Matthew Bootsma, Wendylee Stott, Kim T. Scribner, and Wesley A. Larson

Abstract Genetic and genomic resources are being developed at a rapid pace, offering powerful tools that can help protect and sustain ecologically important fish populations and the valued fisheries that they support. Herein, we discuss recent and ongoing genetic/genomic research in the Great Lakes and how high-throughput sequencing data has informed Walleye (*Stizostedion vitreum*) biology and management. During 2017–2018, RAD-sequencing refined descriptions of population genetic structure in Lake Erie, showing that genomic data can improve assignment accuracy for mixed-stock analysis. During 2018–2019, research demonstrated the

P. T. Euclide (✉)
Department of Biological Sciences, College of Letters and Sciences, University of Wisconsin – Milwaukee, Milwaukee, WI, USA

J. Robinson
New York State Department of Environmental Conservation, Lake Erie Fisheries Research Unit, Dunkirk, NY, USA
e-mail: jason.robinson@dec.ny.gov

M. Faust
Ohio Department of Natural Resources, Sandusky, OH, USA
e-mail: matthew.faust@dnr.state.oh.us

S. A. Ludsin · E. A. Marschall · K.-Y. Chen
Aquatic Ecology Laboratory, Department of Evolution, Ecology and Organismal Biology, The Ohio State University, Columbus, OH, USA
e-mail: ludsin.1@osu.edu; marschall.2@osu.edu; chen.1735@osu.edu

T. M. MacDougall
Lake Erie Management Unit, Ontario Ministry of Natural Resources and Forestry, Port Dover, ON, Canada
e-mail: tom.macdougall@ontario.ca

C. Wilson
Ontario Ministry of Natural Resources and Forestry, Aquatic Research Section, Trent University, Peterborough, ON, Canada
e-mail: chris.wilson@ontario.ca

M. Bootsma
College of Natural Resources, University of Wisconsin-Stevens Point, Stevens Point, WI, USA

J. C. Bruner, R. L. DeBruyne (eds.), *Yellow Perch, Walleye, and Sauger: Aspects of Ecology, Management, and Culture*, Fish & Fisheries Series 41,
https://doi.org/10.1007/978-3-030-80678-1_5

function of Rapture panels to determine the natal origins of Walleye captured in eastern Lake Erie's recreational and commercial mixed-stock fisheries, indicating that both local (eastern basin) and distant (western basin) local spawning populations contribute to these fisheries. In 2019, researchers began to evaluate the hierarchical population structure of Walleye from 31 Great Lakes spawning sites and use a 99,636-bait Rapture panel to identify potential signals of local adaptation within these populations. In addition to highlighting these advances, we discuss how the continued development of these and other molecular tools (e.g., GT-seq panels that can help to reduce the cost and processing time for repeated genetic studies) could allow for 1000s of individuals to be cost-effectively genotyped annually. Such ability would continue to pave the way for researchers and management agencies to identify population structure, estimate relative stock contributions to mixed-stock fisheries, evaluate parentage, inform hatchery practices, or conduct other molecular analyses in support of other management or conservation needs.

Keywords Population genetics · RAD-seq · Molecular ecology · Next-generation sequencing · Applied genetics

1 Introduction

As with other ecologically and economically important species of fish (e.g., *Oncorhynchus* spp., *Salmo salar*, *Gadus morhua*), genetic tools have assisted in the management of Walleye (*Stizostedion vitreum*). For example, in the North American Great Lakes, molecular data have been used to track Walleye movements and migrations (Todd and Haas 1993; Robison and Buchanan 2020), quantify the influence of stocking (Caroffino et al. 2011; Garner et al. 2013), identify recolonization pathways from glacial refugia (Stepien and Faber 1998; Stepien et al. 2015), and estimate harvest of independent stocks and large-scale seasonal movements (McParland et al. 1999; Brenden et al. 2015). This information has in turn helped management agencies evaluate the success of hatchery programs, identify weaknesses in recruitment models, and set harvest quotas (MacDougall et al. 2007; Garner et al. 2013; Zhao et al. 2013). Owing to recent advancements in

W. Stott
Michigan State University/USGS Great Lakes Science Center, Ann Arbor, MI, USA
e-mail: wstott@contractor.usgs.gov

K. T. Scribner
Department of Fish and Wildlife, Michigan State University, East Lansing, MI, USA
e-mail: scribne3@msu.edu

W. A. Larson
Alaska Fisheries Science Center, National Oceanic and Atmospheric Administration, Juneau, AK, USA
e-mail: wes.larson@noaa.gov

molecular methods and bioinformatics, which now allow for genetic and genomic data to be cost-effectively attained and analyzed, we expect the use of molecular information in Walleye fisheries research and management to continue to increase within and outside of the Great Lakes.

Since the last time the genetic contributions to Walleye research were reviewed (Billington et al. 2011; Stepien et al. 2015), extensive advances in genetic/genomic biotechnology have transformed the types and quantity of information available to address complex evolutionary questions (Sect. 1.1). This change from "genetic" to "genomic" techniques (Box 1) has helped to untangle fine-scale patterns of population divergence in Great Lakes Walleye (Sect. 3.1) and estimate the relative contributions of Walleye stocks to some of Lake Erie's most valued recreational and commercial fisheries (Sect. 4.1.1). Ongoing genomic research aims to further aid descriptions of spatial Walleye genetic structure throughout the Great Lakes and identify genes associated with local adaptation (Sect. 4.1.2). Finally, the data generated from these studies are leading to the development of additional molecular resources, such as panels consisting of 100–1000s of informative genetic markers (e.g., Rapture, GT-seq), which can enable more efficient collection of genetic data for stock assessments, quota setting, and hatchery management (Sect. 4.2). In this chapter, we briefly describe and discuss four examples of recent and ongoing research that have used high-throughput sequencing technology to assist with Walleye management and evolutionary research in the North American Great Lakes (hereafter Great Lakes). Our fullest expectation is that during the next 5–10 years, these and other expected advancements (e.g., a fully sequenced Walleye genome) will continue to enhance the value of genomic approaches to fisheries conservation and management applications.

Box 1 Common Genetic and Genomic Research Terms and Their Definitions

Assignment accuracy: The relative confidence (generally out of 100%) that an individual from a given stock or population originated from that stock or population. This metric is often used to assess the predictive performance of genetic markers.

Depth-of-coverage: The number of unique sequences that include a particular nucleotide. High depth-of-coverage is important for accurate genotype calls when using high-throughput sequencing data.

Genetic structure: Sometimes referred to as population structure, this term refers to the degree of genetic divergence or gene flow among subpopulations or stocks within the broader surveyed population.

Genetics: The study of heredity or evolution based on inheritance or DNA, usually through the evaluation of a specific, small number of genes or regions of the entire genome. Common techniques include microsatellite genotyping or mitochondrial DNA sequencing and/or nuclear DNA intron sequencing.

(continued)

Box 1 (continued)

Genomics: The study of heredity or evolution based on an organism's entire genome, or a large subset of the total genome of an organism. Genomics commonly refers to large amounts of genomic sequencing data generated on modern high-throughput sequencing platforms.

Genotyping-by-sequencing: Assignment of genotypes based on observed nucleotide differences rather than variation in fragment size or banding patterns often used in traditional genetic approaches such as microsatellite analysis.

High-throughput sequencing (HTS): This term, also known as massively parallel sequencing or next-generation sequencing, refers to a range of sequencing platforms developed during the early 2000s that greatly increased the number of DNA strands that could be sequenced at one time. The efficiency gained through this development has allowed for much broader applications of DNA sequencing to be applied than was previously possible with traditional Sanger sequencing.

Microhaplotype: A short continuous region of the genome containing multiple SNPs that, when scored together, form a single allele.

Population: A single unit of individuals of the same species located in a particular location that sometimes consists of multiple independent subpopulations (stocks) connected by weak gene flow.

Sequencing: The process of determining the order of nucleotides in DNA or RNA.

Single nucleotide polymorphism (SNP): A mutation or substitution of a single nucleotide at a given locus in the genome.

Stock: A single spawning group of individuals of the same species that spawns in a localized area within a population (i.e., a subpopulation or local spawning populations subunit).

1.1 Transition from Genetics to Genomics

The biggest change in molecular research during the last two decades has been a heightened ability to quantify variation across a larger proportion of an organism's genome (e.g., Allendorf et al. 2010; Helyar et al. 2011; Kumar and Kocour 2017). Genetic studies have historically analyzed subsets of 10–100s of polymorphic sites or markers that are assumed to be representative of the average diversity and differentiation across the genome. This information has helped to describe the evolutionary history of the focal species, population, or individual based on similarities and differences in the diversity or allele frequencies at these markers. Examples of genetic approaches in Walleye research include allozyme variation, mitochondrial and nuclear DNA sequencing, restriction fragment length polymorphisms (RFLPs), and microsatellite DNA loci variation (Todd and Haas 1993;

Stepien and Faber 1998; Gatt et al. 2000; Brenden et al. 2015; Stepien et al. 2015). More recently, molecular studies have scaled up the number of markers such that analysis of thousands to millions of markers is now common. Research using thousands to millions of genetic markers (often referred to as "genomics") becomes especially important when an investigation's objective is to identify regions of the genome that are not well represented by average diversity estimates, when divergence is recent in time, or when more precise estimates are required.

The higher number of markers attained by genomic studies is made possible by high-throughput DNA sequencing (HTS), also known as next-generation sequencing technology. Prior to the development of HTS, determining the exact nucleotide order of DNA (i.e., sequencing) was a slow, expensive process that limited the amount of data that could be produced. For this reason, research was often conducted with approaches such as microsatellite or restriction fragment length polymorphism (RFLP) analysis, which estimate allele frequencies at a small number of loci that do not need to be sequenced. For approaches that relied on the knowledge of nucleotide order, investigations often limited research to small portions of the genome, such as variable regions of the mitochondrial genome (mtDNA), major histocompatibility complex (MHC) genes, or nuclear introns to limit the amount of necessary sequencing (summarized in Billington et al. 2011; Stepien et al. 2015). A key advantage of HTS is its ability to reduce the time necessary to sequence large amounts of DNA, thereby allowing researchers to collect data from a larger portion of the genome. This technological advancement has made it possible to begin to shift from targeted genetic approaches to genome-wide approaches. High-throughput DNA sequencing also has allowed entire genomes to be more easily sequenced and assembled, which in turn has conferred a heightened ability to map species traits onto the genome and identify structural variants such as chromosomal inversion (e.g., Flanagan et al. 2018; Johnson et al. 2018). While mapping the Walleye genome has not yet been accomplished, sequencing and assembly of the Walleye genome is underway.

Genomic data have advantages over traditional genetic data when it comes to understanding the forces shaping a species or population and prioritizing management needs (Table 1; Box 2). First, inclusion of orders of magnitude higher numbers of markers in genomic studies increases discriminatory power, thus enhancing the ability to precisely detect genetic population structure, estimate genetic diversity, and calculate effective population size (Allendorf et al. 2010). Additionally, the investigation of thousands of markers across the genome facilitates identification of neutral regions of the genome and regions that may be under selection. Variation in regions of the genome under selection are outliers and commonly do not reflect the average diversity and differentiation across the genome. In turn, both neutral and non-neutral markers can be evaluated separately and/or together to help differentiate among the evolutionary forces shaping a species or population (Luikart et al. 2003). For example, variation at neutral markers reflects evolutionary processes such as migration and genetic drift, whereas markers under selection reflect adaptive differentiation (Holderegger et al. 2006; Pearse et al. 2014). Furthermore, markers that appear to be under selection can help delineate locally adapted populations that could be important for maintaining long-term species viability (Funk et al. 2012) or for predicting population-specific vulnerability to environmental change (Schindler

Table 1 Optimal applications and conditions of molecular approaches in fisheries research and conservation/management. Although many methods can be used for an application, check marks are limited to applications that we suggest are the best. For example, while it is possible to conduct a mixed-stock analysis using whole-genome resequencing, the benefits of doing so would likely not outweigh the cost and time required to conduct such a study

Application	Microsatellites	GT-seq	RAD-seq	Rapture	Whole-genome resequencing
Parentage-based tagging	✓	✓			
Parentage	✓	✓	✓	✓	
Genetic (population) structure	✓	✓	✓	✓	
Neutral genetic diversity	✓	✓	✓	✓	
Population monitoring		✓			
Close-kin mark-recapture		✓		✓	
Large sample sizes ($N > 1000$)		✓		✓	
Mixed-stock analysis		✓		✓	
Replicated studies		✓		✓	
Fisheries-induced evolution			✓	✓	✓
Local adaptation			✓	✓	✓
Structural variant diversity					✓
Trait mapping					✓

et al. 2010; Harrisson et al. 2017). For example, SNP markers identified through a HTS approach were used to identify Murray Cod (*Maccullochella peelii*) populations that were adapted to warmer, drier climates and thus potentially resistant to climate change (Harrisson et al. 2017). Studies such as this show how genomic approaches can help identify populations and stocks that are vulnerable to environmental change or that might need to be protected to maintain population viability. In turn, this information can be used to prioritize conservation and management efforts to ensure that genetic diversity is maintained to help safeguard against future changes in environmental conditions.

Box 2 Common ways that molecular data can be used to assist with fisheries research and inform fisheries conservation and management efforts. Although traditional genetic approaches (e.g., mtDNA, microsatellites) can be used for many of these applications, their resolution is often limited by low numbers of markers. Thus, genomic approaches, which increase the number and types of markers available for discrimination by orders of magnitude, often offer a better means to assist with these applications

Aquaculture production: Owing to expected food limitations with increasing human population growth, aquaculture has become common worldwide,

(continued)

Box 2 (continued)
including for percids such as Walleye. Genomic data can improve selective breeding programs within aquaculture settings by identifying markers linked to phenotypic or performance traits, enabling marker-assisted selection.

Close-kin mark-recapture: This approach offers a way to estimate the demographics of a population, such as its size, by assessing the relatedness (kinship) among sampled individuals. Genomics has conferred this ability by improving the accuracy parent-offspring and full-sibling relationship identification.

Fisheries-induced selection (rapid evolution): Fisheries harvest can exert selective pressure on populations that results in phenotypic and associated genetic shifts that are not conducive to population persistence and production (e.g., reduced size and age at maturation). In populations where fisheries-induced selection is a concern, genomics offers a way to identify the loci under selection that are associated with these adaptive changes as well as quantify genome-wide shifts in genetic diversity.

Mixed-stock analysis (relative stock contributions): Knowledge of the relative contributions of stocks to a breeding population or fisheries harvest can benefit conservation or management efforts by identifying stocks that are underperforming or that are supporting a population. Molecular data can quantify relative stock contributions to a breeding population or a fishery. In populations with low genetic differentiation, genomic approaches that use lots of markers can improve the ability to accurately discriminate among stocks and assign individuals of unknown origin to their source (natal) population(s).

Parentage-based tagging (hatchery supplementation): Many species of fish, including Walleye, are stocked to enhance fishing opportunities or to keep populations viable. Parentage-based tagging leverages genotypes of hatchery brood stock to assign parentage for individuals of unknown origin, which can be used to assess the success of hatchery programs by tracking the survival and recruitment to the fishery of stocked individuals.

Sex identification: Genetic sex markers differ among species of fish. Genomics provides a method to identify sex-linked genes in fish, and in turn, a means to efficiently sex individuals.

Although collecting data from 1000s of molecular markers is beneficial for many applications, collecting these data remains somewhat costly and time-consuming. Therefore, once differences among stocks or populations have been established, it can be beneficial to choose a subset of the most informative markers (e.g., SNP loci or microhaplotypes) and target only these markers in future studies. Two types of sequence-based genotyping marker panels that are becoming common in fisheries science are RAD-capture (Rapture) panels (Ali et al. 2016) and genotyping-in-thousands by sequencing (GT-seq; Campbell et al. 2015). Both approaches rely on the development of panels that contain a relatively small subset of informative loci compared to other HTS approaches. Because loci can be selected based on predetermined criteria (e.g., the power to distinguish lineages, Zhao et al. 2020; to distinguish parent-offspring relationships, Steele et al. 2019; or to identify established genotype-phenotype associations, Micheletti et al. 2018), their use can lead to better and more cost-effective outcomes than molecular approaches that do not screen for a reduced set of informative markers. Additionally, small marker panels developed by HTS can make data more comparable among laboratories (Larson et al. 2014; Hess et al. 2015; Box 3), as HTS identifies exact nucleotide sequences and are therefore more easily replicated.

Box 3 Two of the most common types of markers used in molecular studies are microsatellites (commonly associated with "genetic" approaches) and single nucleotide polymorphisms (SNPs; commonly associated with "genomic" approaches). However, it is important to understand the differences between these two types of markers when comparing genetic research and that diversity and differentiation metrics (e.g., heterozygosity, allelic richness, F_{ST}) are often not directly comparable between the two

Microsatellites	SNPs
• **Physical structure:** A sequence pattern of highly redundant nucleotides (e.g., ACACAC…AC_n or GCAGCAGCA…GCA_n).	• **Physical structure:** A single nucleotide mutation in the genome (A, T, C, or G).
• **Alleles:** The number of sequence pattern repeats, often defined as the length of the repeated region. Repeated mutations in microsatellite regions can result in many different alleles.	• **Alleles:** The specific inherited nucleotide at the SNP position in the genome (i.e., for an AT SNP in a diploid organism, SNP alleles can only be A or T).
• **Genotyping:** Genotype is determined by estimating size of microsatellite sequences amplified in an individual. Alleles with more repeats of a sequence pattern will appear larger than alleles with fewer repeats.	• **Genotyping:** Genotype is determined by defining the exact nucleotides present, in an individual by sequencing the SNP position.
• **Diversity:** Microsatellites can have many alleles (>10) and therefore often result in higher heterozygosity and allelic richness estimates than analyses relying on SNPs.	• **Diversity:** SNPs only have two alleles. Therefore, allelic richness and heterozygosity are often lower than analyses relying on microsatellites.
• **Neutrality:** Microsatellites most often occur in neutral (noncoding) regions of the genome where mutations leading to repetitive regions can accrue over time without negative selection. For this reason, microsatellites are not good indicators of local adaptation.	• **Neutrality:** SNPs can occur in both coding and noncoding (neutral) regions of the genome. Thus, some SNPs may be classified as "neutral," whereas others may be "non-neutral." Because non-neutral SNPs may be under selection, they can be good indicators of local adaptation.

2 How Genomic Data Are Being Used in the Great Lakes

As with virtually all other ecosystems, the Great Lakes have been experiencing ecosystem change owing to human stressors, including climate change, nutrient pollution, habitat modification/destruction, and invasive species among other stressors (Allan et al. 2013; Bunnell et al. 2014). The resultant rapid changes in environmental conditions are increasing the need to understand the degree to which valued fish species, such as Walleye, and the fisheries that they support will persevere in the face of continued ecosystem change (Dippold et al. 2020). For threatened or recovering species of fishes in the Great Lakes, such as Lake Trout (*Salvelinus namaycush*), coregonines (*Coregonus* spp.), Brook Trout (*Salvelinus fontinalis*), and Lake Sturgeon (*Acipenser fulvescens*), the use of genomics has identified loci associated with ecotypic divergence and populations where this diversity can be conserved (Elias et al. 2018; Ackiss et al. 2020; Smith et al. 2020; Whitaker et al. 2020). For invasive nuisance species such as Sea Lamprey (*Petromyzon marinus*), genomic approaches have discerned loci capable of discriminating populations and conducting pedigree analysis, to aid in tracking colonization, dispersal, and local adaptation (Sard et al. 2020), which in turn can help efforts to control this species. For another invasive species, Round Goby (*Neogobius melanostomus*), genomics estimated effective population size and identified invasion pathways (Sard et al. 2019). In exploited populations, such as Lake Whitefish (*Coregonus clupeaformis*) and Walleye, genomic approaches have refined estimates of population genetic structure (Chen et al. 2020a; Graham et al. 2020) and determined what local spawning subpopulations (i.e., stocks) contribute individuals to the harvested population (Sects. 3.1 and 4.1.1). Collectively, genomic approaches in studies such as these have increased our understanding of the population connectivity in many metapopulations throughout the Great Lakes and have shed insight into how humans have shaped genetic diversity through harvest and environmental change.

To date, the role of genomics-based research for Walleye has primarily centered around understanding stock-specific contributions to recreational and/or commercial fisheries. Because Walleye are intensively fished throughout much of the Great Lakes, the ability to parameterize stock assessments by quantifying their relative contributions is important (Vandergoot et al. 2010; Walleye Task Group 2018). Much of the annually harvested Walleye biomass stems from natural spawning in rivers and reefs around the Great Lakes, which exhibits variable within- and among-year recruitment success (DuFour et al. 2015; Dippold et al. 2020). While assessments on spawning grounds can help to estimate the number of potential recruits to the fishery (Mion et al. 2009; DuFour et al. 2015; Fraker et al. 2015), stock-specific harvest estimates require identifying the natal sources of adult fish (Andvik et al. 2016). Both artificial and natural tags have been developed for this purpose (e.g., microsatellite panels, otolith microchemical composition); however, no single method has been able to consistently discriminate among stocks at the spatial scale necessary for many types of management (Wolfert and Van Meter 1978; Johnson et al. 2004; Stepien et al. 2012, 2015; Chen et al. 2017, 2018). Genomic data and associated analytical tools (e.g., *assignPOP*; Chen et al. 2018) are now beginning to fill this gap and provide the discriminatory power for identifying differences among finely spatially structured stocks (Chen et al. 2020a; Sects. 3.1 and 4.1.1).

3 Reduced-Representation Sequencing

Reduced-representation genomic sequencing methods, such as restriction site-associated DNA sequencing (RAD-seq), also known as genotyping-by-sequencing, can identify and genotype 1000s of genetic markers in non-model organisms (Davey and Blaxter 2010). These methods work by enriching certain regions of the genome, such as those near restriction sites that can then be sequenced at a higher rate than nontarget regions. Due to this enrichment step, less sequencing is needed to reach the depth-of-coverage necessary for genotype calls (Davey et al. 2011), thereby reducing per-sample costs (Narum et al. 2013). However, because reduced-representation sequencing depends on quasi-random enrichment of DNA sequences, each new sequencing run will produce a different subset of markers (Baird et al. 2008; Leigh et al. 2018). This artifact makes genotyping new individuals at the same markers as those used in previous (or in other) investigations difficult. Nevertheless, RAD-seq and similar techniques are particularly suitable for single studies and fisheries applications when comparative datasets are not essential (Harvey et al. 2016).

3.1 RAD-seq Refines Assignment Accuracy and Population Structure

Most evidence suggests that Walleye exhibit spawning-site fidelity and return to the same spawning grounds year after year (Olson and Scidmore 1962; Chen et al. 2017; Hayden et al. 2018). Spawning-site fidelity combined with variable annual and stock-specific recruitment has pushed Walleye researchers to continue to look for natural tags capable of consistently discriminating among stocks to facilitate mixed-stock assessments (Fraker et al. 2015; Chen et al. 2017, 2020b; Faust et al. 2019). In several species, RAD-seq has increased population discriminatory power beyond microsatellite panels (e.g., Jeffries et al. 2016; Whitaker et al. 2020). Results such as these led Chen et al. (2020a) to pursue the use of RAD-seq to improve discrimination among geographically proximal and closely related spawning stocks of Walleye, which could not be reliably distinguished using microsatellite markers (Stepien et al. 2012, 2015; Merker and Woodruff 1996). In their study, Chen et al. (2020a) used RAD-seq to identify 12,264 SNP markers to describe population structure in four western Lake Erie spawning stocks (Sandusky River, Maumee River, Detroit River, and a portion of the Ohio reef complex; Fig. 1). Unfortunately, these authors found genetic differentiation among all western basin stocks to be low (average pairwise $F_{ST} = 0.001$; Chen et al. 2020a), a finding consistent with previous estimates of genetic structure using microsatellites (Stepien et al. 2012, 2015, 2018). Ultimately, while RAD-seq did increase the discriminatory power beyond the previously developed microsatellite panels, reassignment accuracy of individuals to their known source stocks within the western basin remained low (<50%), and not sufficient to facilitate mixed-stock assessments (Chen et al. 2020a). Thus, situations can exist where even the most-promising genomic approaches cannot resolve genetic structure, if it is low to non-existent.

The inability to identify meaningful genetic structure in the western basin, despite there being meaningful biological structure (i.e., natal homing behavior; Chen et al. 2020b), likely emanates from low-level straying of spawning individuals among spawning sites and lack of time for populations to diverge due to genetic drift (Waples and Gaggiotti 2006). The lack of observed genetic difference among stocks may also be the result of differences between male and female Walleye behavior. For example, the degree and timing of natal homing at reproduction likely differs between males and females (Raby et al. 2018), and males may spawn at multiple sites which could obscure patterns that rely on diploid genetic variation (Stepien et al. 2015). While a large number of markers are interrogated by genomic approach such as RAD-seq, this still only constitutes a small proportion of the entire genome (generally <1%). Therefore, neutral patterns of divergence can be obscured even when genetic differences among populations exist. There are biologically meaningful differences in life history (e.g., natal homing behavior; Chen et al. 2020b) and annual recruitment (DuFour et al. 2015) among stocks that indicate that there is the potential for local adaptation that could result in genetic differences among western basin stocks. Genetic markers associated with local adaptation could be identified by (1) sequencing entire Walleye genomes to identify SNPs that appear to be under selection among stocks and (2) by the sequencing of the entire mtDNA genome of a large number of individuals in each stock to document the female lineage and the male Y chromosome. Both of these approaches can be efficiently addressed using HTS approaches.

While both traditional genetic and genomic approaches could not reliably discriminate among western basin stocks to allow for mixed-stock analyses, the ability to discriminate between stocks that spawn in the western and eastern basins of Lake Erie was possible. Reassignment tests showed that Walleye could be assigned to either the western or eastern basin with 95% accuracy (Chen et al. 2020a). This finding indicates that basin-level stock assignment is the best spatial scale to conduct mixed-stock assignments in Lake Erie (also see Gatt et al. 2003; Stepien et al. 2012, 2015). The ability to delineate spawning stocks between lake basins (see Fig. 1) likely emanates from these stocks being regulated by different recruitment mechanisms and demonstrating basin-specific reproductive and migration behaviors (Zhao et al. 2011; Matley et al. 2020). These findings highlight that, if genetic differences exist within or among populations, genomic approaches are likely to identify them;

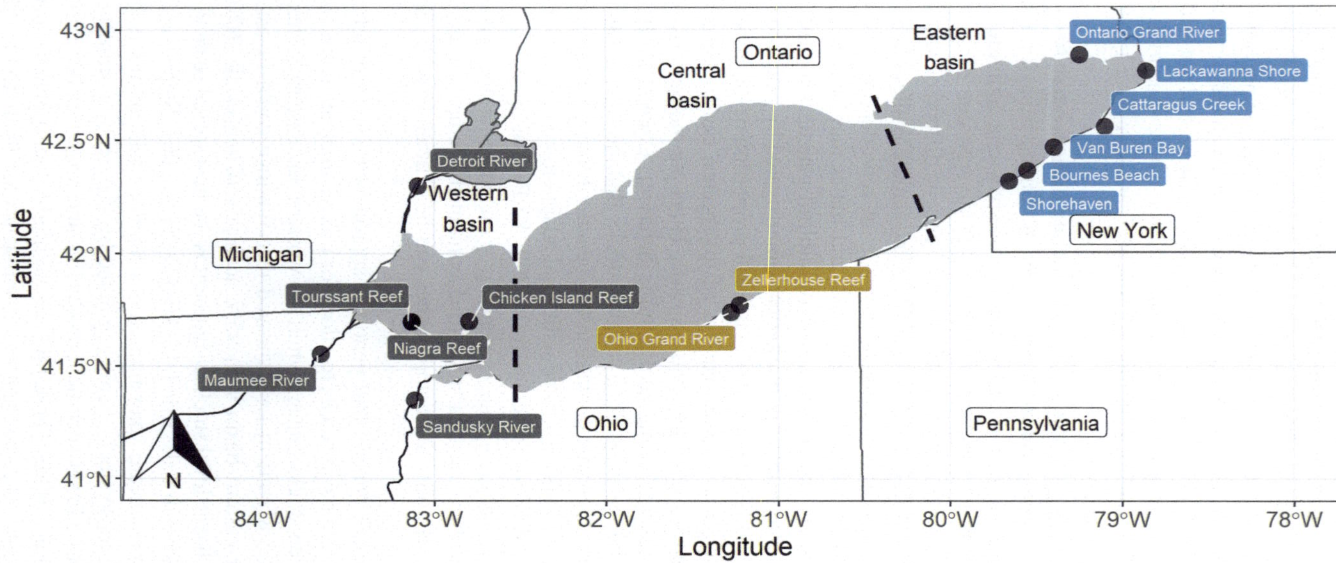

Fig. 1 Major Walleye spawning sites sampled in the Lake Erie genomic studies described in Sects. 3.1 and 4.1.1. Black dashed lines separate the western, central, and eastern basins. The spawning sites of major stocks are color-coded: western basin stocks in gray, central basin stocks in gold, and eastern basin stocks shown in blue. While spawning sites are shown as individual points, the western basin reef complex (e.g., Niagara, Toussaint)—where much of the annual production is believed to occur (DuFour et al. 2015)—is larger and more expansive than the map portrays

however, these findings also indicate that the end-user must be prepared to potentially explore different spatial scales (i.e., broader resolutions) than initially desired.

4 Reproducible Genetic Marker Panels

One of the largest limitations of reduced-representation sequencing is its inability to consistently genotype the same markers among multiple studies (Davey et al. 2011; Harvey et al. 2016). While this limitation might not be problematic for research focused simply on quantifying population structure or that are asking evolutionary questions, for managed species like Walleye, reproducibility of genetic data is essential to facilitate tracking populations through time (Wilson et al. 2008). Below, we discuss Rapture and GT-seq panels, which are being used to leverage HTS to genotype a consistent set of genetic markers in support of Walleye management in the Great Lakes basin.

4.1 RAD-Capture Panels

RAD-capture (Rapture), first described by Ali et al. (2016), is a variation of traditional RAD-seq that provides a more targeted genotyping approach. Rapture contains an additional step that uses a prepared set of capture "baits" (custom oligonucleotides that help to isolate desired genomic regions) to enrich a specific set of markers. In principle, this approach sequences the same set of markers among independent sequencing runs. The markers in Rapture panels must be identified through preliminary RAD-seq or from previously published RAD-seq sequence data. Most Rapture panels contain a few thousand markers selected based on specified criteria, such as diversity or genome alignment position (Sard et al. 2020; Reid et al. 2021). Owing to increased specificity of sequence enrichment during library preparation, the per-sample sequencing cost for Rapture analyses can be greatly reduced when a large number of samples (>1000) are genotyped (Ali et al. 2016; Meek and Larson 2019), making this approach ideal for applications requiring large sample sizes and a large number of genetic markers such as genetic assessments of abundant populations with low genetic structure (see Table 1). The added marker specificity and reduced per-sample cost of Rapture sequencing make this approach an attractive compromise between RAD-seq and more targeted genomic approaches that are often unavailable for non-model organisms. Thus far, two Rapture panels have been designed for Walleye. The first panel was designed to facilitate research in Lake Erie and contains 12,081 markers (Sect. 4.1.1), whereas the second one was designed to facilitate research throughout all 5 of the Great Lakes and contains 99,636 markers (Sect. 4.1.2).

4.1.1 Relative Stock Contributions to Eastern Lake Erie's Fisheries

Owing to the inconsistency of markers and per-sample sequencing cost of RAD-seq, Euclide et al. (2021) developed a Rapture panel to determine the relative contributions of different spawning groups to eastern Lake Erie's recreational and commercial fisheries. The expectation of eastern fisheries being supported by individuals that spawn locally and elsewhere (western and central Lake Erie) was high, given that western Walleye spawners typically migrate eastward into the eastern basin after the spring spawning season (e.g., Wang et al. 2007; Raby et al. 2018; Matley et al. 2020). The use of genetic markers as a natural tag provided a means to estimate the proportional composition of distinct stocks (eastern basin vs. western/central basin) in a novel sampling event without the need for artificial tagging. The use of natural tagging approaches is helpful when population sizes are large and when inter-annual recruitment among spawning stocks is variable, both of which are the case in Lake Erie (DuFour et al. 2015).

In Lake Erie, fishery management agencies use a statistical catch-at-age model to estimate adult (age 2+) Walleye population size and inform the annual total allowable catch and harvest quotas (Vandergoot et al. 2010; Walleye Task Group 2018). However, this model primarily relies on data from the western basin spawning stocks and therefore cannot track the same cohorts of Walleye through time in the eastern basin (Kayle et al. 2015). Using conventional tagging data, Zhao et al. (2011) concluded that western basin Walleye comprised an average of 90% of the eastern basin harvest during 1993–2007. However, until very recently (Euclide et al. 2021), our understanding of relative stock contributions to eastern Lake Erie's recreational fisheries remained unknown as did knowledge of the contributing stocks to both types of fisheries during the past decade. Improved understanding of the harvest composition of eastern Lake Erie's fisheries (i.e., eastern basin vs. western basin Walleye) and associated spatial and temporal variation in stock-specific harvest would facilitate incorporation of the eastern basin into the lake-wide management framework, helping ensure the long-term sustainability of Lake's Erie important Walleye population (DuFour et al. 2015; Kayle et al. 2015).

By developing a Rapture panel, Euclide et al. (2021) were able to genotype individuals of known origin from both eastern and western/central basin spawning sites, identifying distinct allele frequency differences between these two reporting groups (i.e., spawning populations). Stock-population assignment accuracies to a source (natal) basin ranged 85–99% and were found acceptable to facilitate mixed-stock assessments of individuals of unknown origin that were harvested in eastern Lake Erie (Euclide et al. 2021). Assignment of 1274 individuals of unknown origin to their source basin showed that during the summer (July–September), when commercial and recreational harvest was highest, 40–80% of the harvested individuals emanated from the western basin (Fig. 2). However, the contribution of western basin migrants to the eastern basin harvest varied spatially and was consistently higher in the western portion of the eastern basin than in the eastern portion (Fig. 2). While still high, the contribution of western basin Walleye appeared to be

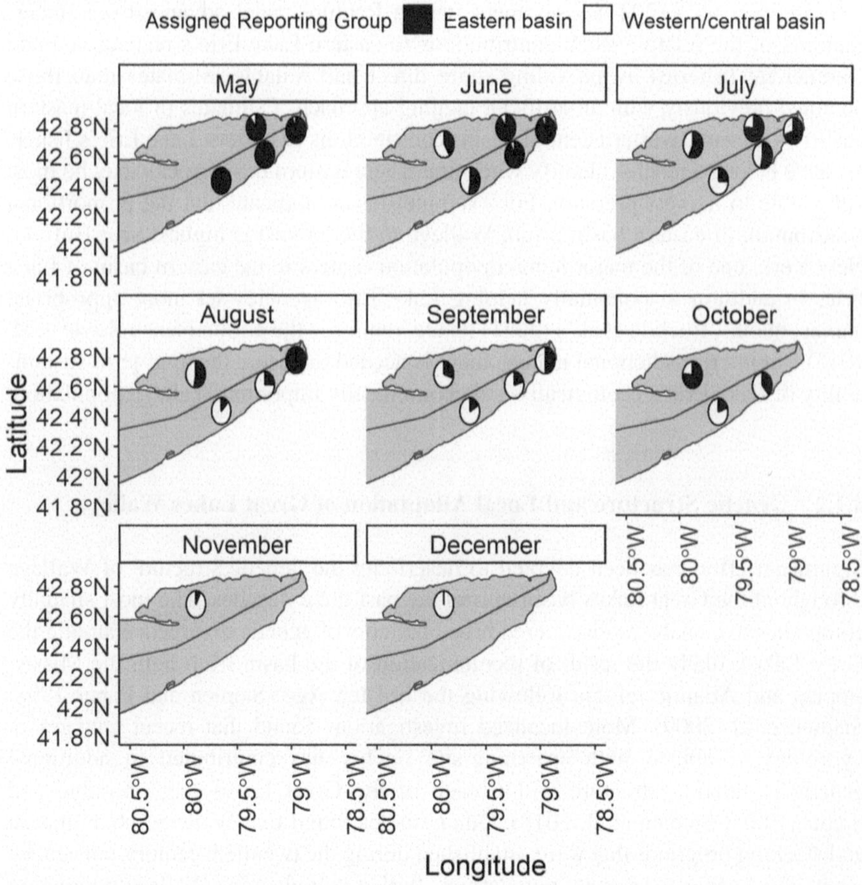

Fig. 2 Proportion of Walleye harvested in eastern Lake Erie's recreational and commercial fisheries during 2017 that were assigned to the eastern basin reporting (spawning) group or the non-eastern basin (i.e., combined western/central basin) reporting group. Pie charts display the proportion of individuals captured in each of the six geographical areas of eastern Lake Erie that were denoted as eastern basin vs. western/central basin fish. Assignment of individuals was conducted using genotypes from markers sequenced with a 12,081 marker Rapture panel, which had a classification accuracy greater than 95% for both basins (Euclide et al. 2021). Pie charts north of the USA-Canada border (distinguished by a solid black line) are based on assignments of individuals captured in the Canadian commercial harvest, whereas pie charts south of the border are based on assignments of individuals captured in New York's recreational harvest. Recreational and commercial harvest data were provided by the New York State Department of Environmental Conservation and Ontario Ministry of Natural Resources and Forestry, respectively. Figure modified from Euclide et al. (2021)

substantially lower than previous estimates, which used artificial tags to estimate that 90% of the annual harvest was comprised of western basin migrants (Zhao et al. 2011; Gatt et al. 2003).

Euclide et al.'s (2021) assessment with a Rapture panel advanced our understanding of the relative stock contributions to eastern Lake Erie's recreational and commercial fisheries by providing more direct and reliable estimates than those obtained previously with an artificial tagging approach. Estimates of local (eastern basin) vs. distant (western/central basin) contributions to eastern Lake Erie's fisheries have helped agencies identify when and where eastern basin stocks may be most vulnerable to harvest pressure. For example, results indicate that the proportional contribution of eastern basin origin Walleye to the harvest is highest near Buffalo, New York, one of the major human population centers in the eastern basin of Lake Erie. In addition to potentially helping Lake Erie agencies set more appropriate harvest quotas, Euclide et al.'s (2021) results reinforce the notion (sensu Kayle et al. 2015) that interjurisdictional management is needed to ensure the long-term sustainability of Lake Erie's ecologically and economically important Walleye population.

4.1.2 Genetic Structure and Local Adaptation of Great Lakes Walleye

Significant effort has been devoted to describing the genetic structure of Walleye throughout the Great Lakes basin during the past three decades. The most spatially comprehensive analyses discovered broad patterns of genetic divergence among the Great Lakes, likely the result of recolonization of the basin from both the Mississippian and Atlantic refugia following the last Ice Age (Stepien and Faber 1998; Stepien et al. 2009). More localized investigations found that recent changes in hydrology combined with spawning-site fidelity has contributed to additional genetic population structure within each of the Great Lakes (e.g., Strange and Stepien 2007; Stepien et al. 2018). Other studies found that Walleye rehabilitation and stocking programs that were established during the twentieth century introduced many novel genes to remnant populations, further complicating Walleye population structure (Wilson et al. 2007; Garner et al. 2013; Haponski et al. 2014). Together, this highlights the complexity of Walleye population genetics in the Great Lakes basin.

One of the major advantages of genomic data is their ability to identify and evaluate the potential role of natural selection on population structure. In large genomic studies, tens of thousands of neutral and adaptive markers are often identified (Seeb et al. 2011; Ackiss et al. 2020), making it possible to evaluate the relative roles of neutral genetic diversity versus local adaptation in shaping population genetic structure (Bernatchez 2016; Reid et al. 2021). While neutral loci and loci under selection need to be analyzed separately in evolutionary models, both are necessary to understand population differences (Morissette et al. 2019; Smith et al. 2020). Previous research established that the neutral genetic structure of Walleye reflects recolonization from refugia maintained by spawning-site fidelity and restricted connectivity among lakes, rivers, and their basins (Stepien et al. 2009, 2015). These studies provide an important baseline of population genetic structure throughout the Great Lakes for genomic investigations into the roles of selection associated with environmental change and stocking on the recent evolutionary history of Walleye.

Currently, a large genomic study involving 1296 Walleye from 31 spawning stocks across the Great Lakes is underway with the objective of describing more recent changes in genetic structure that might be associated with environmentally induced selection (i.e., local adaptation) and stocking. The individuals were genotyped with a Rapture panel of 99,636 SNP markers, which translates to approximately 1 marker per 10,000 base pairs in the Walleye genome. At this density, researchers should be able to identify fine-scale differences among stocks (e.g., within-basin differences that have thus far remained elusive; Chen et al. (2020a)) and identify genetic variants associated with local adaptation (Lowry et al. 2017). These markers can also be used to investigate introgression between stocked and native Walleye populations (Hohenlohe et al. 2011; Ozerov et al. 2016). While final results have yet to be reported, findings should advance our understanding of broad-scale genetic structure and local adaptation of the Great Lake's Walleye population.

4.2 *GT-seq Panels*

Reduced-representation approaches such as RAD-seq and bait-capture panels (Rapture panels) are ideal for assessments where a large number of markers are required to discern genetic structure such as when populations are weakly differentiated. However, data generation using these techniques requires substantial expertise in molecular methods and can be analytically time-consuming, thereby making it difficult to apply RAD-seq or Rapture to research or assessment efforts where large numbers of individuals must be regularly genotyped (e.g., replicated studies, large-scale population monitoring; see Table 1). For these types of studies, other HTS approaches exist that may take more time and cost to develop than RAD-seq and Rapture but provide more replicable and intuitive data and analytical procedures once established (Meek and Larson 2019). These approaches generally involve developing panels of 10–100s of markers (compared to 1000s with Rapture) to address specific applications such as annual fishery assessments using genetic stock assignment or parentage analysis. The markers included in these panels can be chosen to target regions of the genome of interest or selected on their ability to distinguish populations of interest (Hess et al. 2015; McKinney et al. 2020). While these panels sacrifice the genome-wide nature of RAD-seq or Rapture sequencing, their comparatively simplistic preparation and increased specificity make it possible to collect genetic data on many more individuals by substantially lowering the per-sample sequencing cost and analysis time (Meek and Larson 2019). Additionally, when these small marker panels are dependent on sequencing, exact alleles can be identified bioinformatically, making results more easily shared among independent laboratories than results from microsatellite panels which depend on human scorers to call alleles (Wilson et al. 2008; Stott et al. 2010; Scribner et al. 2018).

Genotyping-in-thousands by sequencing panels (GT-seq panels) are becoming one of the most common types of marker panels in fisheries research (Campbell et al. 2015). These panels genotype individuals at 100s of pre-selected SNP and microhaplotype markers using basic polymerase chain reaction (PCR) and standard short-read HTS technology (McKinney et al. 2017; Baetscher et al. 2018). GT-seq is now used by nearly all the fisheries agencies as a method for genetic stock identification and mixed-stock analysis of Pacific salmon (where GT-seq was originally developed; Dahle et al. 2018). The reduced costs of these SNP panels have also allowed management agencies to shift from mass-marking of hatchery outplants with coded wire tags to genetically "marking" all fish based on their parental genotypes (Beacham et al. 2020). The efficiency provided by GT-seq panels, both in sample processing and analysis, allows for large numbers of individuals to be genotyped in a single study (Baetscher et al. 2019). This efficiency means that large-scale research that include 1000s of samples, such as close-kin-mark-recapture of large populations, are becoming feasible to conduct (see Table 1).

As genomic data become more available for Walleye, sets of informative markers are being developed into GT-seq panels to be used as management tools. One such panel was designed with the dual purpose of conducting parentage analysis and genetic stock identification in inland lakes and is currently being used to inform hatchery practices in Wisconsin (Bootsma et al. 2020). The 436-marker panel contains a combination of high F_{ST} SNPs (i.e., markers with alleles that have variable frequencies among lakes) and highly polymorphic microhaplotypes (i.e., markers with a high number of alleles that are helpful when assigning parentage). All markers were identified from a large RAD-seq database of 934 Walleye from 23 inland lakes in Wisconsin and Minnesota (Bootsma et al. 2020). A second dual-purpose panel is now being developed specifically for Walleye populations from the Great Lakes. Although GT-seq panels can be used to evaluate populations other than the ones from which they were developed, the resolution and informativeness often decline due to ascertainment bias (Lachance and Tishkoff 2013). Therefore, genetic markers for the Great Lakes panel are being selected from the Rapture dataset described in Sect. 4.1.2. The findings from this study will create new opportunities to facilitate additional genetic research through the Great Lakes and assist with management decision-making in the basin (e.g., stocking decisions, quota management, habitat restoration efforts, population assessments; see Box 2, Table 1).

5 Conclusions and Future Directions for Walleye Genomics

Given the growing implementation of genomics in fisheries (Ekblom and Galindo 2011) and very recent adoption of genomic approaches by population geneticists in the Great Lakes (e.g., Elias et al. 2018; Chen et al. 2020a; Euclide et al. 2021; Whitaker et al. 2020), much of the research discussed in this chapter is ongoing. The studies discussed herein exemplify how evolutionary approaches and research is advancing to inform fisheries management. Presently, much of the genetic-stock assignment research has centered on Lake Erie's population, which supports the largest recreational and commercial Walleye fisheries in the Great Lakes. Similar approaches, however, could be used to conduct mixed-stock analyses in other regions, such as the Huron-Erie Corridor or Green Bay, Lake Michigan, where relative stock contributions to the fisheries remain unknown (Brenden et al. 2015; Dembkowski et al. 2018). Genomic approaches could also inform Walleye rehabilitation and reintroduction efforts such as those in eastern Lake Erie (MacDougall et al. 2007; Haponski et al. 2014) and Lake Superior's Nipigon Bay and Black Bay (Wilson et al. 2007; Garner et al. 2013). Additionally, because genetic structure is hierarchical, and large-scale research like that discussed in Sect. 4.1.2 can only describe broad-scale genetic structure Great Lakes Walleye, additional targeted studies focusing on localized spawning populations will be required to estimate gene flow and adaptive differences among Walleye populations on a smaller spatial scale. Finally, developing molecular resources such as Rapture and GT-seq panels is just the first step in integrating genetic data into conservation and management. Once panels have been created, additional effort will be necessary to adopt new sampling practices and design experiments that apply these tools effectively.

In this chapter, we covered only a fragment of the ways that genomics is being used to inform fisheries management (Valenzuela-Quiñonez 2016) and our discussion was limited to research on Walleye in the North American Great Lakes basin. However, continued advances in HTS, bioinformatics and other analytical tools, and computing capabilities can be expected to increase the breadth questions related to fisheries that can be addressed with molecular data (Bernatchez et al. 2017; Waples et al. 2020). Outside of the Great Lakes basin, HTS approaches have been used to identify population-level differences in genetic expression and characterize Walleye populations at the southern extent of their North American range (Thorstensen et al. 2020; Zhao et al. 2020). Reference genomes have already been developed for closely related percids (e.g., Yellow Perch *Perca flavescens*) that provide data necessary to understand genotype-phenotype interactions (Feron et al. 2020), and assembly of a Walleye genome will further refine interpretation of genetic data for ecological and evolutionary applications. Thus far, genetic and genomic research has proven to be extremely helpful in fisheries science outside of the Great Lakes basin (Waples et al. 2020), and we are optimistic that continued cooperation between managers and geneticists within the basin will offer new opportunities to apply molecular data to enhance our understanding of Great Lakes Walleye populations and keep the valued fisheries that they support sustainable in the face of continued environmental change.

References

Ackiss AS, Larson WA, Stott W (2020) Genotyping-by-sequencing illuminates high levels of divergence among sympatric forms of coregonines in the Laurentian Great Lakes. Evol Appl 13(5):1037–1054

Ali OA, O'Rourke SM, Amish SJ, Meek MH, Luikart G, Jeffres C, Miller MR (2016) Rad capture (Rapture): flexible and efficient sequence-based genotyping. Genetics 202(2):389–400

Allan JD, McIntyre PB, Smith SD, Halpern BS, Boyer GL, Buchsbaum A, Burton GA Jr, Campbell LM, Chadderton WL, Ciborowski JJH, Doran PJ, Eder T, Infante DM, Johnson LB, Joseph CA,

Marino AL, Prusevich A, Read JG, Rose JB, Rutherford ES, Sowa SP, Steinman AD (2013) Joint analysis of stressors and ecosystem services to enhance restoration effectiveness. Proc Natl Acad Sci U S A 110(1):372–377

Allendorf FW, Hohenlohe PA, Luikart G (2010) Genomics and the future of conservation genetics. Nat Rev Genet 11(10):697–709

Andvik RT, Sloss BL, VanDeHey JA, Claramunt RM, Hansen SP, Isermann DA (2016) Mixed stock analysis of Lake Michigan's lake whitefish *Coregonus clupeaformis* commercial fishery. J Great Lakes Res 42(3):660–667

Baetscher DS, Clemento AJ, Ng TC, Anderson EC, Garza JC (2018) Microhaplotypes provide increased power from short-read DNA sequences for relationship inference. Mol Ecol Resour 18 (2):296–305

Baetscher DS, Anderson EC, Gilbert-Horvath EA, Malone DP, Saarman ET, Carr MH, Garza JC (2019) Dispersal of a nearshore marine fish connects marine reserves and adjacent fished areas along an open coast. Mol Ecol 28(7):1611–1623

Baird NA, Etter PD, Atwood TS, Currey MC, Shiver AL, Lewis ZA, Selker EU, Cresko WA, Johnson EA (2008) Rapid SNP discovery and genetic mapping using sequenced RAD markers. PLoS One 3(10):e3376

Beacham TD, Jonsen K, McIntosh B, Sutherland BJG, Willis D, Lynch C, Wallace C (2020) Large-scale parentage-based tagging and genetic stock identification applied in assessing mixed-stock fisheries and hatchery brood stocks for coho salmon in British Columbia, Canada. Can J Fish Aquat Sci 77(9):1505–1517

Bernatchez L (2016) On the maintenance of genetic variation and adaptation to environmental change: considerations from population genomics in fishes. J Fish Biol 89(6):2519–2556

Bernatchez L, Wellenreuther M, Araneda C, Ashton DT, Barth JMI, Beacham TD, Maes GE, Martinsohn JT, Miller KM, Naish KA, Ovenden JR, Primmer CR, Young Suk H, Therkildsen NO, Withler RE (2017) Harnessing the power of genomics to secure the future of seafood. Trends Ecol Evol 32(9):665–680

Billington N, Wilson C, Sloss B (2011) Distribution and population genetics of walleye and sauger. In: Barton B (ed) Biology, management, and culture of Walleye and Sauger. American Fisheries Society, pp 105–132

Bootsma ML, Gruenthal KM, McKinney GJ, Simmons L, Miller L, Sass GG, Larson WA (2020) A GT-seq panel for walleye (*Sander vitreus*) provides important insights for efficient development and implementation of amplicon panels in non-model organisms. Mol Ecol Resour 20 (6):1706–1722

Brenden TO, Scribner KT, Bence JR, Tsehaye I, Kanefsky J, Vandergoot CS, Fielder DG (2015) Contributions of Lake Erie and Lake St. Clair walleye populations to the Saginaw Bay, Lake Huron, recreational fishery: Evidence from genetic stock identification. N Am J Fish Manag 35 (3):567–577

Bunnell DB, Barbiero RP, Ludsin SA, Madenjian CP, Warren GJ, Dolan DM, Brenden TO, Briland R, Gorman OT, He JX, Johengen TH, Lantry BF, Lesht BM, Nalepa TF, Riley SC, Riseng CM, Treska TJ, Tsehaye I, Maureen G, Walsh MG, Warner DM, Weidel BC (2014) Changing ecosystem dynamics in the Laurentian Great Lakes: bottom-up and top-down regulation. Bioscience 64(1):26–39

Campbell NR, Harmon SA, Narum SR (2015) Genotyping-in-Thousands by sequencing (GT-seq): a cost effective SNP genotyping method based on custom amplicon sequencing. Mol Ecol Resour 15(4):855–867

Caroffino DC, Mwai AM, Evans BI (2011) Population genetics of walleye and yellow perch in the St. Marys River. Journal of Great Lakes Research 37(Suppl. 2):28–34

Chen K-Y, Ludsin SA, Corey MM, Collingsworth PD, Nims MK, Olesik JW, Dabrowski K, Van Tassell JJ, Marschall EA (2017) Experimental and field evaluation of otolith Strontium as a marker to discriminate between river-spawning populations of walleye in Lake Erie. Can J Fish Aquat Sci 74(5):693–701

Chen K-Y, Marschall EA, Sovic MG, Fries AC, Gibbs HL, Ludsin SA (2018) assignPOP: an r package for population assignment using genetic, non-genetic, or integrated data in a machine-learning framework. Methods Ecol Evol 9(2):439–446

Chen K-Y, Euclide PT, Ludsin SA, Larson WA, Sovic MG, Gibbs HL, Marschall EA (2020a) RAD-seq refines previous estimates of genetic structure in Lake Erie walleye. Trans Am Fish Soc 149(2):159–173

Chen K-Y, Ludsin SA, Marcek BJ, Olesik JW, Marschall EA (2020b) Otolith microchemistry shows natal philopatry of walleye in western Lake Erie. J Great Lakes Res 46(5):1349–1357

Dahle G, Johansen T, Westgaard JI, Aglen A, Glover KA (2018) Genetic management of mixed-stock fisheries "real-time": the case of the largest remaining cod fishery operating in the Atlantic in 2007–2017. Fish Res 205:77–85

Davey JW, Blaxter ML (2010) RADSeq: next-generation population genetics. Brief Funct Genomics 9(5–6):416–423

Davey JW, Hohenlohe PA, Etter PD, Boone JQ, Catchen JM, Blaxter ML (2011) Genome-wide genetic marker discovery and genotyping using next-generation sequencing. Nat Rev Genet 12 (7):499–510

Dembkowski DJ, Isermann DA, Hogler SR, Larson WA, Turnquist KN (2018) Stock structure, dynamics, demographics, and movements of walleyes spawning in four tributaries to Green Bay. J Great Lakes Res 44(5):970–978

Dippold DA, Aloysius NR, Keitzer SC, Yen H, Arnold JG, Daggupati P, Fraker ME, Martin JF, Robertson DM, Sowa SP, Johnson M-VV, White MJ, Ludsin SA (2020) Forecasting the combined effects of anticipated climate change and agricultural conservation practices on fish recruitment dynamics in Lake Erie. Freshw Biol 65(9):1487–1508

DuFour MR, May CJ, Roseman EF, Ludsin SA, Vandergoot CS, Pritt JJ, Fraker ME, Davis JJ, Tyson JT, Miner JG, Marschall EA, Mayer CM (2015) Portfolio theory as a management tool to guide conservation and restoration of multi-stock fish populations. Ecosphere 6(12):art296

Ekblom R, Galindo J (2011) Applications of next generation sequencing in molecular ecology of non-model organisms. Heredity 107(1):1–15

Elias A, McLaughlin R, Mackereth R, Wilson C, Nichols KM (2018) Population structure and genomic variation of ecological life history diversity in wild-caught Lake Superior brook trout, *Salvelinus fontinalis*. J Great Lakes Res 44(6):1373–1382

Euclide PT, MacDougall T, Robinson JM, Faust MD, Wilson CC, Chen K-Y, Marschall EA, Larson WA, Ludsin SA (2021) Mixed-stock analysis using Rapture genotyping to evaluate stock-specific exploitation of a walleye population despite weak genetic structure. Evol Appl 14 (5):1403–1420

Faust MD, Vandergoot CS, Brenden TO, Kraus RT, Hartman T, Krueger CC (2019) Acoustic telemetry as a potential tool for mixed-stock analysis of fishery harvest: a feasibility study using Lake Erie walleye. Can J Fish Aquat Sci 76(6):1019–1030

Feron R, Zahm M, Cabau C, Klopp C, Roques C, Bouchez O, Eché C, Valière S, Donnadieu C, Haffray P, Bestin A, Morvezen R, Acloque H, Euclide PT, Wen M, Jouano E, Schartl M, Postlethwait JH, Schraidt C, Christie MR, Larson WA, Herpin A, Guiguen Y (2020) Characterization of a Y-specific duplication/insertion of the anti-Mullerian hormone type II receptor gene based on a chromosome-scale genome assembly of yellow perch, *Perca flavescens*. Mol Ecol Resour 20(2):531–543

Flanagan SP, Forester BR, Latch EK, Aitken SN, Hoban S (2018) Guidelines for planning genomic assessment and monitoring of locally adaptive variation to inform species conservation. Evol Appl 11(7):1035–1052

Fraker ME, Anderson EJ, May CJ, Chen K-Y, Davis JJ, DeVanna KM, DuFour MR, Marschall EA, Mayer CM, Miner JG, Pangle KL, Pritt JJ, Roseman EF, Tyson JT, Zhao Y, Ludsin SA (2015) Stock-specific advection of larval walleye (*Sander vitreus*) in western Lake Erie: implications for larval growth, mixing, and stock discrimination. J Great Lakes Res 41(3):830–845

Funk WC, McKay JK, Hohenlohe PA, Allendorf FW (2012) Harnessing genomics for delineating conservation units. Trends Ecol Evol 27(9):489–496

Garner SR, Bobrowicz SM, Wilson CC (2013) Genetic and ecological assessment of population rehabilitation: walleye in Lake Superior. Ecol Appl 23(3):594–605

Gatt MH, Ferguson MM, Liskauskas AP (2000) Comparison of control region sequencing and fragment RFLP analysis for resolving mitochondrial DNA variation and phylogenetic relationships among Great Lakes walleyes. Trans Am Fish Soc 129(6):1288–1299

Gatt MH, McParland TL, Halyk LC, Ferguson MM (2003) Mitochondrial DNA variation and mixed-stock analysis of recreational and commercial walleye fisheries in Eastern Lake Erie. N Am J Fish Manag 23(2):431–440

Graham CF, Boreham DR, Manzon RG, Stott W, Wilson JY, Somers CM (2020) How "simple" methodological decisions affect interpretation of population structure based on reduced representation library DNA sequencing: a case study using the lake whitefish. PLoS One 15(1): e0226608

Haponski AE, Dean H, Blake BE, Stepien CA (2014) Genetic history of walleyes spawning in Lake Erie's Cattaraugus Creek: a comparison of pre- and poststocking. Trans Am Fish Soc 143 (5):1295–1307

Harrisson KA, Amish SJ, Pavlova A, Narum SR, Telonis-Scott M, Rourke ML, Lyon J, Tonkin Z, Gilligan DM, Ingram BA, Lintermans M, Gan HM, Austin CM, Luikart G, Sunnucks P (2017) Signatures of polygenic adaptation associated with climate across the range of a threatened fish species with high genetic connectivity. Mol Ecol 26(22):6253–6269

Harvey MG, Smith BT, Glenn TC, Faircloth BC, Brumfield RT (2016) Sequence capture versus restriction site associated DNA sequencing for shallow systematics. Syst Biol 65(5):910–924

Hayden TA, Binder TR, Holbrook CM, Vandergoot CS, Fielder DG, Cooke SJ, Dettmers JM, Krueger CC (2018) Spawning site fidelity and apparent annual survival of walleye (*Sander vitreus*) differ between a Lake Huron and Lake Erie tributary. Ecol Freshw Fish 27(1):339–349

Helyar SJ, Hemmer-Hansen J, Bekkevold D, Taylor MI, Ogden R, Limborg MT, Cariani A, Maes GE, Diopere E, Carvalho GR, Nielsen EE (2011) Application of SNPs for population genetics of nonmodel organisms: New opportunities and challenges. Mol Ecol Resour 11(Suppl. 1):123–136

Hess JE, Campbell NR, Docker MF, Baker C, Jackson A, Lampman R, McIlraith B, Moser ML, Statler DP, Young WP, Wildbill AJ, Narum SR (2015) Use of genotyping by sequencing data to develop a high-throughput and multifunctional SNP panel for conservation applications in Pacific lamprey. Mol Ecol Resour 15(1):187–202

Hohenlohe PA, Amish SJ, Catchen JM, Allendorf FW, Luikart G (2011) Next-generation RAD sequencing identifies thousands of SNPs for assessing hybridization between rainbow and westslope cutthroat trout. Mol Ecol Resour 11(Suppl. 1):117–122

Holderegger R, Kamm U, Gugerli F (2006) Adaptive vs. neutral genetic diversity: implications for landscape genetics. Landsc Ecol 21(6):797–807

Jeffries DL, Copp GH, Handley LL, Håkan Olsén K, Sayer CD, Hänfling B (2016) Comparing RADseq and microsatellites to infer complex phylogeographic patterns, an empirical perspective in the Crucian carp, *Carassius carassius*, L. Mol Ecol 25(13):2997–3018

Johnson TB, Dixon B, Stepien CA, Wilson CC (2004) Stock discrimination of Lake Erie walleye: a mixed stock analysis contrasting genetic techniques. Great Lake Fishery Commission Final Report. Great Lakes Fishery Commission, Ann Arbor, MI. Available from http://www.glfc.org/lakecom/lec/WTG_docs/annual_reports/WTG_report_2006.pdf

Johnson RN, O'Meally D, Chen Z, Etherington GJ, Ho SYW, Nash WJ, Grueber CE, Cheng Y, Whittington CM, Dennison S, Peel E, Haerty W, O'Neill RJ, Colgan D, Russell TL, Alquezar-Planas DE, Attenbrow V, Bragg JG, Brandies PA, Chong AYY, Deakin JE, Di Palma F, Duda Z, Eldridge MDB, Ewart KM, Hogg CJ, Frankham GJ, Georges A, Gillett AK, Govendir M, Greenwood AD, Hayakawa T, Helgen KM, Hobbs M, Holleley CE, Heider TN, Jones EA, King A, Madden D, Marshall Graves JA, Morris KM, Neaves LE, Patel HR, Polkinghorne A, Renfree MB, Robin C, Salinas R, Tsangaras K, Waters PD, Waters SA, Wright B, Wilkins MR, Timms P, Belov K (2018) Adaptation and conservation insights from the koala genome. Nat Genet 50(8):1102–1111

Kayle K, Francis J, Murray C, Markham J (2015) Lake Erie Walleye management plan (2015–2019), Ann Arbor, MI

Kumar G, Kocour M (2017) Applications of next-generation sequencing in fisheries research: a review. Fish Res 186:11–22

Lachance J, Tishkoff SA (2013) SNP ascertainment bias in population genetic analyses: why it is important, and how to correct it. BioEssays 35(9):780–786

Larson WA, Seeb LW, Everett MV, Waples RK, Templin WD, Seeb JE (2014) Genotyping by sequencing resolves shallow population structure to inform conservation of Chinook salmon (*Oncorhynchus tshawytscha*). Evol Appl 7(3):355–369

Leigh DM, Lischer HEL, Grossen C, Keller LF (2018) Batch effects in a multiyear sequencing study: false biological trends due to changes in read lengths. Mol Ecol Resour 18(4):778–788

Lowry DB, Hoban S, Kelley JL, Lotterhos KE, Reed LK, Antolin MF, Storfer A (2017) Responsible RAD: striving for best practices in population genomic studies of adaptation. Mol Ecol Resour 17(3):366–369

Luikart G, England PR, Tallmon D, Jordan S, Taberlet P (2003) The power and promise of population genomics: from genotyping to genome typing. Nat Rev Genet 4(12):981–994

MacDougall TM, Wilson CC, Richardson LM, Lavender M, Ryan PA (2007) Walleye in the Grand River, Ontario: an overview of rehabilitation efforts, their effectiveness, and implications for eastern Lake Erie fisheries. J Great Lakes Res 33(1):103–117

Matley JK, Faust MD, Raby GD, Zhao Y, Robinson J, MacDougall TM, Hayden TA, Fisk AT, Vandergoot CS, Krueger CC (2020) Seasonal habitat-use differences among Lake Erie's walleye stocks. J Great Lakes Res 46(3):609–621

McKinney GJ, Seeb JE, Seeb LW (2017) Managing mixed-stock fisheries: genotyping multi-SNP haplotypes increases power for genetic stock identification. Can J Fish Aquat Sci 74(4):429–434

McKinney GJ, Pascal CE, Templin WD, Gilk-Baumer SE, Dann TH, Seeb LW, Seeb JE (2020) Dense SNP panels resolve closely related chinook salmon populations. Can J Fish Aquat Sci 77 (3):451–461

McParland TL, Ferguson MM, Liskauskas AP (1999) Genetic population structure and mixed-stock analysis of walleyes in the Lake Erie–Lake Huron corridor using allozyme and mitochondrial DNA markers. Trans Am Fish Soc 128(6):1055–1067

Meek MH, Larson WA (2019) The future is now: Amplicon sequencing and sequence capture usher in the conservation genomics era. Mol Ecol Resour 19(4):795–803

Merker RJJ, Woodruff RCC (1996) Molecular evidence for divergent breeding groups of walleye (*Stizostedion vitreum*) in tributaries to western Lake Erie. J Great Lakes Res 22(2):280–288

Micheletti SJ, Hess JE, Zendt JS, Narum SR (2018) Selection at a genomic region of major effect is responsible for evolution of complex life histories in anadromous steelhead. BMC Evol Biol 18 (1):140

Mion JB, Stein RA, Marschall EA (2009) River discharge drives survival of larval walleye. Ecol Appl 8(1):88–103

Morissette O, Sirois P, Wilson CC, Laporte M, Bernatchez L (2019) The role of ecotype-environment interactions in intraspecific trophic niche partitioning subsequent to stocking. Ecol Appl 29(3):e01857

Narum SR, Buerkle CA, Davey JW, Miller MR, Hohenlohe PA (2013) Genotyping-by-sequencing in ecological and conservation genomics. Mol Ecol 22(11):2841–2847

Olson DE, Scidmore WJ (1962) Homing behavior of spawning walleyes. Trans Am Fish Soc 91 (4):355–361

Ozerov MY, Gross R, Bruneaux M, Vähä J-P, Burimski O, Pukk L, Vasemägi A (2016) Genome wide introgressive hybridization patterns in wild Atlantic salmon influenced by inadvertent gene flow from hatchery releases. Mol Ecol 25(6):1275–1293

Pearse DE, Miller MR, Abadía-Cardoso A, Garza JC (2014) Rapid parallel evolution of standing variation in a single, complex, genomic region is associated with life history in steelhead/ rainbow trout. Proc R Soc B Biol Sci 281(1783):20140012

Raby GD, Vandergoot CS, Hayden TA, Faust MD, Kraus RT, Dettmers JM, Cooke SJ, Zhao Y, Fisk AT, Krueger CC (2018) Does behavioural thermoregulation underlie seasonal movements in Lake Erie walleye? Can J Fish Aquat Sci 75(3):1–9

Reid BN, Moran RL, Kopack CJ, Fitzpatrick SW (2021) Rapture-ready darters: choice of reference genome and genotyping method (whole-genome or sequence capture) influence population genomic inference in Etheostoma. Mol Ecol Resour 21(2):404–420

Robison HW, Buchanan TM (2020) Fishes of Arkansas, 2nd edn. The University of Arkansas Press, Fayetteville, AR, 959 p

Sard NM, Robinson J, Kanefsky J, Herbst S, Scribner K (2019) Coalescent models characterize sources and demographic history of recent round goby colonization of Great Lakes and inland waters. Evol Appl 12(5):1034–1049

Sard NM, Smith SR, Homola JJ, Kanefsky J, Bravener G, Adams JV, Holbrook CM, Hrodey PJ, Tallon K, Scribner KT (2020) RAPTURE (RAD capture) panel facilitates analyses characterizing sea lamprey reproductive ecology and movement dynamics. Ecol Evol 10(3):1469–1488

Schindler DE, Hilborn R, Chasco B, Boatright CP, Quinn TP, Rogers LA, Webster MS (2010) Population diversity and the portfolio effect in an exploited species. Nature 465(7298):609–612

Scribner K, Tsehaye I, Brenden T, Stott W, Kanefsky J, Bence J (2018) Hatchery strain contributions to emerging wild lake trout populations in Lake Huron. J Hered 109(6):675–688

Seeb JE, Carvalho G, Hauser L, Naish K, Roberts S, Seeb LW (2011) Single-nucleotide polymorphism (SNP) discovery and applications of SNP genotyping in nonmodel organisms. Mol Ecol Resour 11(Suppl. 1):1–8

Smith SR, Amish SJ, Bernatchez L, Le Luyer J, Wilson C, Boeberitz O, Luikart G, Scribner KT (2020) Mapping of adaptive traits enabled by a high-density linkage map for lake trout. G3 (Bethesda) 10(6):1929–1947

Steele CA, Hess M, Narum S, Campbell M (2019) Parentage-based tagging: reviewing the implementation of a new tool for an old problem. Fisheries 44(9):412–422

Stepien CA, Faber JE (1998) Population genetic structure, phylogeography and spawning philopatry in walleye (*Stizostedion vitreum*) from mitochondrial DNA control region sequences. Mol Ecol 7(12):1757–1769

Stepien CA, Murphy DJ, Lohner RN, Sepulveda-Villet OJ, Haponski AE (2009) Signatures of vicariance, postglacial dispersal and spawning philopatry: population genetics of the walleye *Sander vitreus*. Mol Ecol 18(16):3411–3428

Stepien CA, Banda JA, Murphy DM, Haponski AE (2012) Temporal and spatial genetic consistency of walleye spawning groups. Trans Am Fish Soc 141(3):660–672

Stepien CA, Sepulveda-Villet OJ, Haponski AE (2015) Comparative genetic diversity, population structure, and adaptations of Walleye and Yellow Perch across North America. In: Kestemont P, Dabrowski K, Summerfelt RC (eds) Biology and culture of percid fishes: principles and practices. Springer, Dordrecht, pp 643–689, 901 p

Stepien CA, Snyder MR, Knight CT (2018) Genetic divergence of nearby walleye spawning groups in central Lake Erie: implications for management. N Am J Fish Manag 38(4):783–793

Stott W, VanDeHey JA, Justin JA (2010) Genetic diversity of lake whitefish in lakes Michigan and Huron; sampling, standardization, and research priorities. J Great Lakes Res 36(Suppl. 1):59–65

Strange RM, Stepien CA (2007) Genetic divergence and connectivity among river and reef spawning groups of walleye (*Sander vitreus vitreus*) in Lake Erie. Can J Fish Aquat Sci 64 (3):437–448

Thorstensen MJ, Jeffrey JD, Treberg JR, Watkinson DA, Enders EC, Jeffries KM (2020) Genomic signals found using RNA sequencing show signatures of selection and subtle population differentiation in walleye (*Sander vitreus*) in a large freshwater ecosystem. Ecol Evol 10 (14):7173–7188

Todd TN, Haas RC (1993) Genetic and tagging evidence for movement of walleyes between Lake Erie and Lake St. Clair. J Great Lakes Res 19(2):445–452

Valenzuela-Quiñonez F (2016) How fisheries management can benefit from genomics? Brief Funct Genomics 15(5):352–357

Vandergoot CS, Cook HA, Thomas MV, Einhouse DW, Murray C (2010) Status of walleye in western Lake Erie, 1985-2006. In: Status of walleye in the Great Lakes: proceedings of the 2006 Symposium. Great Lakes Fishery Commission Technical Report 69, pp 123–150

Walleye Task Group (2018) Report of the Lake Erie Walleye Task Group to the Standing Technical Committee. Great Lakes Fishery Commission, Ann Arbor, MI

Wang H-Y, Rutherford ES, Cook HA, Einhouse DW, Haas RC, Johnson TB, Kenyon R, Locke B, Turner MW (2007) Movement of walleyes in Lakes Erie and St. Clair inferred from tag return and fisheries data. Trans Am Fish Soc 136(2):539–551

Waples RS, Gaggiotti O (2006) What is a population? An empirical evaluation of some genetic methods for identifying the number of gene pools and their degree of connectivity. Mol Ecol 15 (6):1419–1439

Waples RS, Naish KA, Primmer CR (2020) Conservation and management of salmon in the age of genomics. Annu Rev Anim Biosci 8(1):117–143

Whitaker JM, Price LE, Boase JC, Bernatchez L, Welsh AB (2020) Detecting fine-scale population structure in the age of genomics: a case study of lake sturgeon in the Great Lakes. Fish Res 230: 105646

Wilson CC, Lavender M, Black J (2007) Genetic assessment of walleye (*Sander vitreus*) restoration efforts and options in Nipigon Bay and black Bay, Lake Superior. J Great Lakes Res 33 (1):133–144

Wilson CC, Stott W, Miller L, D'Amelio S, Jennings MJ, Cooper AM (2008) Conservation genetics of Lake Superior brook trout: issues, questions, and directions. N Am J Fish Manag 28 (4):1307–1320

Wolfert DR, Van Meter HD (1978) Movements of walleyes tagged in eastern Lake Erie. N Y Fish Game J 25(6):16–22

Zhao Y, Einhouse DW, MacDougall TM (2011) Resolving some of the complexity of a mixed-origin walleye population in the east basin of Lake Erie using a mark–recapture study. N Am J Fish Manag 31(2):379–389

Zhao Y, Kocovsky PM, Madenjian CP (2013) Development of a stock-recruitment model and assessment of biological reference points for the Lake Erie walleye fishery. N Am J Fish Manag 33(5):956–964

Zhao H, Silliman K, Lewis M, Johnson S, Kratina G, Rider SJ, Stepien CA, Hallerman EM, Beck B, Fuller A, Peatman E (2020) SNP analyses highlight a unique, imperiled southern walleye (*Sander vitreus*) in the Mobile River Basin. Can J Fish Aquat Sci 77(8):1366–1378

Walleye Larviculture in Water Reuse Aquaculture Systems

J. Alan Johnson, Kevin Kelsey, and Robert Summerfelt

Abstract Meeting of the three authors during the 2011 Midwest Fish and Wildlife Conference in Des Moines, Iowa, led to frequent communication that formed the basis for collaborating on the status of their respective production facilities. The operation techniques of the water reuse aquaculture system (RAS) facilities in Vermont and Iowa are compared and contrasted. Kelsey and Johnson present detailed descriptions of Walleye (*Stizostedion vitreum*) larviculture in innovative RASs at the fish culture facilities in Vermont and Iowa, respectively. Since 2011, intensive culture of Walleye fry/fingerlings has been conducted at the Ed Weed Fish Culture Station in Grand Isle, Vermont, with the goal of large-scale production from the facility's program inception to supplement existing extensive pond culture efforts of fingerlings that are used for sports fishing restoration. Tank volumes of 1940 L are now used in a RAS dedicated exclusively for intensive Walleye culture. Proof-of-concept techniques have been applied with successive production years to duplicate identified advances related to feed and feeding rates as well as various rearing environment conditions. After two successive years of trials in four self-cleaning tanks (2018 and 2019), as of 2020 all tanks within the system are now self-cleaning, providing optimum rearing conditions. The feeding of blended feeds through the entire culture run has also been applied since 2017. Larviculture survivals from day 1 post hatch (1 dph) through 34 dph in excess of 60% are being achieved averaging 50 mm in length, providing recruitment to the fishery that can be documented.

J. A. Johnson (✉)
Iowa Department of Natural Resources, Moravia, IA, USA
e-mail: alan.johnson@dnr.iowa.gov

K. Kelsey
Vermont Fish and Wildlife Department, Ed Weed Fish Culture Station, Grand Isle, VT, USA
e-mail: Kevin.Kelsey@vermont.gov

R. Summerfelt
Department of Natural Resource Ecology and Management, Iowa State University of Science and Technology, Ames, IA, USA

J. C. Bruner, R. L. DeBruyne (eds.), *Yellow Perch, Walleye, and Sauger: Aspects of Ecology, Management, and Culture*, Fish & Fisheries Series 41,
https://doi.org/10.1007/978-3-030-80678-1_6

The Iowa Department of Natural Resources fish hatcheries rely on surface water sources for Walleye advanced fingerling production in single-pass systems. Aquatic invasive species as well as some pathogens are present in these water sources. RAS technology with secure water sources is one solution to these challenges. A pilot-scale larviculture RAS was built at the Rathbun Fish Culture Research Facility, and fingerling production began in 2019. The larviculture RAS produced 107,800 to 139,284 fingerlings in each crop to 1.0-g size with a 77.2% survival rate over the 2019–2020 trials. Except for an outbreak of bacterial gill disease, none of the several bacterial and protozoan pathogens that frequently infect Walleye during intensive culture using traditional surface water source were observed on fish reared in RAS during the 2019–2020 trials.

Keywords Walleye · *Stizostedion vitreum* · RAS · Larviculture · Cannibalism · Vermont · Iowa

1 Introduction

Robert C. Summerfelt

1.1 Objectives

This chapter presents detailed descriptions by Kelsey (Sect. 2) and Johnson (Sect. 3) of Walleye (*Stizostedion vitreum*) larviculture in innovative reuse aquaculture systems (RASs) at the fish culture facilities in Vermont (Ed Weed Fish Culture Station (EWFCS)) and Iowa (Rathbun Fish Culture Research Facility (RFCRF)), respectively. They achieved noteworthy success for intensive larviculture of Walleye based on their many years of experience, use of science-based protocol, and application of modern RAS engineering. They address issues specific to larviculture of Walleye that are applicable to facilities using single-pass as well as water reuse systems. Their descriptions of RAS technology and system management will be immeasurably helpful for others that are seeking to transition from intensive culture using single-use flowing to reuse.

The survival and production of fingerling Walleye improved following the adoption of RAS technology at both sites. At EWFCS, larval survival in the RAS was 52–61% from hatch to 33–35 days post hatch (dph) compared with 32% average survival in ponds. In 2019, at RFCRF, a total of 107,804 fingerlings (75.2% survival rate) were produced in the RAS in production-scale tanks. At both sites, the culture systems were designed with surface sprays to overcome the ubiquitous and serious problem of noninflation of the gas bladder (NGB) in the tank culture of many physoclistous fishes (Summerfelt 2013).

A literature review of Walleye larviculture was not a purpose of this report; therefore, citations have been intentionally limited to only serve as an aid for comprehension or to give credit to the most relevant previous work. Comprehensive literature reviews are available elsewhere (Summerfelt 2005, 2013; Summerfelt et al. 2011; Johnson and Summerfelt 2015). Likewise, technical details on water reuse technology are described elsewhere by many others (Summerfelt 1996a; Summerfelt et al. 2001; Timmons et al. 2001).

1.2 RAS Technology

The engineering technology for a reuse system includes an organized set of complementary unit processes that are needed to maintain water quality and to maximize the reuse of water from the fish culture system, with a minimum addition of freshwater to replace evaporation and water loss from clarification. Backwashing the drum filter or flushing water to prevent the accumulation of high nitrate accounts for the major water loss. The unit processes include (from Summerfelt 1996a) the following:

- Clarification for removing settleable and suspended solids
- Biofiltration for removing dissolved organics and ammonia
- Stripping carbon dioxide
- Oxygen addition to levels generally greater than saturation
- Disinfection using ozone or ultraviolet (UV) filters
- Automatic adjustment of pH

There are many options for managing each of the unit processes involved using RAS, for example, as a means for accomplishing biofiltration, the process of ammonia removal by nitrification in the filter. Alternatives include submerged biofilters, trickling biofilters, rotating biological contactors (RBC), bead filters, fluidized bed biofilters, or a moving bed bioreactor (MBBR). The latter was selected at both EWFCS (Sect. 2) and RFCRF (Sect. 3).

Major Benefits of RAS Technology

- There is reduced water consumption because a high percentage—greater than 95–99.5% in advanced systems—of water is reused after passing through the culture tanks. This allows expansion in production with the same water supply or opportunity for a facility to be sited where water supply is more limited. RAS is adaptable to small and large facilities.
- There are reduced costs for the heating and cooling of the water because only the replacement water needs treatment, whereas it is generally cost-prohibitive to control the temperature in single-pass systems.

- There are more options and reduced costs for biosecurity, such as preventing the intake of indigenous parasites such as Ich (*Ichthyophthirius multifiliis*) and microbial pathogens that cause morbidity or mortality to the cultured fish or that serve as a threat to environments that are included in the stocking plans (e.g., viral hemorrhagic septicemia virus (VHSV)). At the Rathbun Fish Hatchery, formalin treatment to control Ich is greater than 20% of the variable cost for Walleye production (Johnson and Summerfelt 2015). Another potential savings of RAS is to substantially reduce the volume of water needed, thus requiring smaller size and more effective filtration and disinfection system to treat the use of water from a surface source. The reduced water requirement can sometimes even make an economically viable use of municipal water sources for the water supply.
- Given the reduced need for large volumes of water in facilities using RAS, effective treatment of the intake water can reduce problems with aquatic invasive species (AIS). Facilities with a single-pass or partial reuse require exceptional efforts to prevent intake from surface water sources and the subsequent distribution of nonindigenous plants and animals in the water of the distribution unit. A prime example is the special effort and added expenses to prevent the distribution of Zebra Mussel (*Dreissena polymorpha*) veligers in the water used to transport fish for stocking (Edwards et al. 2002).

Disadvantages of RAS Technology

- Substantial capital investment for equipment is required for water treatment processes that allow the water to be reused in the same tank or other tanks. Although out-of-date with 1996 dollars O'Rourke (1996) has itemized investment-related costs for equipment for Walleye larviculture and fingerlings, which is not specific for larviculture in a RAS; however, exclusive of the building and land, the costs in 1996 dollars for "fixed equipment" was $33,440. A hatchery going from a flow-through system to RAS will find that some increased costs may be compensated by reduced costs for disease treatments, as well as by eliminating hatchery modifications that are often needed to prevent the intake and distribution of AIS. Also, increased production at an existing site will reduce the need for a substantial expense for expanding the current facility or for building a new one.
- RAS technology can be challenging for those with limited experience with the equipment, plumbing, and electrical service that is required. A new installation may require assistance from an engineering specialist; fortunately, however, commercial sources for the equipment are available to offer suggestions, and there are educational opportunities for hatchery personnel to gain a fundamental understanding of the technology, e.g., the short course on RAS technology taught by the staff of the Freshwater Institute, Shepherdstown, WV. That short course includes pumping and piping, fish health and biosecurity, monitoring RAS water quality, and supervisory control and data acquisition (SCADA) hardware and software. The SCADA system is used to continuously monitor and/or control water pumps, level, flow rate, and quality parameters such as dissolved oxygen,

carbon dioxide, pH, total gas pressure, oxidation-reduction potential, and temperature.
- There are many options for managing the major unit processes involved in a RAS. For example, biofiltration, the process of ammonia removal by nitrification in the filter, may be done with submerged biofilters, trickling biofilters, RBC, bead filters, fluidized bed biofilters (Timmons et al. 2001), or a MBBR. The latter was selected at both EWFCS (Sect. 6.2) and RFCRF (Sect. 6.3).
- There is an increased operating cost for the electrical power needed for pumping water, heating or cooling, UV disinfection, and providing a source of oxygen, whether produced on-site or purchased as liquid oxygen from a commercial source.
- If ever an opportunistic pathogen establishes in the RAS, it can be extremely challenging to control the pathogen. Sometimes, the best recourse has been to depopulate, disinfect, and restock, which is extremely expensive.
- Backup power, pumps, oxygen, and other critical infrastructure are often required to reduce the risk of catastrophic losses when operating RAS.

1.3 Larviculture of Walleye

The present chapter focuses on the intensive (i.e., tank) culture of Walleye from hatch to 35 dph. The first 35 dph encompasses three larval stages (prolarva, postlarva I and II), and early juvenile at about 15–18 days and a length of about 20 mm (Summerfelt 1996c). Early development of scales is visible at 24 dph but not completed until 45 days (Priegel 1964), which is an important factor affecting handling as they are more susceptible to injury before scale development.

Critical elements for Walleye larviculture were described in the Walleye Culture Manual (Summerfelt 1996c) and updated with a comprehensive literature review by Summerfelt and Johnson (2015). A detailed presentation of gas bladder inflation and noninflation of the gas bladder is described by Summerfelt (2013). An expansive body of relevant experience by hatchery personnel is difficult to access by internet search, but when available, the proceedings of the Annual Meetings of the Coolwater Fish Culture Workshop provide insight into issues and practical solutions to cultural problems reported by hatchery biologists from a cross-section of North America. Nevertheless, substantial scientific literature already exists on the culture of Walleye that spans the twentieth century and has been growing ever since.

Although the configuration of the systems at EWFCS (Sect. 2) and RFCRF (Sect. 3) differ in many details, such as the means to achieve turbid water and their feeding strategies, both systems incorporate practices essential to overcome NGB, clinging behavior, and cannibalism, which are critically important problems affecting the success of Walleye culture. The interplay of NGB and clinging behavior strongly influence the incidence of cannibalism. Summerfelt (1996c) described the methodology used to overcome these problems. Briefly, the attraction of larval Walleye to direct and reflected light has a major influence on design criteria for larviculture. A comparison of larval Walleye behavior in laboratory aquaria from hatch to 17 dph showed that in clear water, larvae had a strong association with the sides of the aquaria, but in turbid water, larvae avoided the sides of the aquaria (Bristow and Summerfelt 1994, 1996). Importantly, larvae in turbid water had greater average swimming speeds, faster growth rates, and improved gas bladder inflation (GBI) than larvae cultured in tanks with clear water (Rieger and Summerfelt 1998). The improved performance and viability in turbid water are attributed to the changes in larval distribution as a consequence of larval reaction to diffused light in turbid water.

The other problem was NGB, dependent on the ability of the larvae to penetrate the water surface to gulp air for the first filling of their gas bladder. Walleye, other percids, and nearly all spiny-rayed fishes are physoclists, meaning that air gulped at the water surface is able to pass through the pneumatic duct for only a short interval after yolk sac absorption. Inflation cannot occur if they cannot penetrate surface tension (Rieger and Summerfelt 1998). The problem was resolved in Iowa by equipping tanks with a surface spray to homogenize the oil droplets to a size that will pass through the standpipe screen (Moore et al. 1994). The method removes oil from the surface of the culture tank during the critical period when the larvae must inflate their gas bladder. For larval Walleye, GBI takes place from the 6th to the 12th day post hatch (Marty et al. 1995). A high percentage of larval mortality occurs in this interval for fish that do not achieve GBI. Both Vermont and Iowa use surface sprays.

2 Hatchery-Scale Production of Walleye Fingerlings in a Water Reuse Aquaculture System at the Ed Weed Fish Culture Station, Grand Isle, VT

Kevin Kelsey

2.1 Walleye Culture in Vermont

Walleye are native to Lake Champlain, Vermont, and its tributaries where the species served a commercial seining fishery from the late 1800s through the early 1900s. When catches declined, culture for the fishery restoration of Walleye on Lake Champlain was undertaken beginning in 1899 when up to 100 million eggs were collected at a stripping station/hatchery established in Sandy Point on Missisquoi Bay. The resulting fry were stocked into Lake Champlain, throughout Vermont, and

Fig. 1 The Ed Weed Fish Culture Station is located on the west shore of Grand Isle in Lake Champlain (Photo courtesy of Vermont Agency of Natural Resources Engineering)

other New England states, New York, and Pennsylvania (McKenzie personal communication, Vermont Fish and Wildlife Department, retired). This site continued to provide fry for stocking into Lake Champlain until 1954. Due to the decline of Walleye harvest numbers, commercial fishing in Missisquoi Bay in Canada for Walleye ceased in 1971 (Marsden and Langdon 2012), and with recreational catch rates dropping by more than 50%, stricter regulations were being put into effect and management plans were considered and implemented.

In 1986, the Vermont Fish and Wildlife Department (VFWD) and the Lake Champlain Walleye Association (LCWA) formed an agreement to produce Walleye fingerlings using extensive rearing ponds stocked with newly hatched fry and later using intensive culture techniques to stock ponds with advanced fry. This effort was started at the Bald Hill Fish Culture Station (BHFCS) in 1992, where newly hatched fry were reared in tanks and fed *Artemia nauplii* for 5 days and then distributed to the LCWA ponds as well as ponds on the hatchery site in an effort to improve survivals. Prior to the adoption and expansion of the RAS technology, success was inconsistent with a range from no fish harvested, due to water quality or food shortage problems, to years with good harvest. The average survival in cooperative extensive ponds over the last 30 years in Vermont has been 32%.

The EWFCS came online in Grand Isle, Vermont, in 1992 (Fig. 1). The need for this facility was recognized as the fishery division's management request for cultured fish was consistently not being met and reliance on surplus from federal facilities and neighboring states was frequently unreliable. For 16 years, production was focused on catchable Brook (*Salvelinus fontinalis*), Brown (*Salmo trutta*), Lake (*Salvelinus namaycush*), and Rainbow (*Oncorhynchus mykiss*) Trout for distribution throughout the waters of Vermont. Also, a portion of the annual production was dedicated to Lake Champlain, providing a percentage of the landlocked Atlantic Salmon (*Salmo salar*) smolts required for the lake, as well as Steelhead Trout (*Oncorhynchus mykiss*), Brown Trout, and Lake Trout.

Since 2005, when VHSV caused massive fish kills in the Great Lakes (Spickler 2007), Vermont decided to use a basin management approach to fishery efforts on Lake Champlain as it is connected to the Great Lakes drainage from the Richelieu River, a tributary of the St. Lawrence. In 2009, the production at EWFCS had become exclusive for Lake Champlain. The Walleye program was transferred from BHFCS to EWFCS in 2011; BHFCS continues to produce Walleye for Vermont waters outside the Champlain basin. The current program production objectives for EWFCS are as follows:

- 145,000 landlocked Atlantic Salmon smolts
- 58,000 Steelhead Trout smolts
- 57,000 Lake Trout yearlings
- 49,000 Brown Trout yearlings
- 165,000 advanced Walleye fry for LCWA ponds
- 200–250,000 Walleye fingerlings

The early-life stage rearing (sac fry to parr/fingerling) for all these species is done in production-scale systems using RAS technology and artificial diets exclusively. Techniques for solving problems were developed using a proof-of-concept approach wherein technology innovation in one year is validated by duplication in the following season. A detailed account of the system and techniques encompasses a culture run from the introduction of fry into tanks to fingerling harvest. The success of the system has allowed the EWFCS Vermont facility to achieve a consistent output of quality fingerlings needed to achieve program goals to enhance Sport Fish Restoration and angling opportunities for Walleye in Lake Champlain. The RAS technology allows for the control of environmental parameters critical for successful hatchery-scale production of Walleye fry, advanced fry, and fingerlings. The technique of using RAS to produce fingerlings for stocking extensive ponds managed by LCWA is carried out at EWFCS using and expanding on the same techniques that had been applied at BHFCS. The resulting fingerlings from this production are harvested in cooperation with the VFWD staff. This partnership extends beyond the ponds that the LCWA manages. Their involvement has been significant to our success. They have advocated politically, as well as financially supporting the development of intensive larviculture of Walleye in Vermont at both EWFCS and BHFCS.

2.2 System Design/Unit Processes

The system used to intensively rear Walleye fingerlings was designed to support biomass at harvest of 250–300,000, 45–50 mm Walleye fingerlings (250–300 kg) (Fig. 2). The maximum carrying capacity was estimated by an analysis of system production based on expected biomass and projected feed use to establish the requirements of the unit processes within the system (Summerfelt 1996a). Monitoring for alarm conditions such as system water levels, oxygen, temperature, and electrical function is done through the system control panel that is linked to the SCADA system that alerts day staff of alarm conditions; the three staff that live on-site 24/7 monitor overnight alarm outputs.

2.2.1 Solid Removal

General process flow begins with water gravity flowing from the tank center drain standpipe (standard tanks) or from the sidewall drain (self-cleaning tanks) to the drum filter for solid removal (Fig. 3). Tank depth is maintained by the height of the drains. The drum filter has an external bypass incorporated into the plumbing to permit water to continue flowing in the event of mechanical failure. The drum filter is sized to process 475 Lpm. Filter panels are 60 μm screen (Fig. 4).

Conducting Walleye larviculture in RAS with the addition of a turbidity agent (algae or clay) results in environmental conditions that mimic an enriched pond. This requires the "acceptance" of the presence of suspended material by both the fish and the unit processes of the system. Some solid accumulation occurs in the tanks, and

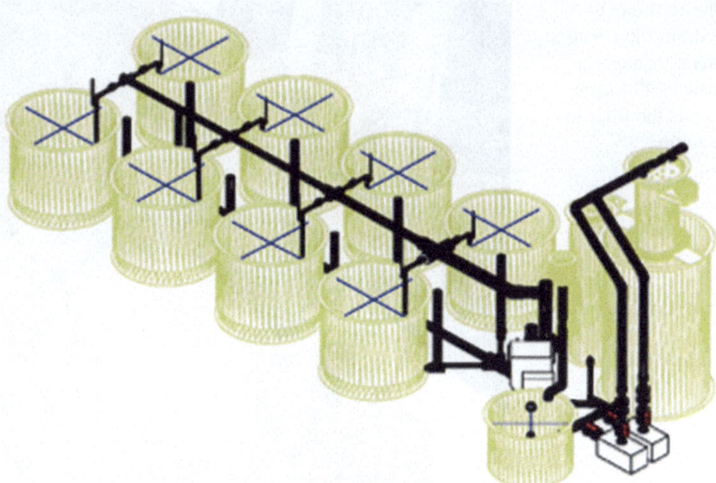

Fig. 2 A CAD drawing of a conceptual design of the intensive Walleye system at the Ed Weed Fish Culture Station (Photo courtesy of INNOVASEA)

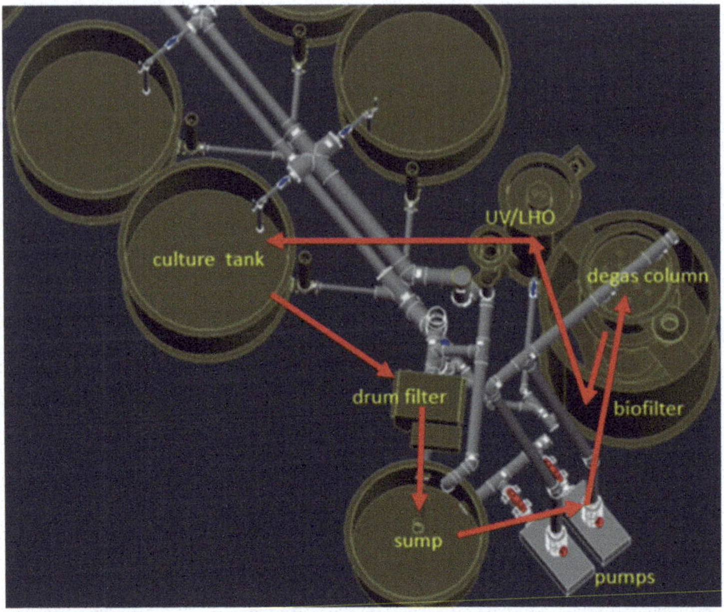

Fig. 3 Major components and process flow (red arrows) of the Walleye RAS at EWFCS. Water exits the culture tanks via surface drains to the rotary drum filter. Filtered water enters the sump where a 1-horsepower pump lifts water to the gas-stripping tower that is mounted on top of the MBBR. Water then passes by gravity through the aerated biomedia to the UV/LHO vessel and back to the culture tanks. Makeup water for the system (4–8 Lpm) is introduced to either the biofilter or sump. The overall system volume is approximately 20,000 L (CAD drawing courtesy INNOVASEA)

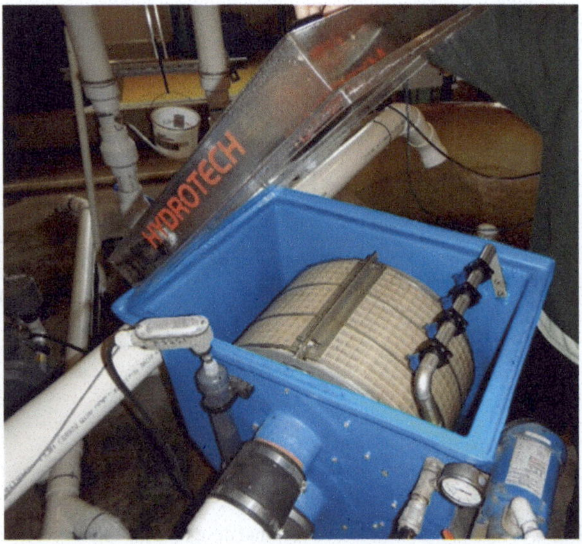

Fig. 4 The Hydrotech HDF-501 drum filter with a 60-μm screen manages a process flow of 475 Lpm. The flow exits the filter to the pump sump

the water has a higher quantity of both dissolved and suspended solids than typical of a single-pass system. The conditions require monitoring to maintain environmental conditions within acceptable target margins for desired growth rate and fish health. Settleable solids within the culture tanks that are greater than 100 μm settle to the tank floor and are removed manually. Finer suspended solids pass through the surface screens on the tanks and are removed by the drum filter. Algae is applied to the system (1–20 dph) to disperse the larvae and avoid clinging behavior. About 10 days after the application of algae, a biofilm of organic matter (biofloc) develops on the screens, which we try to remove as much as possible, but it dissipates completely as the system is cleared later in the culture run. Changing sizes on surface screens allows for more of this to be handled at the rotary drum as it passes through the culture tank screens. The rotary drum filter panels are 60-μm mesh. A 0.5-horsepower (hp) high-pressure pump rinses off accumulated solids from the panels to a waste trough that diverts the solids from the system.

The processed water from the drum filter discharges to the pump sump; the pump sump is the low point in the system and has an overflow standpipe to discharge exchange water that is continuously added to the system (4–8 Lpm). A 1.0-hp. pump draws water from the sump and delivers it to the top of the degassing column above the biofilter. A second pump is in standby backup with a selector switch. The backup pump will start in the event of a motor protection trip of the primary pump. A wide-angle float switch starts and stops the pump in the event of a low system water level.

2.2.2 Stripping Carbon Dioxide

Carbon dioxide was removed using a stripping column, which is designed to force large volumes of air through cascading water within an enclosed column. The stripping column is 0.9 m in diameter and 1.2 m high and is mounted on the top of the biofilter (Fig. 5). The column can process a flow of up to 475 Lpm. Water entering the top of the degassing column is uniformly distributed through a plate with antivortex crown nozzles. An inline continuous duty fan draws air upward and runs counter to the water dropping through the media. The system removes 70% of the generated carbon dioxide with each pass-through. Timmons et al. (2018) gave a recommendation for a system carbon dioxide concentration of 30–60 mg/L for warm water species and 15–20 mg/L for cool water species. We used a 15-mg/L concentration as our target limit with the recognition that levels could be as much as 30 mg/L or higher when approaching a maximum carrying capacity.

2.2.3 Biofiltration

An MBBR is used for biofiltration in the RAS at EWFCS. This type of biofilter was selected because it provides the necessary removal rate of total ammonia nitrogen (TAN) for the expected carrying capacity of the system while affording ease in maintenance and operational energy use.

Fig. 5 The gas-stripping tower is mounted on the top of the moving bed bioreactor

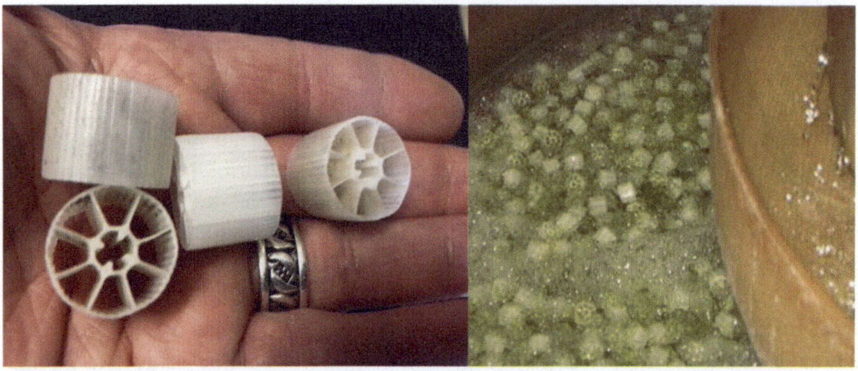

Fig. 6 MB3 media (left) was used as a surface for nitrifying bacteria in the MBBR (right)

The MBBR diameter is 1.5 m, and the height is 2.75 m. Processed water enters the top of the unit and flows downward through the media zone. A bowl below the degassing column collects and directs the water into the center of the biofilter to aid the movement of the 1.5 m^3 of MB3 polyethylene media (INNOVASEA Boston, Massachusetts) (Fig. 6). The biofilter aeration grid is powered by a 1.5-hp. regenerative blower, rolling the media up the exterior walls and down the center for thorough mixing. Aeration provides sufficient oxygen to the microbial biofilm within the biofilter to oxidize ammonia to nitrate. A shunt valve on the supply line to the biofilter diverts air to the tanks to be used for feeders to disperse feed and box screens that have bubble curtains to discourage fry from congregating in the box area. A bottom screen retains the biofilter media. The outlet is plumbed to allow for the gravity flow of water to the vertical UV/low head oxygenator (LHO) vessel.

2.2.4 Ultraviolet Disinfection and Oxygenation

Disinfection, oxygenation, and carbon dioxide removal are achieved by three unit processes. Water from the biofilter passes to the center core of the LHO. An integral open channel vertical design UV unit in the LHO center core passes the full system flow past the UV lamp field for treatment. A minimum of 30,000 μw/cm^2 is achieved at the maximum process flow of 475 Lpm. This level of UV disinfection is effective for most pathogens of concern. Undoubtedly, some loss of effectiveness of the UV light occurs with the use of algae as a turbidity agent, which is a necessity for Walleye larviculture. However, UV dose is inversely proportional to the flow rate (Losordo and Conwell 2014); thus, in the 1–14-day interval when turbidity is being applied to the system, we assume that the system flow during that period of 208–285 Lpm is sufficient to offset the reduced efficacy caused by the turbidity. As evidenced, disease events have occurred only once in 10 years; a severe bacterial gill disease in 2013 attributed to the excessive feeding of a microdiet.

Water flow entering the LHO is dispersed by the distribution plate, thereby enhancing oxygenation by the air-water contact as it passes through the LHO. A side box standpipe prevents overflowing the LHO in the event the distribution plate is blocked and directs the water below the LHO to maintain water flow to the culture tanks. The LHO serves as a head tank for water distribution to the rearing tanks. An excess water flow overflow standpipe directs any water that is not being used in the tank gallery back to the pump sump.

2.3 *Culture Tanks*

The rearing units for the intensive culture of fish have been rectangular (Colesante 1996), called raceways, and circular tanks (Moore 1996; Summerfelt 1996c). Circular tanks are the predominant tank shape used for the larviculture of Walleye. The advantages of circular tanks have been described by Summerfelt (1996b), but a noteworthy feature is that they operate with a rotating flow about the center drain that concentrates solids at the bottom, and they maintain uniform water quality throughout the tank.

Two types of circular tanks are used within the system at EWFCS. Originally, eight tanks with a skirted exterior bottom and center drain sump arrangement were installed (Fig. 7a). The tank operating depth was controlled with an external

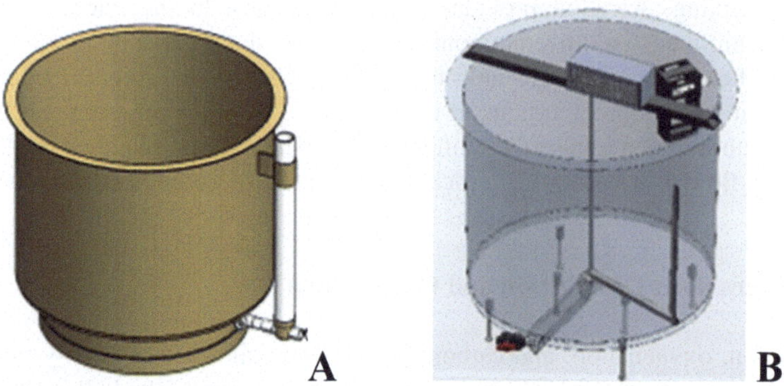

Fig. 7 Structure of tanks utilized within the system, (**a**) INNOVASEA and (**b**) Oceans Design. The dimensions of both styles of tanks were 1.5 m × 1.2 m

standpipe to facilitate draining from the sump or by the insertion of a solid wall 5-cm center standpipe to drain water from the surface. The later configuration was chosen to allow for the deposition of solids to remain in the tank bottom until they were removed with a siphon (Fig. 7a). Currently, a 1.5-m diameter tank is used with an operating depth set at 1.1 m. The tank walls have been painted black and buff-sanded to a flat finish to minimize the reflection of light that would be attractive to the photopositive fry. The floors of the tanks are light blue in color. The standpipes are retrofitted with custom surface screened drains constructed from modified plastic buckets modified to install mesh drain screens (Fig. 8b). The total surface area of drain screens provided is 0.13 m^2. The screen hole size employed is determined by the size of feed being applied and based on the developmental size of the larvae as well. Daily cleaning of the tank walls is done manually using a modified squeegee design developed at RFCRF. Floors and sumps are manually cleaned with a siphon twice daily.

In 2018 and 2019, self-cleaning tanks (Oceans Design Colorado Springs, Colorado) replaced four of the standard tanks (Fig. 7b). The self-cleaning tanks have a gear motor that moves a squeegee on the bottom and a brush on the sidewall at one rotation per hour. The squeegee moves solids to a trough in the tank bottom. The twice-daily siphoning is the same as with the standard tanks. Tank dimensions are identical to the standard tanks, and to ensure proper tank elevation, adjustable legs are used instead of tank skirts. Tank walls are a flat black finish with the tank bottom being light gray. The surface drain is an Easy Slide Larval Screen Box (Oceans Design Colorado Springs, Colorado) with a screen surface area of 0.18 m^2 mounted on the sidewall of the tank. Screen sizes are changed based on feed size and the size of the larvae (Table 1).

The operating volume of both tank styles is 1940 L. Both style tanks require attention to screen cleaning and changing screen sizes with larval development. Water exits the tanks through a surface drain—cylindrical center for standard tanks and box sidewall for self-cleaning tanks. In the last 2 years, the monitored cleaning events indicated that self-cleaning tanks take on average one-fourth of the time to

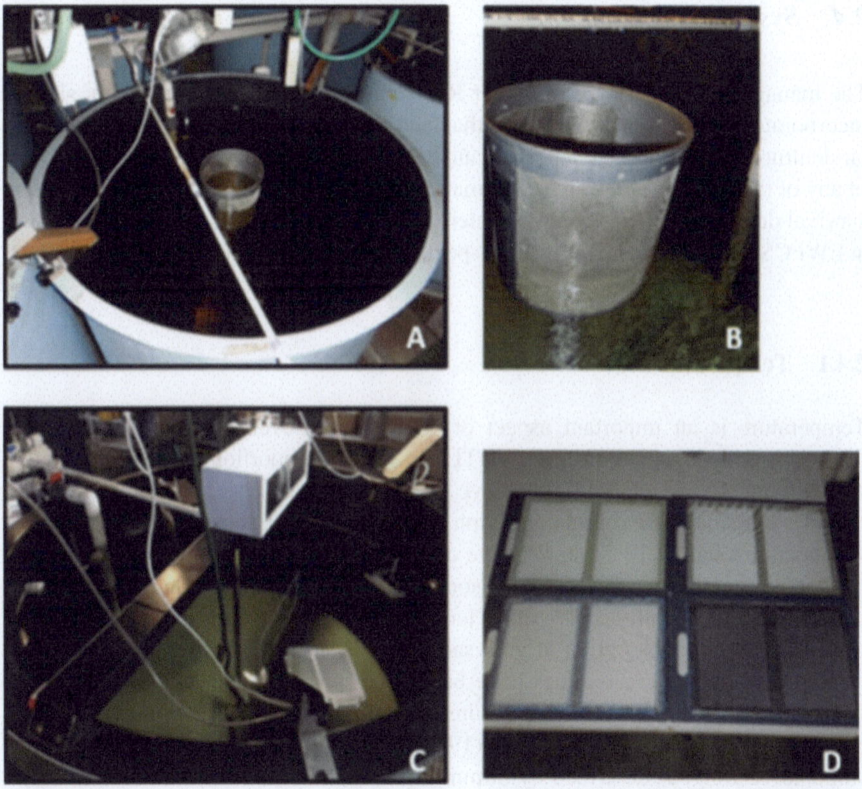

Fig. 8 Overhead photos of tanks and screens: (**a**) standard tank, (**b**) standard tank center screen, (**c**) self-cleaning tank, (**d**) screen panels of various mesh size for self-cleaning tank sidewall box

Table 1 Screen sizes used on standard and self-cleaning tanks in relation to the ranges of fry length and increased feed size

Days post hatch (dph)	Fry length range (mm)	Feed size range (μm)	Screen hole size (mm)
1–11	8–13	360–910	0.8
12–17	14–20	910–1410	1.65
18–25	21–34	910–1410	3.0
26–35	34–50	910–1800	4.0

clean than standard tanks, which is comparable to observations made by Rotman et al. (2017). Based on the increased efficiency and improved survival in self-cleaning tanks, the remaining standard tanks will be replaced with self-cleaning tanks for the 2020 culture season.

2.4 System Management

The management of systems, whether RAS, partial reuse, or flow-through, should incorporate practices and measures that address three major hurdles in Walleye larviculture: NGB, clinging behavior, and cannibalism. The presence and persistence of any of these issues alone are problematic, and in combination they will cause poor survival during and after the culture interval. Procedures utilized during culture runs at EWFCS that inhibit or reduce these specific issues are described in this subsection.

2.4.1 Temperature

Temperature is an important aspect of growth and development in larviculture. Sufficient daily temperature units (DTU) in proper proportions for critical stages of development are required to achieve optimum growth. The availability of DTU directly influences the number of dph needed to reach the target harvest size. Temperatures below 10 °C in Walleye culture can inhibit development and reduce metabolism, resulting in mortalities associated with nonfeeding behavior. During the early stages of larviculture (1–5 dph), temperatures greater than 20 °C may result in causing premature absorption of yolk sac and bacterial diseases.

Varying temperature regimes have been used in the past, with recognition that feed acceptance is greater at temperatures above 16 °C. An ideal temperature range of 16–18 °C was described by Moore (1996) to include a temperature rise at 5 dph to stimulate feeding, as described by Summerfelt (1996c). Our earlier culture runs fell within these ranges and would reach a maximum of 22 °C, similar to temperatures used by Moodie and Mathias (1996). Temperature in our RAS was controlled using a 5-cm diameter 12,000-watt in-line titanium heater that is threaded into the return plumbing line from the UV/LHO unit to the culture tanks. Temperature is monitored in the system sump, and transmission to the control panel activates and deactivates the heater based on the programmed settings. The temperature is manipulated to rise, coinciding with progressive larval stage transitions to optimize feeding and growth potential. Beginning in 2014, temperature started at 19–20 °C but was increased to 20–21 °C at 18 dph with positive results. Thereafter, a temperature of 20–22 °C was selected to reach a target size for the stocking of 1-g fingerlings, 50 mm in length in 35 days or less (Table 2).

Table 2 Temperature settings for the RAS during the 36-dph culture interval

Days post hatch (dph)	Temperature °C
1–4	19–20
5–17	20–21
18–36	21–21

2.4.2 Light

In 2011–2014, which were the first 4 years of Walleye larviculture, we used fluorescent shop lights to illuminate the tanks with a light intensity of 100–700 lux. Since 2015, light intensity has been maintained at 25 lux or less for the entire culture run. Each tank had a hooded lamp mounted 0.75 m above the water surface. The entire system was surrounded by a black curtain that eliminates external light. The subdued lighting provided in the otherwise dark environment was a strong enough attractant for the photopositive phase of the larvae but gentle enough to reduce excessive excitability observed with higher levels of light intensity. Fry began to exhibit photonegative behavior at 16–18 dph, during which point the hooded lamps were tilted at a 45°angle to reduce light intensity until the time of harvest (Fig. 9).

2.4.3 Turbidity

Turbidity is a necessity for Walleye larviculture (Bristow and Summerfelt 1994, 1996). Turbidity reduces reflected light within the culture tank and disperses the larvae, which in turn increases survival, reduces cannibalism, and enhances feed utilization and growth. The turbid environment also aids in the acceptance of artificial diets for initiating exogenous feeding, along with reducing cannibalistic tendencies as well. Except for developments reported by Johnson in this publication, an intensive culture of Walleye fry has been conducted in flow-through systems using clay to produce turbid water. However, the use of clay in a RAS was considered problematic because clay may adhere to and cover over (smother) the biofilm in the biofilter. Although clay solution has been the traditional method for producing turbidity for the larviculture of Walleye, microalgae were used to

Fig. 9 Hooded lamps are used to provide lighting to each tank (left). The hooded lamp is turned at a 45°angle as the fry transitions from photopositive to photonegative (right)

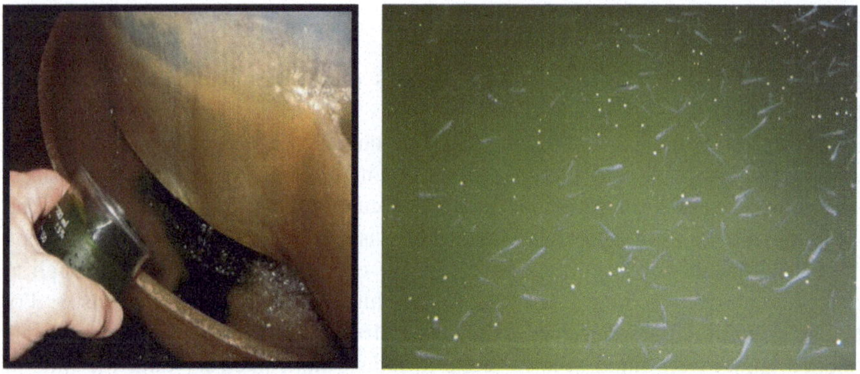

Fig. 10 Batch dosing is applied to the bowl below the system degassing column at the top of the biofilter (left). Example of turbidity created by the green water showing well-dispersed fry reacting to feed (right)

encourage the initiation of first feeding by several species of marine fish larvae (Cutts and Batty 2005). Houde (1975) described positive nutritional properties of algae by direct ingestion during the feeding of Sea Bream (*Archosargus rhomboidalis*). There is the prospect of nutritional and probiotic benefits to using algae as a turbidity agent for Walleye.

The use of algae instead of clay as the turbidity agent is a major difference between the protocol at EWFCS and that followed at RFCRF. Obtaining the concentration of algae to provide the desired turbidity is essential. Rønfeldt and Nielsen (2010) used *Chlorella* (*Chlorella pyrenoidosa*) algae paste in a turbidity experiment with Pikeperch (*Stizostedion lucioperca*). In comparison with clay, turbidity using algae was about 17 NTU (nephelometric turbidity units), while clay was close to 60 NTU. An experimental comparison of clay and algae with Walleye has not been undertaken.

In our system, an algal concentrate of *Nannochloropsis*, Nanno 3600 (Reed Mariculture Campbell, California), is employed by batch dosing directly to the system biofilter and sump (Fig. 10). Batch dosing of 800 mL of algae at 8-h intervals is applied between the biofilter and sump through 1–14 dph. Beginning at 15 dph, algae dosing is reduced by 58% with the use of a green water substitute concentrate, Sanolife GWS (Inve Aquaculture Salt Lake City, Utah), which is introduced at a batch dosing rate of 1 L every 8 h. Sanolife GWS is used exclusively for one day prior to clearing the system at 18 dph.

2.4.4 Surface Spray

The use of surface spray is required to reduce surface tension for successful GBI. Walleye larvae and that of other physoclistous fishes initially fill their gas bladder by penetrating the water surface and gulping air. NGB occurs due to the presence of a

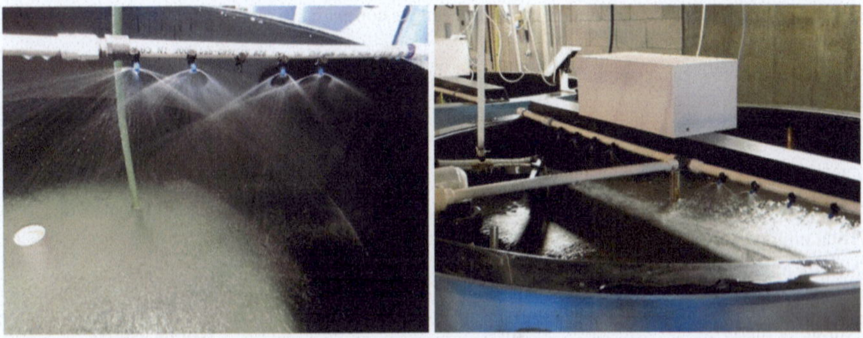

Fig. 11 Surface spray on the radial arm of the standard tank with bucket screen removed (left), spray bar spanning a self-cleaning tank (right)

surface film that larvae cannot penetrate (Rieger and Summerfelt 1998). El Gamal (2015) noted that in the culture of White Sea Bass (*Dicentrarchus labrax*), if an oil film is allowed to accumulate within tanks, consequences such as reduced feeding and growth occur, as well as irregular swimming behavior due to negative buoyancy. Colesante et al. (1986) opined that feed is a source for surface oil, and Boggs and Summerfelt (1996, 2003) identified larval mortality as an additional source for oil accumulation. Surface skimmers are more commonly used in marine larviculture to enhance GBI by blowing low-pressure air laterally along the surface of the water to trap oil in a floating containment area for removal. High-pressure water spray that can penetrate through the surface has been the general application of choice for Walleye fry culture. Several studies have been conducted to address the hurdle of NGB. Higher percentages of GBI were documented by Barrows et al. (1993) and Clayton and Summerfelt (2010) with the use of surface spray. Johnson et al. (2008) obtained GBI rates of 93–100% at Rathbun Fish Hatchery.

In the system at EWFCS, spray bars are mounted on the tanks to span the full diameter of the tank (Fig. 11). A 3.2-cm water supply line is run directly from the biofilter driven by a 0.125-hp. submersible pump. Each arm of the spray bar is threaded to allow for slight changes in angle that can be made to control surface directional flow if desired. Five adjustable spray nozzles (Rain Drip/NDS Woodland Hills, California) are on each arm of the bar providing overlapping semi-circle coverage. The spray bars are operated for the entire culture run through harvest to keep any feed or fines from accumulating on the surface. GBI is monitored in every tank throughout the culture run and at harvest. The 5-year GBI average is 95.6%.

2.4.5 Water Quality and Quantity

Temperature, oxygen, pH, carbon dioxide, ammonia, nitrite, and alkalinity are measured. The frequency of water quality measurements within the system ranges from weekly for carbon dioxide, TAN, alkalinity, nitrite, and pH to multiple times

Table 3 EWFCS RAS water quality parameter ranges compared to recommended values for warm water species (Timmons et al. 2018)

Parameter	Recommended values	EWFCS Walleye RAS
Temperature °C	24–30	19–22
Oxygen mg/L	4–6	6–10
CO2 mg/L	30–50	15–35
TSS mg/L	20–30	10–20 (post 18 dph)
Total ammonia-N mg/L	<3	0.2–1.3
NH_3-N mg/L	<0.06	0.01–0.04
Nitrite-N mg/L	<1	0.01–0.07
Nitrate-N mg/L	'high'	Not measured
Chloride mg/L	>200	Not measured

Table 4 Tank flow rate adjustments based on developmental stages of fry. Values do not include flow from spray bar ~4 Lpm

Days post hatch (dph)	Flow Lpm
1–7	30
8–11	34
12–20	38
21–35	42

per day for in-tank oxygen levels and temperature, as well as system-wide oxygen saturation (Table 3). More frequent monitoring and measurements were required when levels of pH and in-tank oxygen began to decrease, and carbon dioxide increased with rising biomass during the final week (28–35 dph). Although suspended solids should be considered to be reduced to the lowest levels achievable in RAS, that concern had to be counterbalanced by the necessary use of turbid water culture with algae (i.e., green water) for Walleye larviculture. At EWFCS, green water was cleared at 18 dph.

Values for pH within the system range from 6.5 to 8.2 with the lower values occurring toward the end of the culture run just prior to harvest as the system reaches maximum carrying capacity. Alkalinity values range between 136 and 188 mg/L. Makeup water for the system is derived from facility processed water that is filtered to 20 μm, disinfected with UV at 100,000 mw/cm^2, and heated to 10–11 °C.

Water is continuously added to maintain adequate water quality (Table 4). Hydraulic retention time (HRT) rates between 30–60 min are considered suitable for maintaining water quality parameters with properly functioning unit processes within the system. The HRT range is under an hour at system inoculation and just over 40 min for the last 2 weeks before harvest.

Direct flows at these rates through a 5-cm pipe can exceed 30 cm/s, a velocity that would easily overwhelm fry and disrupt many processes of development due to expending extra energy to combat the current. High exchange rates cause higher current velocity; therefore, to achieve acceptable HRT and counteract current velocity, a 3.8-cm vertical inlet pipe is used with 15 holes that are 1 cm in diameter. We made 10 holes, 10 cm apart on the front side of the pipe and five holes 12 cm apart on the back side of the pipe. The holes are parallel to the tank wall, velocity is subdued

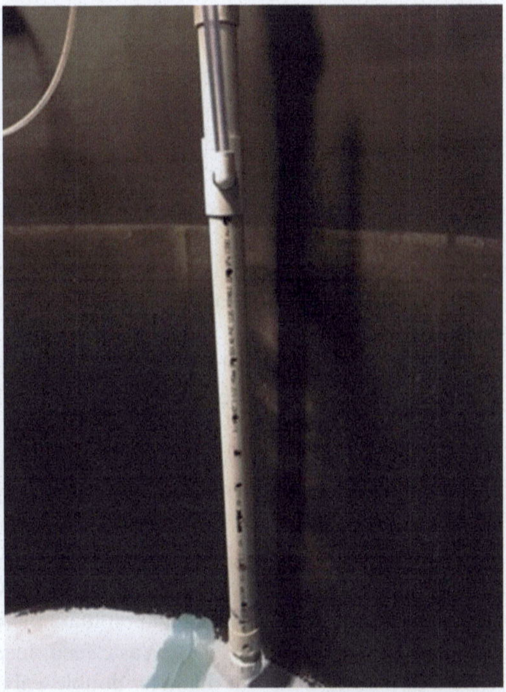

Fig. 12 A veritcal inflow pipe with 10 holes and clear tube manometer. The other side had five holes to dampen velocity within the tank

significantly with the backflow counteracting and dampening the main flow. Flow was measured by using a clear tube manometer (Fig. 12).

2.4.6 Stocking Density

In laboratory research, stocking densities of larvae were in the range of 3 to 100/L of tank volume (Summerfelt et al. 2011). Although it is important to use high density to maximize output, the higher density incurs risks of reduced survival, growth, and cannibalism. For rearing advanced fry at EWFCS to stock in LCWA cooperative ponds, a stocking density of 60 fry/L was used, with a survival rate average of 70% to 8 dph. Densities between 30 and 40 fry/L have been used for successful dry diet investigations at the RFCRF (A. Johnson, Iowa Department of Natural Resources (IDNR) personal communication). At EWFCS, from 2011 to 2014, a stocking density of 30–35 fry/L was used, but survival was poor (2–28%), perhaps due to other cultural conditions as well. Thereafter, fry stocking densities were lowered to 22–24 fry/L, which allowed for a comfortable margin for system inoculation at ~45,000 fry/tank with a potential to achieve production targets when a high level of survival to harvest was achieved.

2.4.7 Tank Hygiene

Consistent cleaning of culture tanks is critical in accomplishing positive results during larviculture. The removal of any accumulation of biofloc and fungus is necessary to maintain optimum environmental factors for both the fish and unit processes within the system. Matter accumulates from using algae for turbidity and microdiets, as well as from feces, waste feed, and mortalities. The EWFCS is staffed 16 h/day with three staff living on-site to maintain 24-h coverage. During the Walleye culture run, which occurs from mid-April to mid-June, the Walleye culture systems are staffed throughout the day, ending at midnight with an operational check of the feeders and RAS.

The screens are cleaned twice daily. Standard tank bucket screens are scrubbed in place and siphoned on the interior portion. Self-cleaning box screens have two screen panels that can be removed (Fig. 13). Parallel slots are designed so that a fresh clean screen can be inserted prior to removing the screen to be cleaned, which is brought to a sink to be hosed off and set aside for the following day. The interior of the box is cleaned with a small 1.5-cm diameter siphon.

Tank walls are cleaned continuously in self-cleaning tanks and once daily in standard tanks with a squeegee design similar to what has been used for culture tanks at RFCRF (Fig. 14). As feeding and biomass increase, inlet pipes are removed every other day for cleaning (Fig. 14). Waste feed, feces, and mortality are also removed twice a day using a 2.5-cm diameter double-valved wand (Fig. 15) designed with a clear tube section at eye level to observe waste being drawn through. The discharge of the wand is cam locked into a manifold that deposits waste and mortalities into a screen insert that holds back mortalities that are enumerated, and the waste falls through the screen into a discharge collection sump.

Fig. 13 The box screens were removed for cleaning in a self-cleaning tank. Note the clean screen that has been placed in the parallel slot (left). A CAD drawing of a box screen with parallel slots for screen exchanges (right) (Photo courtesy Oceans Design Inc)

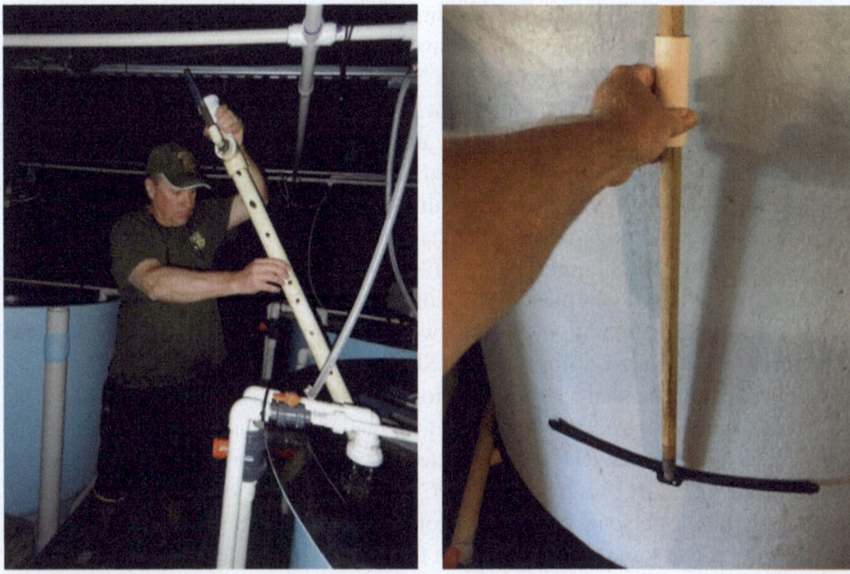

Fig. 14 The inlet pipe is removed for cleaning (left). The manual squeegee device used to clean the interior walls of standard tanks (right)

Fig. 15 A double-valved cleaning wand (left) with the vacuum head removed is used to clean tank sumps (center), and the discharge cleaning manifold and sump (right) are used for the vacuuming of standard tanks to process waste and collect mortalities

2.4.8 Feeds, Feeding Rates, and Feeders

In early research on intensive Walleye larviculture, fish were initially fed *Artemia* nauplii and zooplankton then weaned to a dry-formulated diet (Howey et al. 1980). Colesante (1996) used the protocol for large-scale production: *Artemia* were fed for one month, followed by 2 weeks of offering *Artemia* and a dry diet, after which a dry diet was fed exclusively. A modification of this technique is practiced at EWFCS.

The separate RAS that is used for producing advanced fry (8 dph) fed with *Artemia* often has surplus fry remaining after fry have been counted out for distribution to the LCWA cooperative rearing ponds. The remaining surplus of advanced fry is inventoried. To evaluate survival after stocking, fish were marked with oxytetracycline (OTC); details of this procedure are described in Sect. 2.6. Survival to fingerling harvest (33–35 dph) has ranged from 25% to 45% post OTC marking, with the bulk of mortality happening within 48 h of the mark application.

An extensive chronology of work that has been done feeding Walleye fry entirely on a dry diet appears in Summerfelt et al. (2011). Ongoing efforts in assessing diets, ratios, and rates have continued with work done by Johnson and colleagues at RFCRF. To date, comparative trials between Otohime (Japan, sourced by Reed Mariculture, Campbell, California) and Skretting Gemma (Fontaine-lès-Vervins, France) feeds indicate that Otohime performs better than Skretting Gemma. The choice to use Otohime exclusively as a diet for the EWFCS culture was based on the multiyear feeding trials at RFCRF conducted in smaller culture tanks. The rates and ratios applied from 2011 to 2014 at EWFCS were based on the findings at RFCRF (A. Johnson, IDNR personal communication). Variable results were achieved during these years because some critical feed and feeding issues needed further evaluation:

- Ratio of feed (g) to the number of fry/L—Was the feeding rate too high, thus impacting tank hygiene? Or should it be increased to enhance growth and survival?
- Diet blending—Feed trials have indicated the inconsistent performance of one diet versus another. Could there be a benefit to combining two diets that would be accepted by fry and provide desired performance?

During these earlier culture runs at EWFCS, fry were exclusively given a feed between 250 and 360 μm in size, until 7 dph, then and in combination with the larger feed until 14 dph. Though fry will consume this size feed, observations of larvae being first fed to *Artemia* in the system that is used to grow advanced fry for extensive ponds were easily consuming nauplii sizes of 350–550 μm. The smaller size dry diet created tank hygiene issues and, in the instance of the culture runs in 2013 when increases in rates were attempted, likely contributed to a bacterial gill disease outbreak, which caused significant mortality. Since 2015, the use of this feed size range was reduced and by 2018 eliminated.

When fed separately, Otohime has shown to have increased palatability in the RFCRF trials compared to Gemma; therefore, in 2015, the blending of two diets (Otohime and Skretting Gemma) was considered for first-feeding larvae. Otohime has a variable size range of crumbles and fines compared to the uniform pelleted extrusion of Gemma, and the profile of Gemma diets is more nutritionally dense compared to Otohime (Table 5). During a routine tank sampling of fry conducted through the culture runs, both diets were seen in the stomach and digestive tract by a visual examination of the transparent fry. This observation is evidence that fry will consume both diets when presented in equal proportion. Perhaps, the attractiveness of one diet may encourage the consumption of the other diet with higher nutritional

Table 5 Comparative size and nutritional values of Skretting Gemma and Otohime diets

Diet	Size range µm	% protein	% fat	% fiber	% ash
Gemma					
Wean 0.3	350–500	62	14	0.5	9
Wean Diamond 0.5	500–800	62	14	0.5	9
Diamond 0.8	800	57	14	0.2	10
Diamond 1.0	1000	57	15	0.2	10
Diamond 1.2	1200	57	15	0.2	10
Otohime					
B-2	360–650	51	11	3.0	15
C-1	580–840	51	11	3.5	15
C-2	840–1410	51	11	3.5	15
S-2	920–1800	52	14	3.5	15

Table 6 Diets, diet size, and ratios fed and the feeding rates used in 2014 and 2019 by age (dph): B-, C-, and S- are Otohime; G is Skretting Gemma; SC is Silver Cup

2014					2019								
Dph	B-1	B-2	C-1	SC 1.0	Dph	B-2	G 0.3	C-1	G 0.5	C-2	G 0.8	S-2	G 1.0
2–7	100				3	50	50						
8–9	75	25			7	35	35	15	15				
10–12	50	50			8	25	25	25	25				
13–14	25	75			10	10	10	40	40				
15–16		100			11			50	50				
17–18		75	25		13			25	25	25	25		
19–20		50	50		14			20	20	30	30		
21–22		25	75		17			15	15	35	35		
23–25			100		19			20		40	40		
26–30+			75	25	22			10		45	45		
					24					40	60		
					26					30	50	10	10
					28					15	45	20	20
					30						35	15	50
					31						20	20	60
					32–35						15	15	70

value. The blending of both diets in equal amounts through the entire culture run has been established since 2017.

Along with increases in feed size and truncating ratio transition times, feed rates were also increased well beyond rates described by Summerfelt (1996c) and diet trials conducted from 1999 to 2003 by Johnson and Rudacille (2009) (Table 6). Moving through the feed sizes and ratios faster and increasing feeding rates reduced

Fig. 16 (Left) Cannibalism in Walleye showing that size differential between the cannibal and prey can promote cannibalism. (Right) A random grab of fingerlings from a harvest net illustrating uniformity in fish size

the proportion of the presence of smaller size fry. These smaller fish have the potential to exacerbate cannibalistic behavior due to the substantial size differentiation (Fig. 16). Reduction of these fish from the population is attributed to the harvested population being more robust and uniform in size while still maintaining potential levels of survival to exceed 60%.

Various feeders have been described for intensive culture, such as belt feeders (McCauley 1970), vibratory (Loadman et al. 1989), and augers (Summerfelt 1996c). Moodie and Mathias (1996) had "custom built" feeders made because belt feeders lacked precision. At EWFCS from 2011 to 2013, belt feeders were mounted on tanks to deliver feed; however, small feed sizes adhered to the belt, and feed of all sizes would fall into the tank with limited dispersal. There was an obvious need for better feeders.

In recent years, feeding systems have been refined for a variety of applications for different species focusing on delivering microdiets used for dry diet exogenous weaning and feeding. In 2014, Hatchery Feeding System (HFS) feeders (Nutrakol Pty. Ltd Perth, Australia) were evaluated (Fig. 17). These feeders deliver a microdiet in doses determined by the plate size selection (single or double slot) and by the number of shots in a 24-h period that occurs as a solenoid is actuated based on selected preprogramming on the controller which can operate up to 24 feeders. The feeders are 24 volts and can be adapted to utilize interchangeable hoppers of variable capacity (250 g and 1 kg). There is also an air induction port that can be throttled to use an air current to loft and disperse feed on the tank surface.

In the event of a malfunction, backup feeders have been purchased to prevent the interruption of the 24-h feeding and so that maintenance and cleaning of the units can be accomplished. Continuous feeding is critical to avoid events of cannibalism. For instance, in 2016, at 22 dph, 50% of the feeders were inoperable for 8–12 h during the night due to an unmonitored electronic malfunction. At this stage of development, the expected daily system mortality for the eight tanks combined would be 1200–2000. The loss of these feeders resulted in a mortality of 25,000 fish 25 mm in

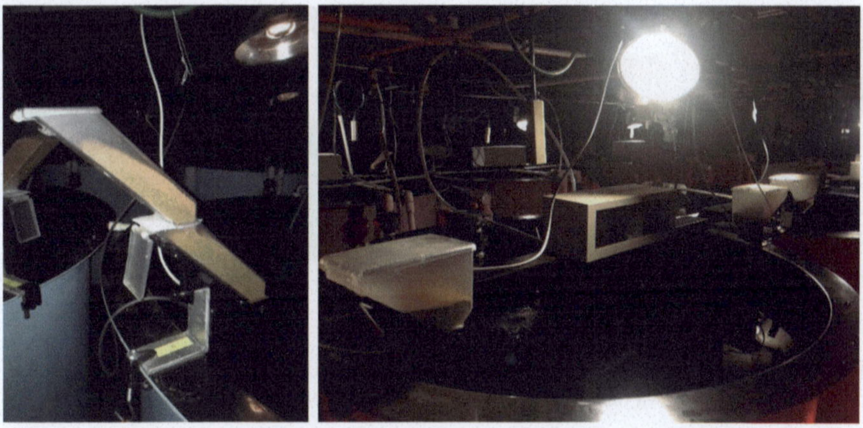

Fig. 17 Position of mounted HFS feeders on tanks: (left) a custom design modification of joining two 250-g hoppers together for increased capacity, (right) photo of 1-kg-size hoppers

length. Almost all mortalities had evidence of bite marks on the isthmus and caudal areas.

The rate of feed dispensed from the HFS feeders has now been elevated to closely numerically correlate to the days post hatch. For instance, in 2019, at 19–20 dph, 19 g/1000 was programmed. It should be noted that from 14 dph to harvest, hand-feeding events are also applied to the tanks 8–12 times across 16 h. The total daily amount of hand feeding applied correlates to an additional 30–35% of the calculated daily program amount of the automatic feeders. Final feed application rates toward the time of harvest exceed 45 g/1000. Feed conversion ratio (FCR) from 2017 to 2019 has averaged 0.9. Low FCR values are influenced by the large differential in water content of the feed and fish.

Data from earlier culture runs identified ranges of sizes at the time of harvest in excess of 25 mm. The propensity of Walleye fingerlings to be piscivorous at an early stage of development is key to their success in the natural setting. Creating an environment that promotes this tendency during intensive larviculture could lead to increased cannibalistic behavior that could potentially reduce harvest numbers. At EWFCS, feeding regimes have been developed to hurry the inclusion of larger size feed as that improves tank hygiene and focuses on feeding the top of the population bell curve, dropping out poor performers earlier to help reduce counterproductive cannibalistic behavior. The application of this approach may have contributed to an increase in overall growth and survival.

Our feeding rate (g feed/1000 fish) is determined by “feeding to the number.” Daily mortality assessments of individual tanks are required to ensure that proper feed rates are being applied to the existing population within the tank (Fig. 18). From 1 to 7 dph, an estimate is made of 800–1000 mortalities/tank/day. By 7–9 dph, fully formed body mortalities can be identified in the waste stream and counts are begun. After rinsing the mortalities from the waste sump tray screen insert, mortalities are enumerated. Either of the following method of preference is used: shallow tray

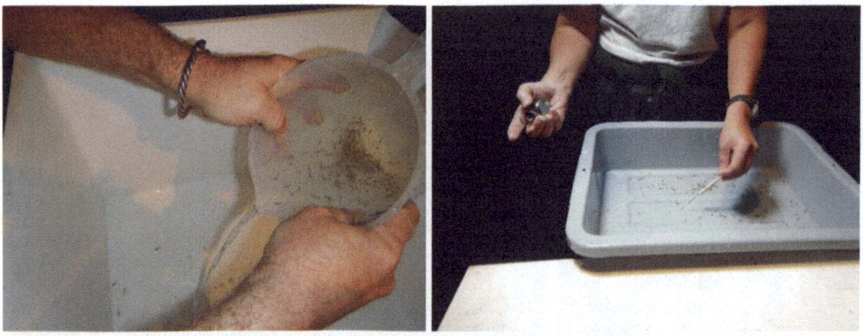

Fig. 18 Accurate accounting of tank mortalities is needed to adjust feeding rates. The beaker pour counting method is shown (left), and shallow tray is used to count mortality (right)

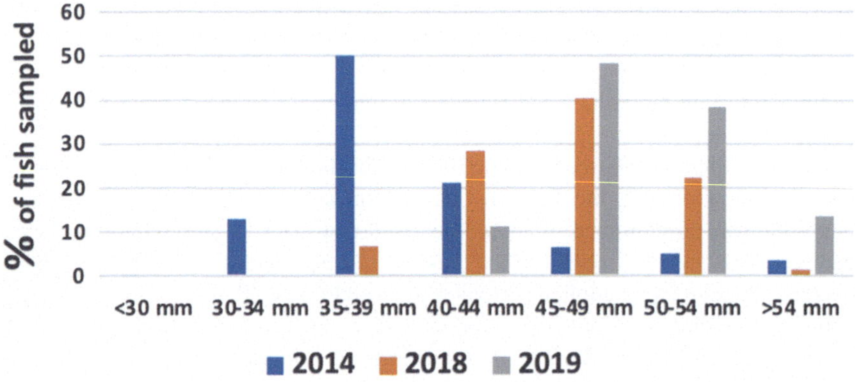

Fig. 19 Lengths of samples of 60 fish per tank measured during harvest. The stronger bimodality represented in 2014 can promote cannibalism due to the expanded ranges of size and potentially reduce survival due to the strikes and bite wounding that can be created by this size differential in the population in contrast to the greater size uniformity exhibited in 2018 and 2019

counting or slow beaker pouring (Fig. 19). With the shallow tray method, mortality from a tank is placed into a polyethene tray 40 cm × 28 cm × 12 cm with 1.2–2.5 cm of water. Mortalities are spread out on the floor of the tray, and the counting process is conducted. The beaker pour method uses a 2-L beaker filled to the halfway point with water, at which point the rinsed fry mortality from a tank is introduced. Water is added to the fill mark and then slowly poured out toward the technician to enumerate the fry. A small utensil or dissecting needle can be used to keep the fry from binding together. Additional water is added as needed to keep the fry flowing at a rate that allows each individual to be counted.

It is important to get a careful accounting of mortalities and avoid estimation so that as accurate as possible feeding rates can be applied. In dealing with cannibalism or any other unknown loss that may transpire during the first few days of the culture run, unaccounted mortality can occur. Though there can be a ranging variability based on a number of factors, minimizing the guesswork with precise mortality counts has provided inventory accuracy from tank inoculation to harvest that averages +/− 10–12%.

2.5 *Growth and Survival*

Growth in fish length was monitored primarily using lengths at 5–7-day intervals. Vermont Fish and Wildlife Department fish health biologists sample each tank (20 fish/tank) and conduct measurements of fish, along with observations related to GBI/NGB, the prevalence of deformities, and feed utilization. The method used for samples is rapidly "trawling" two large aquarium nets toward each other. This method provides a reasonable snapshot, though it is likely that larger, faster swimming individuals avoid sampling during the last 7–10 days, indicated by length samples (60/tank) conducted at final harvest (Fig. 19). Fish weight was determined during the later stages of the culture run and at harvest (Table 7).

Survival from the initial stocking of the culture tanks to harvest is a measure of the success of larviculture. The goal at EWFCS has been to engage in intensive culture techniques using RAS to attain a consistent level of fingerling production that would result from survivals surpassing 50%. After several years of trial and error, the quality and quantity of fingerlings improved significantly beginning in 2017, which has led to consecutive years of production that has provided a substantial number of fingerlings for contribution to the Walleye program for Lake Champlain (Table 8) (Fig. 20).

2.6 *Performance Assessment of Stocked Fingerlings*

Measuring the contribution of cultured fish to the fishery is important data collected by district fisheries field staff. Marking the otoliths of cultured Walleye using OTC began in 1996 to establish the level of contribution and recruitment from the cultured fish program to the Walleye fishery. The source of fry for the RAS at EWFCS comes

Table 7 Average length and weight of 20 fish sampled from each tank during the 2019 culture run at EWFCS. At harvest (34 dph), 60 fish are measured and weighed from each tank

Days post hatch (dph)	Length (mm)	Weight (g)
6	10.52	NA
10	12.78	NA
15	18.13	NA
21	26.10	0.22
27	35.10	0.42
34	49.62	0.91

Table 8 EWFCS Walleye fingerling harvest data from 2011 to 2019

	EWFCS Walleye fingerling harvest trends				
Year	# of initial fry in system	Fingerlings harvested	% survival	Dph	Length mm
2011	344,000	20,043	5.8	41	38.1
2012	178,500	49,866	27.9	33	39.12
2013	175,190	3762	2.2[a]	35	34.29
2014	202,125	15,004	7.4	31	40.39
2015	152,500	56,344	36.9	31	37.34
2016	222,000	55,209	24.9[b]	30	37.85
2017	280,000	146,024	52.2	35	45.72
2018	320,000	194,914	60.9	33	46.1
2019	360,000	204,628	56.8	34	49.62

[a]Bacterial gill disease caused major mortality
[b]At 22 dph, 50% of tank feeders malfunctioned through the night (8–12 h), resulting in a mortality of 25,000 due to cannibalistic behavior

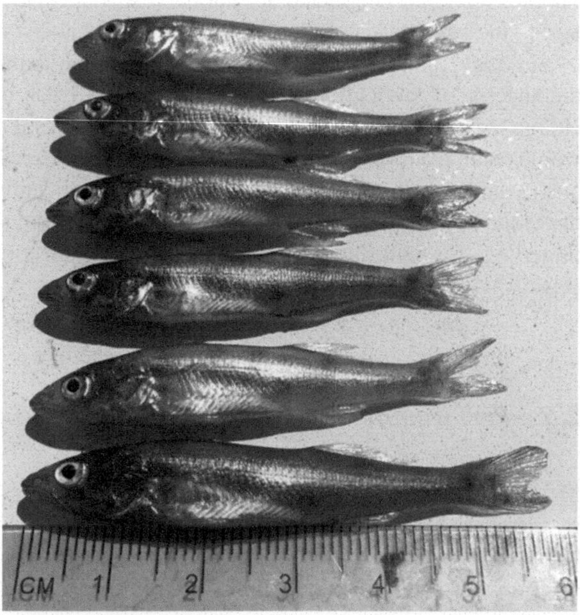

Fig. 20 Illustrated in this photo are uniform robust fingerlings from 2019 harvest samples

from broodstock obtained from Lake Champlain. Three tributaries in Vermont are used rotationally on an annual basis: the Poultney, Winooski, and Missisquoi rivers.

Three OTC marks are applied using a bath at a 6-h exposure (Fig. 21): a fry mark at 3 dph and an advanced fry mark at 8 dph, both of which are done at a concentration of 700 ppm, and a fingerling mark at 500 ppm is applied at the time of harvest at 34–36 dph for intensively reared fingerlings and 50–55 dph for fingerlings reared extensively in ponds. Walleye stocked at the unfed fry and advanced fry stages have shown returns that are statistically insignificant over 20 years, which is why efforts to

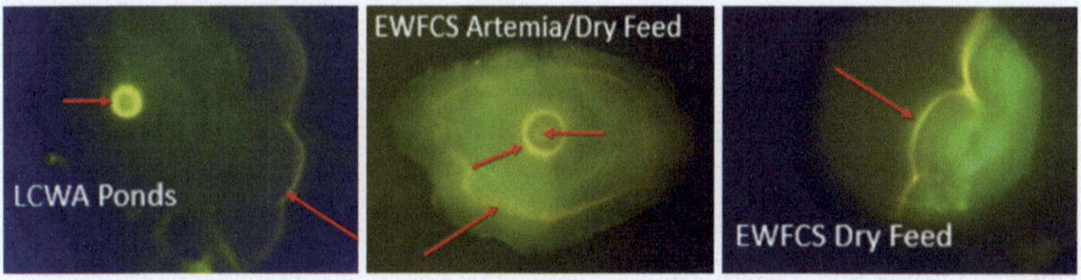

Fig. 21 Fluorescing OTC marks are observed using a Nikon E-400 microscope. LCWA extensively reared pond fingerlings have a fry and fingerling mark (left). Surplus advanced fry weaned from *Artemia* to a dry diet have a triple mark (center). Fry that are raised to fingerlings on a dry diet exclusively receive only a fingerling mark (right)

rear fingerlings have been established. A separate RAS is used for supplying the LCWA ponds with advanced fry that have been fed *Artemia nauplii* for 5 days. If there are any surplus fish in this system after the ponds have been stocked, a mark at the advanced fry stage is applied to differentiate the group having a triple mark. These fish are weaned from *Artemia nauplii* to dry feed (over a 3-day transition period) and grown to the fingerling stage.

Good (2020) describes the process of preparing and analyzing sagittal otoliths in this manner: otoliths are extracted from Walleye, wiped clean and dried, mounted concave side down to glass slides with Super Glue (Ontario, California), and allowed to harden for at least 24 h. Once the glue is cured, otoliths are gently ground on wetted 1500-grit automotive sandpaper along the sagittal plane to remove the top layers of glue and establish a level polishing surface. Specimens are sanded until the entire surface of the otolith is revealed then covered in mineral oil and viewed under a compound microscope at ×10 magnification with a reflected fiber-optic light source. Annuli are counted to estimate age.

After age estimation, otoliths are polished further using a wetted 30-μn lapping film (Precision Surfaces International, Houston, Texas), stopping occasionally throughout the process to examine the otoliths for the OTC marks until presence or absence is determined (Secor et al. 1991; Brooks et al. 1994; Fielder 2002).

Otolith inspection for OTC marks is conducted using a Nikon Eclipse E-400 epi-fluorescent microscope with fluorescent lighting and filter blocks designed to fluoresce the tetracycline marks. The Nikon E-400 microscope was outfitted with a B-3A filter cube (505-nm dichroic mirror, 420–490-nm exciter filter, and 520-nm barrier filter) and a 100-W mercury UV light source, as described by Bumguardner (1991) and Logsdon (2006). Otoliths are viewed through ×100 and ×200 magnification.

Up to 50 additional 3-year-old males are collected every year in excess of the fish that are utilized for broodstock. These males are being collected to assess otoliths for the presence of OTC marks that designate what part of the culture program contributed using the process described above. If there is no mark detected, this indicates that the fish has been naturally reproduced.

Table 9 The performance of cultured Walleye fingerlings in the Poultney River to document the contribution of intensively reared fingerlings

River	Culture method	OTC mark applied	# stocked	Return ratio (%)	*n* (OTC marked fish)	*n* unmarked fish
Poultney '20	LCWA Ponds	Fry/fingerling	78,178	18	9	0
(2017 stocking)	EWFCS Intensive	Fingerling Fry/Ad.Fr/Fing	167,745	82	42	
Poultney '17	LCWA Ponds	Fry/fingerling	69,109	66	19	10
(2014 stocking)	EWFCS Intensive	Fingerling	15,004	34	10	
Poultney '14	LCWA Ponds	Fry/fingerling	41,200	94.60	35	3
(2011 stocking)	EWFCS Intensive	Fingerling	20,100	5.40	2	

Highlights of the OTC mark analysis for 3-year-old males sampled on the Poultney River for the last three rotations are shown in Table 9. Sample years are assessing the stocking that occurred 3 years prior. The Lake Champlain Walleye program transferred to EWFCS in 2011, and efforts to work toward continued improvements were applied with each culture run. The analysis and assessment of the 2017 stocked fingerlings have just been completed. The performance of intensively reared fingerlings has developed progressively. There were three naturally reproduced fish present in sampling in 2014 and 10 present in 2017. In 2020, all fish sampled exhibited OTC marks. The reduced presence of naturally produced fish is a trend that is also being recognized on other tributaries by division biologists and could be potentially related to increased predation from invasive species such as Alewife (*Alosa pseudoharengus*) and White Perch (*Morone americana*). As investigation continues, cultured fingerlings will play an important role in the efforts to maintain a quality Walleye fishery in Lake Champlain.

3 Walleye Fingerling Production in a Reuse Aquaculture System at the Rathbun Fish Culture Research Facility, Moravia, IA

J. Alan Johnson

3.1 Walleye Culture in Iowa

Walleye are spawned and eggs incubated at three IDNR hatcheries: Fairport Fish Hatchery (FFH), Rathbun Fish Hatchery (RFH), and Spirit Lake Fish Hatchery (SLFH). In 2019, the hatcheries distributed 148 million fry (1–3 day-old prolarvae) to enhance Iowa's fishery resources. Pond fingerlings are produced at FFH and RFH while advanced fall fingerings are produced at SLFH and RFH. In 2019, the FFH located by the Mississippi River produced 472,639 pond-reared fingerlings (ca. 35–40 mm). In 2019, the RFH, located below Rathbun Lake Dam, distributed 91,240,988 fry; 748,248 pond fingerlings (ca. 40–45 mm); and 176,567 advanced fall fingerlings (mean size 225 mm). Spirit Lake Fish Hatchery, adjacent to Spirit Lake, produced 59,309,340 larvae and 21,093 tank-reared fall fingerlings (152 mm). All advanced fall fingerlings were reared from larvae stocked in a pond, harvested at about 42 mm, transferred to tanks for habituation to pelleted feed, and then grown to about 225 mm at either RFH or SLFH.

Rathbun Fish Hatchery began fish production in 1975, but intensive culture of Walleye did not begin until 1985 with the habituation of pond-reared fingerlings to pelleted feeds. Hatchery research investigations into intensive Walleye larviculture began in 1991. Larger fingerlings were desired, and the early spawning of broodstock coupled with intensive larviculture could extend the growing season to consistently produce fall fingerlings at 204 mm. The Iowa DNR's Fisheries Bureau recognized the need for a dedicated research facility and staff, providing the rationale for the development of the Rathbun Fish Culture Research Facility (RFCRF), which began operation in 1996. In 2000, six 0.04-ha research ponds and 10 0.4-ha production ponds were added to RFCRF and RFH, respectively. Production ponds are double-cropped by the production of Channel Catfish (*Ictalurus punctatus*) following Walleye harvest. All culture operations use surface water (Rathbun Lake) in single-pass culture or pond culture. Water supply to the Hatchery and Facility is from Rathbun Lake and passes through a drum screen (300 μn) before entering outdoor culture units or passing through sand filtration and UV disinfection before filling indoor tanks. In 2015, a RAS was constructed at RFCRF to evaluate pilot-scale a growout of Walleye fingerlings from 100 mm to 225 mm. The growout RAS is the largest system at RFCRF with three 10.9 m^3 tanks. Additional RASs were added for Walleye egg incubation and intensive larviculture in 2018 and 2019, respectively. The incubation RAS is a partial RAS because freshwater exchanges are needed in the absence of a biofilter and has 20 jars with 66 L of egg capacity. All RASs utilize the municipal water supply for filling, replacement, backwashing, and equipment washing; surface water is not used. Municipal water is dechlorinated by activated carbon filtration before addition to the system with one exception: drum screen backwash.

Using RAS, Walleye can be cultured from eggs to large fingerlings using municipal water that is free from AIS contamination. Other sport fish, such as Rainbow Trout and Muskellunge (*Esox masquinongy*) or yearling Walleye for future broodstock, have been cultured in the same system when the RAS tanks were not utilized for Walleye culture. All three Iowa DNR fish hatcheries that produce Walleye utilize surface water sources (Rathbun Lake, Spirit Lake, Mississippi River) that are considered infested with Zebra Mussels and threatened by other AIS as well as surface waters that are known sources of many parasites (e.g., Ich) and pathogens. While disinfecting surface water for single-pass systems can eliminate sources of infectious pathogens or AIS, a single-pass system replaces 100% of the tank volume per hour. Comparatively, RASs rely on a small amount of makeup water volume of 5–20% of the total system volume per day. Compared to single-pass systems, the greatly reduced water requirement of a RAS substantially reduces costs and equipment to improve water quality or eliminate the ingress of pathogens or AIS into the hatchery. RAS use can overcome current limits to the production capacity of the Iowa DNR fish hatcheries that are at or near maximum capacity based on space, water quantity, and/or water quality.

The use of RAS for sport fish production is an emerging trend among government agencies. Egg incubation to food size fish production in RAS is common for many food fish species. A few studies have evaluated RAS for Walleye production (Summerfelt 1996a; Summerfelt and Penne 2007; Zarnock et al. 2010; Davidson et al. 2016), and two focused on Walleye production for stock enhancement (Aneshansley et al. 2001; Harder et al. 2014). Limitations to the successful application of RAS technology to Walleye in future hatchery renovations require applied research to establish optimal production densities, determine swimming speed tolerances, feeding regimes, and other issues. This section will describe Walleye larviculture in pilot-scale RAS at RFCRF.

3.2 *System Design/Unit Process*

The larviculture RAS was added to RFCRF by aquisition of a turn-key system (Fig. 22) manufactured by Oceans Design Inc. (Colorado Springs, CO). Critical system conditions such as dissolved oxygen, water levels, and electrical status are monitored by a SCADA system that alerts staff.

3.2.1 Solids Removal

Solids are removed from the tank by siphoning the waste trough once daily. Suspended solids in the tank effluent are removed by a drum screen filter (Trome, 60 microns, Londerzeel, Belgium). Culture water passing the drum filter passes on to the horizontal MBBR.

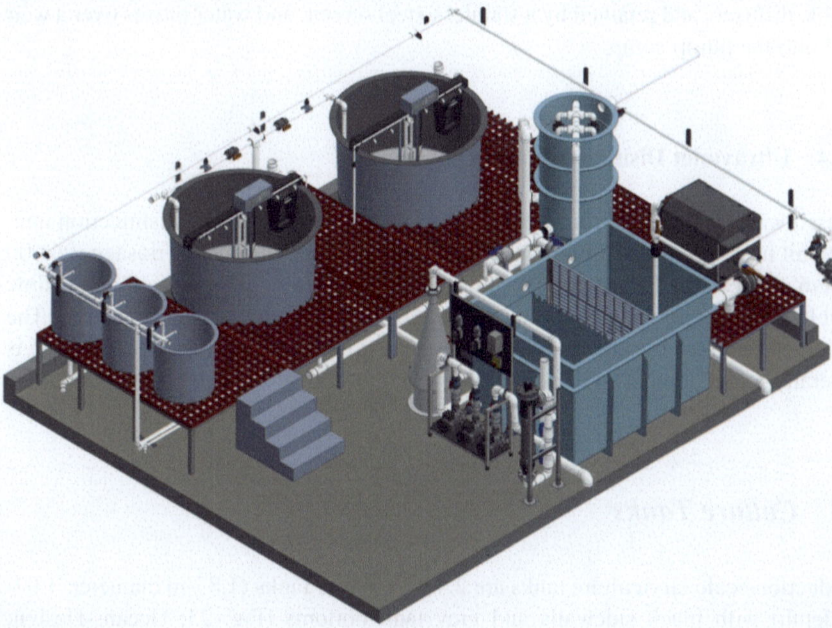

Fig. 22 Designed view of the larviculture water reuse aquaculture system at Rathbun Fish Culture Research Facility, Moravia, IA (provided by manufacturer Oceans-Design, Colorado Spring, CO). In the sequence of water flow, the components of the system are five culture tanks (three 275-L tanks and two 2400-L tanks), microscreen drum filter, biofilter and pump sump, pumps and control panel, CO_2 stripper, UV lamp array, and oxygenation cone

3.2.2 Degassing

The biofilter and pump sump are in one unit, with a media retention screen and weir wall separating the biofilter and the pump sump. A separate circulation loop pumps water from the sump through the chiller heat exchanger into the CO_2 stripper then returns water to the biofilter. A portion of the flow from the pump sump may be diverted through two inline heaters then discharged into the biofilter sump. The CO_2 stripper is filled with tube-form Nor-Pac degassing media (Oceans-Design, Colorado Springs, CO) to prevent biofouling from the combination of clay particles and biofilms that could impede water flow through the CO_2 stripper. Other types of random pack media may be more prone to plugging water flow due to biofilm and clay accumulation.

3.2.3 Biofiltration

The horizontal flow MBBR contains a 1.0-m^3 K3 media (Oceans Design, Colorado Springs, CO) that provides a total surface area of 500 m^2. Media is fluidized by air bubble diffusers and retained by a stainless-steel screen, and water passes over a weir wall into the pump sump.

3.2.4 Ultraviolet Disinfection and Oxygenation

Reused water is pumped from the sump and through an ultraviolet disinfection unit. The unit has three lamps (390 watts total) that provide a 96-mj/cm^2 dosage (96,000 μw/cm^2), 90% UVT, and a 189-Lpm maximum system flow rate. A downflow bubble contactor oxygenates water with LOX-derived pure oxygen injection. The cone is pressurized by a modulating knife valve downstream of the cone. Pressure is generally maintained at 7–10 psi.

3.3 Culture Tanks

Production-scale larviculture tanks are 2.4 m^3 circular tanks (1.83-m diameter, 1.07-m depth) with black sidewalls and grey tank bottoms (Fig. 23; Oceans-Design; Colorado Springs, CO). Each tank has a wiper arm that pivots around the center drain driven by a gear reduction motor. The arm has a bottom squeegee wiper and a sidewall brush wiper. The wiper arm rotates at 1 rpm to sweep waste into a trough at the tank bottom extending the radius of the tank. The trough has a 1-inch drain that is only used to completely drain water from the tank during harvest. Tanks are fitted with a center drain screen and a screened box insert mounted to the tank wall and connected to a waste return line. The center screen has three parts: a 101.6-mm Sch 40 PVC pipe with cutout sections for a frame that inserts into a fitting in the bottom center of the tank, a removable plastic mesh (3.2 mm opening) to provide screen support over the frame openings, and a nylon mesh sock fitting over the plastic mesh. The nylon mesh socks used in Walleye larviculture are 0.7-, 1.0-, 1.5-, 2.0-, and 3.2-mm square openings.

3.4 System Management

3.4.1 Temperature

The water temperature for newly hatched larvae is typically maintained at 15.0–16.7 °C for 2 days post hatch. On day 5 post hatch, when larvae begin to feed exogenously, temperature is increased by 1.5–2.0 °C, which is thought to stimulate uniform feeding behavior. Temperature is gradually increased to up to 22.0 °C prior to 35 dph (Table 10).

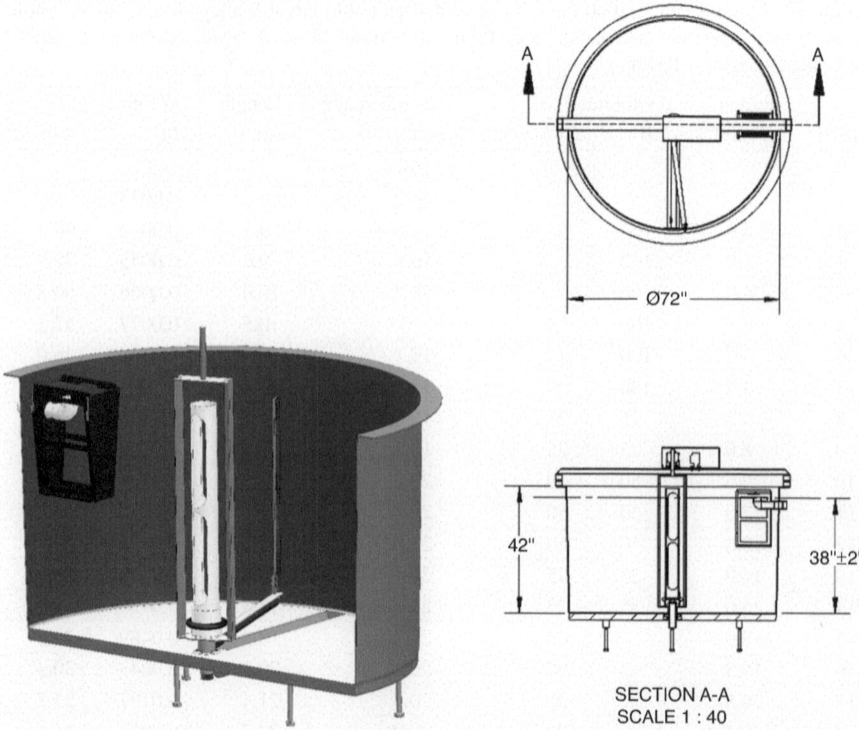

Fig. 23 Schematic cross-sectional, overhead, and side drawings of the 2.4-m^3 self-cleaning larviculture tank with dual effluents fitted with a side-wall screen box and center drain screen

3.4.2 Light

Each tank is equipped with an overhead light fixture with a full spectrum LED lamp suspended over the center of the circular tank. Light levels are measured with a lux meter and set by rheostat to 75 lux at the water surface.

3.4.3 Turbidity

Turbidity is artificially increased by the addition of a clay slurry. The clay used is the OM-4 clay (Imerys, Paris, France), a powdered clay that is mixed with water on-site (25 g/L) in a stock solution in a 113-L cone-bottom tank, aerated to keep the clay in suspension, and pumped to the culture system by a peristaltic pump. The frequency of clay addition is controlled to provide 50 NTU of turbidity during the first 21 days post hatch, and then turbidity addition is reduced over the next 3 days. The clay dissipates from the system, and turbidity levels are about 5 NTU for the remaining culture period. Turbidity levels of culture water are sampled once daily in each culture tank.

Table 10 Feeding rate (g/1000 larvae) and feed sizes fed to larval Walleye from 2 to 35 dph in relation to temperature and larval size. Feed rate expressed as % bw/d calculated as ration/(estimated weight × 1000)

	Ration	Otohime			Temperature	Length	Weight	
DPH	g 1000^{-1}	B2	C1	C2	(°C)	(mm)	(g)	%bw/d
0					16.1			
1					16.2	8.6	0.0033	
2	4.0	100			16.6	9.1	0.0044	90.7
3	4.0	100			16.9	9.6	0.0055	72.8
4	4.0	100			18.2	10.0	0.0066	60.8
5	4.0	100			19.0	10.5	0.0077	52.2
6	4.0	100			19.1	10.9	0.0088	46.0
7	4.5	100			19.2	11.4	0.0099	45.7
8	6.5	75	25		19.5	11.7	0.0105	62.1
9	8.0	75	25		19.7	12.7	0.0160	50.1
10	9.0	50	50		20.0	13.7	0.0214	42.0
11	12.0	50	50		20.3	14.7	0.0269	44.6
12	13.0	50	50		20.3	15.7	0.0324	40.1
13	14.0	25	75		20.3	16.8	0.0379	36.9
14	15.0	25	75		20.2	17.8	0.0434	34.6
15	17.0		100		20.1	18.8	0.0489	34.8
16	19.5		100		20.1	20.1	0.0643	30.3
17	20.5		100		20.4	21.4	0.0797	25.7
18	20.5		100		20.9	22.7	0.0951	21.6
19	22.0		75	25	21.2	24.0	0.1105	19.9
20	22.0		75	25	21.5	25.3	0.1259	17.5
21	23.0		75	25	21.8	26.6	0.1413	16.3
22	23.0		50	50	22.0	27.9	0.1567	14.7
23	24.0		50	50	22.0	29.4	0.1967	12.2
24	26.0		50	50	22.0	30.8	0.2366	11.0
25	28.0		50	50	22.0	32.3	0.2766	10.1
26	33.0		25	75	22.0	33.8	0.3166	10.4
27	41.0		25	75	22.0	35.2	0.3565	11.5
28	50.0			100	22.0	36.7	0.3965	12.6
29	62.0			100	22.0	38.1	0.4365	14.2
30	70.0			100	22.0	40.0	0.5214	13.4
31	80.0			100	22.0	41.9	0.6063	13.2
32	90.0			100	22.0	43.8	0.6912	13.0
33	100.0			100	22.0	45.6	0.7761	12.9
34	110.0			100	22.0	47.5	0.8611	12.8
35						49.4	0.9460	

3.4.4 Surface Spray

Surface spray nozzles provide an open area for GBI and assist in clearing feed debris trapped in the surface tension of the water. The surface spray enhances GBI (Clayton and Summerfelt 2010). Tanks have surface spray nozzles (model 73–501, Hydro-Gardens, Colorado Spring, CO) spaced evenly across the diameter of the tank by mounting to the wiper motor support frame and a second spray bar perpendicular to this spray bar.

In general, single-pass larviculture has resulted in acceptable GBI rates using one spray nozzle for a 0.76-m diameter tank. In 2019, samples of fry showed variable rates of GBI between samples obtained during culture, but the final GBI rate was 98%. In 2020, we installed a second spray bar perpendicular to the existing spray bar for a total of 12 nozzles. Rates of GBI were consistently high in 2020. The surface spray bars are supplied with pressurized system water from the oxygen cone.

3.4.5 Water Quality and Quantity

A municipal water supply is dechlorinated by passing through two activated carbon filters before entering the larviculture system as either system replacement water or drum filter backwash spray. Freshwater replacement water is 0.5–1.0 Lpm. Additional freshwater is added to the system during drum filter backwashing, but the rate of addition has not been quantified. Rathbun Lake is the water source for the local municipal system and has a low level of alkalinity (ca. 90 mg/L). Reused water inflow rates to tanks are initially set at 65% tank exchange per hour and increased to 200% tank exchange per hour at the end of the culture period. Water quality (NH_3 and pH) is measured in each tank and in the pump sump (alkalinity, NH_3, NO_3 and pH) once per week. Alkalinity is maintained above 150 mg/L with a goal of 200 mg/L by the addition of feed grade sodium bicarbonate at a rate of 25–30% of the total daily feed ration. System water is periodically tested for alkalinity and replenished to 200 mg/L.

3.4.6 Stocking Density

Eggs are incubated with water from a RAS that is supplied with dechlorinated tap water. Fry for larviculture are collected over about a 24–30-h period to minimize variation in cohort size that could lead to cannibalism. Larvae are enumerated at 2 days post hatch using a commercial fry counter (Model FCM, Jensorter; Hillsboro, OR) and stocked into tanks. The counter is calibrated according to the instruction manual, then a sample of about 500 fry is enumerated by hand and passed through the counter three times to obtain an average error estimate, and the stocking number is adjusted by the error rate to achieve the desired quantity.

Fry are concentrated in the catch tank, collected in beakers, and placed into the fry counter hopper. Fry exiting the counter are captured in a 20-L bucket with a 500-μm screen-covered window for water release. The bucket receives about 10,000 fry before larvae are transferred to the larviculture tank. In previous single-pass experiments, a stocking density of 40 fry/L was used. Currently, density has been reduced to 30 fry/L in order to produce larger (>0.6 g) size fingerlings at harvest. In the RFCRF RAS, stocking density was 30–35 fry/L in 2019 and 2020.

The offspring of Rathbun Lake broodstock, originally stocked with eggs from Spirit Lake broodstock, have been reared in single-pass larviculture with success at RFCRF. In 2019, tanks were stocked with larvae obtained from Rathbun Lake broodstock. Production, however, was limited with only two 2.4 m^3 tanks; therefore, in 2020, production was increased by rearing three back-to-back crops of fingerlings by obtaining eggs earlier and later than the normal Rathbun spawn. First crop eggs were obtained from Kerwin Reservoir, KS. The eggs were incubated and hatched at the RFCRF and cultured to 35 dph then harvested. The second crop of larvae eggs from Rathbun Lake was incubated and hatched for culture immediately following the first crop. The third crop was produced using eggs from Lake Sakakawea, ND, broodstock incubated and hatched to follow the harvest of the 35 dph Rathbun fingerlings.

3.4.7 Tank Hygiene

Solids accumulate in the waste trough in the tank bottom and are removed once per day by a siphon tube. Solids are collected in a screened basket to retain mortalities; some wastes require removal with a water spray to leave mortalities cleaned for counting. Siphoning the small trough area reduces the amount of time to clean compared to siphoning the entire tank bottom and improves the completeness of siphoning in a turbid water tank with no visibility of the tank bottom.

Center drain and sidewall box drain screens are removed and replaced with cleaned screens once daily. Typically, a few mortalities are impinged on the screen and are counted prior to cleaning. Tank water levels are elevated due to screen or effluent pipe biofilms impeding effluent flow occasionally toward the end of the culture period.

3.4.8 Feeds, Feeding Rates, and Feeders

Fish are fed Otohime B2, C1, and C2 diet sizes (Reed Mariculture, Campbell, CA), starting with B2 and progressed to C2 (Table 10). Feed size is increased by blending progressively larger feed sizes until the C2 diet is fed exclusively. Feed rates are expressed in g/1000 larvae and increased from 4 g/1000 up to 110 g/1000 fry at 35 dph.

Feed is dispensed from an ARVOTec T-drum 2000 (Huutokoski, Finland) using an industrial computer control system to continuously monitor and control the feeders. Each 2.4 m^3 tank has two feeders mounted to the wiper frame spanning the diameter of the tank. Feed is offered at 5-min intervals 24 h/day. The feed dispensing rate is measured during a set interval (100 s), and the PLC program delivers the desired ration (total g/d) at determined intervals (300 s) over a 24-h/day. Feed drops from the feeders directly into the tank without the use of a spreader.

3.5 *Growth and Survival: Culture Period*

Four culture runs, one in 2019 and three in 2020, were completed in the larviculture RAS at RFCRF with an average survival rate of 77.2%, resulting in 470,050 total fingerlings produced in the two 2400-L tanks. In 2019, a total of 107,802 fingerlings were produced in the production scale tanks with a 75.2% survival rate (Table 11). In 2020, the two production tanks were stocked and harvested successively three times, identified here by the source (number and % survival): 118,252 (75.6%) from Kerwin Lake; 139,284 (88.5%) from Rathbun Lake; and 104,712 (67.5%) from Lake Sakakawea. Over the three culture runs, GBI ranged from 96.5% to 100.0% and averaged 98.4%.

In 2019, growth rates ranged from 1.02 mm/d to 1.23 mm/d with a lower temperature of 18.7 °C resulting in slower growth rates than fish cultured at 20.3 °C average temperature in 2020. The feed conversion ratio was 1.65 in 2019. In 2020, feed conversions were 0.81–1.00 and there were observations of very little waste feed. The 2020 temperature regime increased, and the feed regime may have

Table 11 Summary of Walleye fingerling production by year and egg source in a water reuse aquaculture system in 2.4-m^3 culture tanks at Rathbun Fish Culture Research Facility, 2019 and 2020

	2019 Rathbun	2020 Kerwin	2020 Rathbun	2020 Sakakawea
Age at harvest (dph)	42	35	35	35
Final number/tank	53,901	59,126	69,642	52,356
Survival (%)	75.2	75.6	88.5	67.5
Final length (mm)	47.1	48.7	50.7	48.8
Final weight (g)	0.86	0.94	0.99	0.91
Gas bladder inflation rate (%)	98.0	99.0	100.0	96.5
Harvest density (kg/m^3)	16.0	25.4	32.3	21.1
Feed conversion ratio	1.65	0.84	0.81	1.00
Specific growth rate	15.4	16.9	16.6	16.8
Daily length gain (mm/d)	1.02	1.21	1.23	1.19
Mean temperature (°C)	18.7	20.0	20.4	20.5

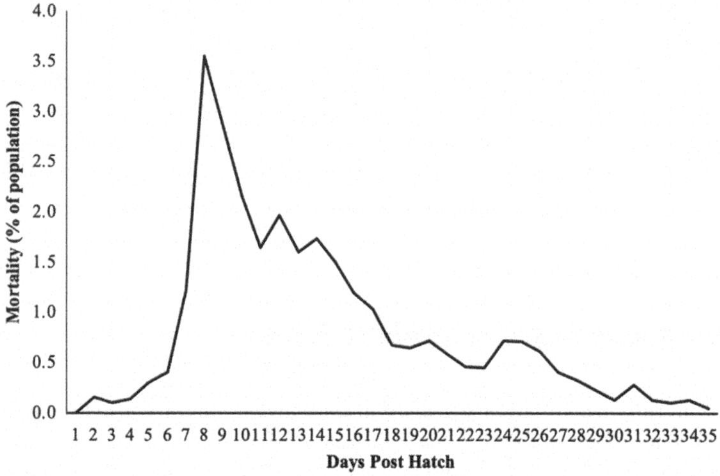

Fig. 24 Mean daily mortality rate (percent of tank population) during the 2020 three larviculture runs

been the minimum feeding rate in two runs: with 0.81 and 0.84 FCR; increased feed rates should be evaluated to determine if such a low FCR indicates underfeeding.

Mortality during the culture interval peaked at 8 dph (Fig. 24). A similar mortality peak was observed in 2019 and is commonly observed in single-pass culture experiments at RFCRF prior to RAS culture. Observations of food in the gut during the 8 dph measurement indicated high rates of feed present in the gut of sampled fish; therefore, cannibalism attempts may be the cause of mortality. Larvae were observed attacking other fish in the opercula region in an attempt to overcome the victim (Fig. 25), and this attack likely resulted in death. Trunk attacks are a greater source of mortality compared to tail attacks (Loadman et al. 1986). The main culture techniques used to minimize cannibalism are feed amount, frequency, and palatability. Some feed was observed in the waste siphoned from tanks; however, it is not known if higher feed rates may help curtail this instinctive behavior.

Feeding rates and temperature regimes have changed at RFCRF in the past 20 years based on work in single-pass larviculture experiments. In the late 1990s and early 2000s, temperature was held constant at 18.3 °C throughout a 28-day period for continuity between years of open formula diet studies. At this temperature, fingerlings reached a length of 25–28 mm. Subsequently, the focus of intensive larviculture moved toward production scale with an emphasis on growth and postlarviculture survival rates. To be competitive with Walleye pond culture growth rates at RFH, temperature regimes for larviculture were increased to obtain higher growth rates to match pond culture growth rates. Additionally, a larger fingerling size is desirable to improve survival immediately after the larviculture period when transferring to grow-out tanks. Other research at the facility demonstrated improved survival of pond-reared fingerlings of at least 0.56 g that began habituation to formulated diets in tanks (Johnson and Rudacille 2010). A predominant factor

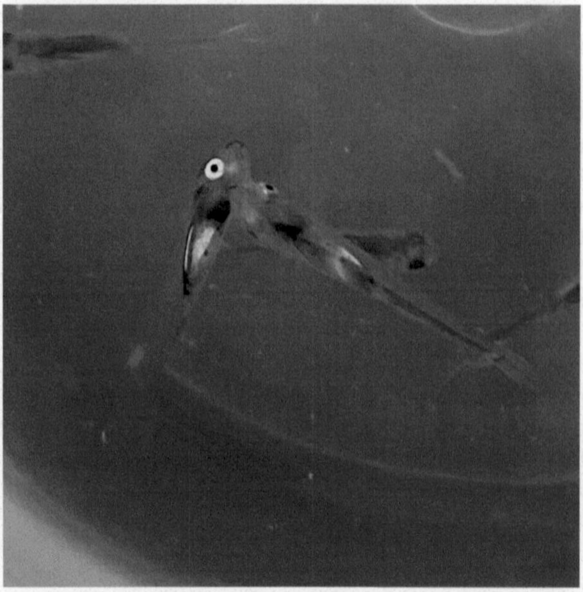

Fig. 25 Cannibalistic behavior of 15 dph larval Walleye, shown ventral side upward, holding the victim by the throat, an attack that differs from an attempt by the predator to swallow its siblings head-on (Fig. 16)

contributing to the higher survival rates of larger fingerlings in surface water single-pass culture is likely the presence of a full complement of scales that develop at 42 mm (Priegel 1964) and that Columnaris disease (CD) is a result of mechanical damage during handling (Hussain and Summerfelt 1991).

Feeding rates utilized from 1999 to 2003 at RFCRF with BioKyowa FFK and reared at 18.3 °C began with 3-g/1000 larvae and increased to 7-g/1000 larvae by 23 mm length and fed 10% body weight per day (bw/d) at 25–40 mm length (Johnson and Rudacille 2009). In 2006 and 2007, using the Otohime diet, larvae were fed 4–7-g/1000 larvae from first feeding to 25 mm (Johnson and Rudacille 2009). Later, feeding rates were increased to match growth rates as water temperature increased. In 2012, the mean temperature in a 35-day study was 19.7 °C and resulted in 43.5-mm and 0.67-g fingerlings with a 75.8% survival rate. In that study, feed rates increased from 4-g/1000 to 6-g/1000 larvae and from 2 dph to 20 dph then changed to a 10% bw/d rate. Feed conversion ratios were not monitored in previous single-pass larviculture studies until RAS larviculture began in 2019.

Past research trials (2006–2007) at RFCRF fed Otohime B1 diet (250–360 μ; Johnson and Rudacille 2009). That size is similar to the BioKyowa FFK B400 size (250–400 μ) fed by Barrows et al. (1993). Through gradual experimentation and collaboration (Kevin Kelsey, Personal Communication, EWFCS), Otohime B1 was phased out and only Otohime B2 (360–650 μ) is offered as the first feed without observed reduction in survival rates or feed acceptance.

The temperature regime used in 2020 was modeled after a temperature regime used at the EWFCS (Lake Champlain, Vermont, Kevin Kelsey personal communication). The feeding rates used in 2020 were based on previous feed rate and growth rate observations at RFCRF and were increased substantially compared to rates used between 1999 and 2012, particularly after seven days post hatch when the 2020 rates increased rapidly and ended with about 12.8% bw/d (110-g/1000 larvae; Table 11) rather than 10% bw/d used in previous studies. Feeding rates expressed as g/1000 larvae are used to determine the daily feed fed at RFCRF. However, these feeding rates, converted to percent body weight per day, compared with rapid larval growth rates, result in initial rates of 90.7% bw/d, decline to 45%, then rebound to 62% when the g/1000 larvae feeding rate is increased. Further research could optimize feed rates if related to survival; however, total feed costs (including shipping) were $0.016 per fish at 35 dph. Therefore, minor adjustments to the feed rate of small larvae are not likely to affect cost per fish.

These trials demonstrated that larviculture protocols and techniques previously established in single-pass culture systems could be successfully implemented on production-scale tanks in a RAS. Three geographic sources of Walleye larvae were reared in the same RAS under similar conditions with results that demonstrate that techniques for Walleye larviculture are successful in RAS regardless of larvae source.

3.6 Performance Assessment of Stocked Fingerlings

Currently, the Iowa DNR has two fishery evaluations of intensive-culture Walleye. One study is evaluating single-pass intensive-larviculture-reared fingerlings to extensive pond culture fingerlings for stocking interior rivers. Walleye fingerlings (~50 mm) cultured in a RAS may be a replacement for pond-reared fingerlings for stocking in interior rivers when pond-reared fingerlings are typically stocked. In June, the rivers are often at flood stage, and stocking at that time results in poor recruitment. Fingerlings reared on pelleted diets in tanks can be held and fed in the hatchery for a short period until river levels decline. However, natural zooplankton and benthic insect larvae in extensive pond production have a limited production capacity that requires the harvest of fingerlings when food sources in the ponds are exhausted, regardless of river level.

Poor recruitment of pond-reared Walleye fingerlings (~50 mm) stocked in constructed or natural lakes in June commonly occurs due to predation. For this reason, size has been identified as a factor in previous IDNR fishery research evaluations. Many impoundments >202.3 ha in Iowa generally receive 225-mm fall fingerlings. Walleye ≥203 mm (70 g) produced at RFH and stocked into some Iowa impoundments had mean survival rates of 45% (Mitzner 1995), and overwinter survival of these large fingerlings was twice that of fingerlings from extensive nursery lakes, which averaged 125 mm. Lower and more variable survival of 4.9–31.7% was reported for 150–178-mm average size Walleye produced at SLFH (Larscheid 1995). The IDNR Fisheries Bureau is currently comparing Walleye (225 mm) produced in RAS and fingerling fish produced in a single-pass system.

4 Discussion: Similarities, Contrasts, and Collaboration

J. Alan Johnson, Kevin Kelsey, and Robert Summerfelt

4.1 Similarities

The RASs at EWFCS and RFCRF employ similar but not identical unit processes for clarification, biofiltration, gas stripping, oxygenation, and UV treatment of the water. Both RASs employ the same commercial self-cleaning tanks, drum filters, and moving bed bioreactors for biofiltration. Both use surface sprays to eliminate NGB and turbid water and culture tanks with black sidewalls to prevent clinging behavior and to disperse the fish to reduce cannibalism. Otohime is fed at both facilities as the first feed using the same commercial feeders at both sites.

Both systems are obtaining better than 60% survival rate in the first 35 days with 90% or better gas bladder inflation. It is important to note that both systems have not experienced frequent outbreaks of bacterial gill disease (BGD), CD, or parasites. In single-pass culture, RFCRF Walleye fingerlings are frequently treated for the control of BGD and CD, which typically cause mortality when fish are >20 mm. During the 10 years of using a RAS at EWFCS, only one outbreak of BGD occurred—in 2013, which was due to significantly increasing the feeding rate in fish at 250–360-μm size range. The smaller dry diet had consistently created tank hygiene issues, which prompted reduced use, thereafter followed by complete elimination in 2018.

In the past, Walleye larviculture turbidity was maintained for the entire production interval, but now both EWFCS and RFCRF start with turbidity regimes then clear the system water of turbidity at 18–24 dph, respectively with no negative effects on production.

4.2 Contrasts

It is apparent that although both EWFCS and RFCRF use RASs, turbid water, and surface sprays, the details of their respective systems were developed independently, resulting in differences in the aspects of system design and in how culture techniques are carried out. The most substantial difference between facilities is how turbid water is achieved, with RFCRF using clay and EWFCS using commercial algae. Turbidity at RFCRF is developed by the addition of a slurry of inert clay, which has been practiced at the facility since its development in small-scale laboratory tanks (Bristow and Summerfelt 1994, 1996). The Blue Jay Creek Fish Hatchery (Paul

Methner, personal communications, Ontario Ministry of Natural Resources) was an early adopter of clay for use in a RAS for Walleye larviculture system, which was prior to the development of the larviculture RAS at RFCRF. Thus far, clay turbidity has proven to work with all RAS component processes at RFCRF. At EWFCS, turbidity is developed by the addition of a commercial microalgae product, which is used in the larviculture of marine species.

The biofilter at RFCRF has a horizontal tank style with an open top for accessibility to allow the clay to be siphoned from the bottom, but thus far, this has not been necessary until the system is annually disinfected. RFCRF uses K3 media, which has a similar diameter but shorter length compared to the MB3 media used at EWFCS. It is not known if using a longer cut of media with clay turbidity would clog the biofilter with biofilm and clay and prevent aeration from mixing. The system at EWFCS utilizes a vertical biofilter configuration with microalgae turbidity being directly batch dosed into the filter three times per day. Microalgae do not accumulate to create fouling in the biofilter as the cellular structure of the algae breaks down beyond 24 h.

Tank sizes are larger at RFCRF, 2.4 m^3, compared to 1.94 m^3 at EWFCS. Decisions for tank size evolved separately; the 1.83-m diameter tanks were chosen at RFCRF as the largest tank that could allow the center drain screen to be accessed by workers and because larger tanks could not fit within the space available. Center standpipe screens were desirable at RFCRF because this was similar to previous experiences on a smaller tank in single-pass culture. Additional rationale for the use of the center screen was the need for overflow points to allow surface oils to be moved out of the tank with the assistance of the surface spray bar. One side box was added to ensure that effluent capacity is adequate to prevent overflows when screen flow was reduced. The 1.52-m diameter tanks at EWFCS were chosen to match the tank diameter already in use on the system. Also, in contrast, the center screen was undesirable for EWFCS staff as cleaning had the potential to result in fish loss. Self-cleaning tanks at EWFCS were engineered to have two boxes for each tank, but only one has proven to be necessary. Tank hygiene practices vary between systems. Daily tank care at RFCRF is done with one cleaning per day. At EWFCS, two cleanings are conducted as the facility is staffed 16 h/day.

4.3 Collaboration

All three authors began communicating on Walleye larviculture techniques at the Midwest Fish and Wildlife Conference in Des Moines, Iowa, in 2011. EWFCS first reported the use of RAS for Walleye larviculture at this conference, which encouraged the collaboration between EWFCS and RFCRF. Since then, continual communication has continued before, during, and after culture runs to understand the focus and approach that each facility will conduct. The sharing of findings and results has led to advances such as the elimination of the use of B1 feed size, the application of optimum temperature regime, and determining the elimination of turbidity to coincide with the transition of larvae from photopositive to photonegative. Communicating with each other during the culture run to discuss challenges while they are unfolding, by going back to the basics of what we are seeing and how it could be addressed with an understanding of existing biology knowledge, has been invaluable. Our partnership has resulted in advancing techniques that produce Walleye fingerlings of consistent quantity and size. This alliance has extended throughout both North America and Europe through the sharing of the knowledge and success gained with these techniques.

It is noteworthy that success has been achieved with the use of science-based protocols and modern RAS engineering for the intensive larviculture of Walleye. At the EWFCS, larval survival in the RAS was 52–61% from hatch to 33–35 dph compared with 32% average survival in cooperative extensive ponds over the last 30 years. At RFCRF, larval survival in the RAS over 2 years was 77.2% to a size of 1.0 g with gas bladder inflation of 96.5% to 100.0%.

Acknowledgments Kevin Kelsey would like to thank the entire technical staff at EWFCS, particularly the current staff on station: Paige Blaker, Jim Bellinghiri, Mark Roche, Michelle Ayer, and John Talbot from BHFCS. Their hard work was immeasurably important to achieving the results being presented in this chapter. An extended thank you is given to Ben Rooks, facility engineer at EWFCS, for leading the staff in the construction, design modifications, and overall operation and maintenance of the RASs on-site. I would like to thank Diana Arteaga Alvarez for her multifaceted support throughout every culture run. Thanks are also due to the Lake Champlain Walleye Association for all of their support and the Fisheries Division and Fish Culture Section of the Vermont Fish and Wildlife Department for efforts related to the Walleye program, particularly Shawn Good for his assistance in preparing some of this chapter content and Tom Jones for his backing related to all things fish health. Glenn Snapp and the team at INNOVASEA provided RAS design expertise and support. I am grateful to Alan Johnson for the many years of collaboration that has elevated the techniques that are now being applied. Finally, I would like to thank Dr. Summerfelt for sharing his extensive knowledge of Walleye culture, for his guidance in the preparation of this manuscript, and for supporting my enthusiasm for the application of RAS for the production-scale culture of Walleye.

Alan Johnson wishes to thank the RFCRF culture teams in 2019–2020 for their work on this project: Coy Blair, Randall Esser, Conner Johnson, Rachel McDonnell, Jacob Miller, Steve Pecinovsky, and Adam Todd. I would like to thank Kevin Kelsey for a continued discussion of the challenges and successes of larviculture. I recognize Robert Summerfelt for his mentorship and shared enthusiasm for the science of Walleye culture. I would like to express my gratitude to the agencies that provided eggs for 2020: Kansas Wildlife, Parks, and Tourism for contributing Kerwin eggs and the USFWS and North Dakota Game & Fish for providing North Dakota eggs. Funding for this research was provided by the Federal Aid in Sport Fish Restoration.

Robert Summerfelt thanks Kevin Kelsey and Alan Johnson for their invitation to collaborate in the preparation of this report. Personally, this has been an integrative experience of the laboratory research I have had on this topic. They (Kelsey, Sect. 2, and Johnson, Sect. 3) describe successful achievements that substantially advance prospects for hatchery-scale production methods for the intensive larviculture of Walleye, with an obvious application for on-growing Walleye and other species. Their accomplishments are a product of exceptional dedication over many years of research and development activities that have shown extraordinary resilience to accept, learn, and respond from setbacks that I know to be common with any effort to control living systems in the manner that we would like. Of course, their starting point and progress benefited from the experience of others that are cited in the text and also from unpublished studies by many hatchery persons who presented talks at the annual meetings of the Coolwater Workshop, Midcontinent Workshop, and World

Aquaculture Society. It is as in the metaphor by Issac Newton, "If I have seen further it is by standing on the shoulders of Giants" (Chen 2013). Their success, however, has come not only from what they have learned from others but also from their own innovation in combining what was known of critical early life stages of this species and cultural systems using the latest RAS technology. Yet, all of that said, they recognize that more needs to be done.

The reference of trade names stated is to illustrate the application of these products to the culture process and does not suggest endorsement by the authors or their agency.

References

Aneshansley ED, Timmons MB, Colesante RT (2001) Survival and growth of walleye *Stizostedion vitreum* fry as affected by water recirculation, feed transition age, and stocking density. J World Aquac Soc 32:89–95

Barrows FT, Zitzow RE, Kindschi GA (1993) Effects of surface water spray, diet, and phase feeding on swim bladder inflation, survival, and cost of production of intensively reared larval walleyes. Prog Fish Cult 55:224–228

Boggs CT, Summerfelt RC (1996) Origin of oil film on the water surface of tanks used for the mass culture of larval walleye. In: Libey GS, Timmons MB (eds) Success and failures in commercial recirculating aquaculture. Northeast Regional Agricultural Engineering Service (NREAS), NREAS-98, vol 2. Cornell University, Ithaca, NY, pp 631–636

Boggs CT, Summerfelt RC (2003) Enhancing gas bladder inflation in larval walleye: comparison of two methods for removing an oily film from the water surface of culture tanks. In: Barry TP, Malison J (eds) Proceedings of Percis III, the third international percid symposium, University of Wisconsin, Madison, Wisconsin, US, July 20–24, 2003. University of Wisconsin Sea Grant Institute, Madison, WI, pp 19–20

Bristow BT, Summerfelt RC (1994) Performance of larval walleye cultured intensively in clear and turbid water. J World Aquac Soc 25:454–464

Bristow BT, Summerfelt RC (1996) Comparative performance of intensively cultured larval walleye in clear, turbid, and colored water. Prog Fish Cult 58:1–10

Brooks RC, Heidinger RC, Kohler CC (1994) Mass-marking otoliths of larval and juvenile walleye by immersion in oxytetracycline, calcein, or calcein blue. NA J Fish Manag 14:143–150

Bumguardner BW (1991) Marking subadult red drums with oxytetracycline. Trans Am Fish Soc 120:537–540

Chen C (2013) Mapping scientific frontiers: The quest for knowledge visualization. Springer-Verlag, London, 344 p

Clayton RD, Summerfelt RC (2010) Gas bladder inflation in walleye fry cultured in turbid water with and without a surface spray. N Am J Aquac 72:338–342

Colesante RT (1996) Intensive culture of walleye using brine shrimp and formulated diets. In: Summerfelt RC (ed) Walleye culture manual, NCRAC culture series, vol 101. North Central Regional Aquaculture Center Publications Office, Iowa State University, Ames, IA, pp 191–194

Colesante RT, Youmans NB, Ziolkoski B (1986) Intensive culture of walleye fry with live foods and formulated diets. Prog Fish Cult 48:33–37

Cutts CJ, Batty RS (2005) Progress report on Algal-mediated turbidity and larval performance in marine fish: primary milestone 1. SAMS*ardtoe*, Ardtoe Marine Laboratory, Ardtoe, Acharacle, Argyll, PH36 4LD. Report number: SR 502.1, 11 p

Davidson J, Summerfelt R, Barrows F, Gottsacker B, Good C, Fischer G, Summerfelt S (2016) Walleye *Sander vitreus* performance, water quality, and waste production in replicated recirculation aquaculture systems when feeding a low phosphorus diet without fish meal versus a traditional fishmeal-based diet. Aquac Eng 75:1–13

Edwards WJ, Babcock-Jackson L, Culver DA (2002) Field testing of protocols to prevent the spread of zebra mussels *Dreissena polymorpha* during fish hatchery and aquaculture activities. N Am J Aquac 64:220–223

El Gamal AR (2015) Oil trapping and swim bladder inflation of marine fish. Fish Consulting Group. Available at: http://fishconsult.org/?p=12433

Fielder DG (2002) Methodology for immersion marking walleye fry and fingerlings in oxytetracycline hydrochloride and its detection with fluorescence microscopy. Michigan Department of Natural Resources, Fisheries Technical Report 2002-1, Ann Arbor, MI, 11 p

Good SP (2020) Walleye oxytetracycline mark assessment. Federal Aid in Sport Fish Restoration, Project F-35-R-22, Study Number 1, Montpelier, Vermont, pp 1–18

Harder TM, Gotch GG, Summerfelt RC (2014) Growth response of juvenile Walleyes after cold banking. N Am J Aquac 76:255–260

Houde ED (1975) Effects of stocking density and food density on survival, growth and yield of laboratory-reared larvae of seabream *Archosargus rhomboidalis* (L.) (Sparidae). J Fish Biol 7: 115–127

Howey RG, Theis GL, Haines PB (1980) Intensive culture of walleye (*Stizostedion vitreum vitreum*). US Fish and Wildlife Service, Lamar Fish Cultural Development Center Information Leaflet 80-05, Lamar, Pennsylvania 11 p

Hussain M, Summerfelt RC (1991) The role of mechanical injury in an experimental transmission of *Flexibacter columnaris* to fingerling walleye. J Iowa Acad Sci 98:93–98

Johnson JA, Rudacille JB (2009) Production scale testing of formulated diet for the intensive culture of walleye and muskellunge. Iowa Department of Natural Resources. Federal Aid in Sport Fish Restoration, Project F-160-R, Completion Report, Des Moines, IA, 145 p

Johnson JA, Rudacille JB (2010) Intensive culture of walleye converted to a diet as small fingerlings. Iowa Department of Natural Resources. Federal Aid in Sport Fish Restoration, Project F-160-R, Completion Report, Des Moines, IA, 91 p

Johnson JA, Summerfelt RC (2015) Intensive culture methods of Walleye during ongrowing. In: Kestemont P, Dabrowski K, Summerfelt RC (eds) Biology and culture of percid fishes. Springer, New York, pp 437–468

Johnson JA, Summerfelt RC, Clayton RD (2008) Evaluation of replacement feeds for Fry Feed Kyowa for larval walleyes. N Am J Aquac 70:445–451

Larscheid JG (1995) Contribution of stocked walleye and population dynamics of adult walleye in Spirit, East and West Okoboji Lakes. Iowa Department of Natural Resources. Federal Aid in Sport Fish Restoration, Project F-135-R, Completion Report, Des Moines, IA, 69 p

Loadman NL, Moodie GE, Mathias JA (1986) Significance of cannibalism in larval walleye (*Stizostedion vitreum*). Can J Fish Aqua Sci 43:613–618

Loadman NL, Mathias JA, Moodie GE (1989) Method for the intensive culture of walleye. Prog Fish Cult 51:1–9

Logsdon DE (2006) Contribution of fry stocking to the recovery of the walleye population in the Red Lakes. Minnesota Department of Natural Resources Investigational Report 535, 41 p

Losordo TM, Conwell D (2014) Unit processes in RAS systems, ultraviolet radiation. Global Aquaculture Alliance technical article

Marsden JE, Langdon RW (2012) The history and future of Lake Champlain's fishes and fisheries. J Great Lakes Res 38:19–34

Marty GD, Hinton DE, Summerfelt RC (1995) Histopathology of swimbladder noninflation in walleye (*Stizostedion vitreum*) larvae: role of development and inflammation. Aquaculture 138: 35–48

McCauley RW (1970) Automatic food pellet dispenser for walleyes. Prog Fish Cult 32:42

Mitzner L (1995) Man-made lake investigations. Iowa Department of Natural Resources. Federal Aid in Sport Fish Restoration, Project F-134-R, Completion Report, Des Moines, IA 11 p

Moodie GE, Mathias JA (1996) Intensive culture of larval walleye on formulated food. In: Summerfelt RC (ed) Walleye culture manual. NCRAC Culture Series 101. North Central

Regional Aquaculture Center, Publications Office, Iowa State University, Ames, IA, pp 187–190
Moore AA (1996) Intensive culture of walleye fry on formulated feed. In: Summerfelt RC (ed) Walleye culture manual, NCRAC Culture Series, vol 101. Ames, IA, North Central Regional Aquaculture Center, Publications Office, Iowa State University, pp 195–197
Moore A, Prange MA, Summerfelt RC, Bushman RP (1994) Evaluation of tank shape and a surface spray for rearing larval walleye on formulated feed. Prog Fish Cult 56:100–110
O'Rourke PA (1996) Economic analysis for walleye aquaculture enterprises. In: Summerfelt RC (ed) Walleye culture manual, NCRAC Culture Series, vol 101. North Central Regional Aquaculture Center, Publications Office, Iowa State University, Ames, IA, pp 371–384
Priegel GR (1964) Early scale development in the walleye. Trans Am Fish Soc 93:199–200
Rieger PW, Summerfelt RC (1998) Microvideography of gas bladder inflation in larval walleye. J Fish Biol 53:93–99
Rønfeldt JL, Nielsen JW (2010) Effect of turbidity on rearing of pikeperch larvae (*Sander lucioperca* L.). In: Biologispeciale udarbejdet, Detnaturvidenskabelige Fakultet Københavns Universitet, pp 19–27
Rotman F, Stuart K, Drawbridge M, Paquette M (2017) Self-cleaning larval rearing tanks to improve larval production of California yellowtail *Seriola doralis* and white sea bass *Atractoscion nobilis*. Presented at Aquaculture America 2017, International Conference and Exposition, February 19–22, 2017, San Antonio, TX
Secor DH, White MG, Dean JM (1991) Immersion marking of larval and juvenile hatchery produced striped bass with oxytetracycline. Trans Am Fish Soc 120:261–266
Spickler AR (2007) Viral hemorrhagic septicemia. Retrieved from http://www.cfsph.iastate.edu/DiseaseInfo/factsheets.php
Summerfelt ST (1996a) Engineering design of a water reuse system. In: Summerfelt RC (ed) Walleye culture manual. NCRAC Culture Series 101. North Central Regional Aquaculture Center Publications Office, Iowa State University, Ames, IA, pp 277–309
Summerfelt ST (1996b) Aquaculture of walleye as a food fish. In: Summerfelt RC (ed) Walleye culture manual. North Central Regional Aquaculture Center Publications Office, Iowa State University, Ames, IA, pp 215–230
Summerfelt RC (1996c) Intensive culture of walleye fry. In: Summerfelt RC (ed) Walleye culture manual. NCRAC Culture Series 101. North Central Regional Aquaculture Center, Publications Office, Iowa State University, Ames, IA, pp 161–185
Summerfelt RC (2005) The culture of walleye *Sander vitreus* (Mitchill 1818). In: Kelly A, Silverstein J (eds) Aquaculture in the 21st century: manual of fish culture III. American Fisheries Society Symposium 46, Bethesda, MD, pp 373–412
Summerfelt RC (2013) Gas bladder inflation in larval fish aquaculture. In: Qin JG (ed) Larval fish aquaculture. Nova Science, Hauppauge, NY, pp 125–142
Summerfelt RC, Johnson JA (2015) Intensive culture of Walleye from egg incubation to juvenile. In: Kestemont P, Dabrowski K, Summerfelt RC (eds) Biology and culture of percid fishes. Springer, New York, pp 313–335
Summerfelt RC, Penne CR (2007) Nutrient retention by fish in a multispecies recirculation aquaculture facility. Int J Recirculating Aquac 8:43–64
Summerfelt ST, Bebak-Williams J, Tsukuda S (2001) Controlled systems: water reuse and recirculation. In: Wedermeyer G (ed) Fish hatchery management, 2nd edn. Bethesda, MD, American Fisheries Society, pp 285–395
Summerfelt RC, Johnson JA, Clouse CP (2011) Culture of walleye, sauger, and hybrid walleye. In: Barton BA (ed) Biology, management, and culture of walleye and sauger. Bethesda, MD, American Fisheries Society, pp 451–570
Timmons MB, Ebeling JM, Wheaton FW, Summerfelt ST, Vinci BJ (2001) Recirculating aquaculture systems. Northeastern Regional Aquaculture Center Publication No 01–002, 650 p
Timmons MB, Guerdat T, Vinci BJ (2018) Recirculating aquaculture, 4th edn. Ithaca Publishing Co LLC, 775 p
Zarnock CB, Schreibman CRT, Timmons MB (2010) Growth performance of walleye, *Sander vitreus*, in recirculating aquaculture systems. J Appl Aquac 22:285–296

Effects of Parasiticidal Hydrogen Peroxide Treatments on Walleye Hatching Success in a Recirculating System

Guy D. Eroh, Robert B. Bringolf, Alvin C. Camus, Jean L. Williams-Woodward, and Cecil A. Jennings

Abstract Infections of fish eggs by organisms of the family Saprolegniaceae are a common problem in hatcheries. Recently, these infections have been implicated in instances of low Walleye hatching success experienced by the Georgia Department of Natural Resources (GADNR). Unlike most Walleye hatcheries, which use flow-through systems, GADNR uses a recirculating system to incubate its Walleye eggs. In 2018 and 2019, the effectiveness of various hydrogen peroxide concentrations and dosing frequencies on the hatching success of Walleye eggs infected with *Saprolegnia* spp. and incubated in recirculating systems were evaluated experimentally. Further, attempts were made to identify the pathogen causing mortalities in the experimental systems. Each combination of three hydrogen peroxide concentrations (100, 250, or 500 mg/L) and two exposure frequencies (once or twice daily) was tested in triplicate. A two-way analysis of variance revealed that treatment concentration significantly affected the hatching success in 2018 but not in 2019. Specifically in 2018, eggs treated with 100-mg/L hydrogen peroxide (at both frequencies) had greater mean percent viability (39%) than every other treatment concentration (0–9%). Results suggest that lower treatment concentrations may improve the hatching success in recirculating systems. Treatment frequency and the interaction

G. D. Eroh (✉) · R. B. Bringolf
Warnell School of Forestry and Natural Resources, University of Georgia, Athens, GA, USA
e-mail: gderoh@gmail.com

A. C. Camus
Department of Pathology, University of Georgia, College of Veterinary Medicine, Athens, GA, USA

J. L. Williams-Woodward
Department of Plant Pathology, University of Georgia, College of Agricultural and Environmental Sciences, Athens, GA, USA

C. A. Jennings
Warnell School of Forestry and Natural Resources, University of Georgia, Athens, GA, USA

U.S. Geological Survey, Georgia Cooperative Fish and Wildlife Research Unit, Athens, GA, USA

J. C. Bruner, R. L. DeBruyne (eds.), *Yellow Perch, Walleye, and Sauger: Aspects of Ecology, Management, and Culture*, Fish & Fisheries Series 41,
https://doi.org/10.1007/978-3-030-80678-1_7

between concentration and frequency did not affect the hatching success in either year. The use of quantitative polymerase chain reaction (qPCR) methodologies to quantify the pathogen was unsuccessful during both experiments. Estimated zoospore concentrations did not align with observed hyphal growth and were unaffected by hydrogen peroxide treatments. Further, the quantification process revealed that published *Saprolegnia* spp.-specific PCR primers also amplified deoxyribonucleic acid (DNA) from another genus. Specifically, the DNA sequencing of hyphae revealed that *Aphanomyces laevis*, though not previously associated with Walleye, may cause the mortality of fertilized eggs in hatchery conditions and may represent a naturally occurring pathogen of Walleye eggs. This pathogen may respond differently to parasiticidal treatments than *Saprolegnia* spp. and may help explain the overall lower hatching success in all treatment combinations during 2019.

Keywords *Aphanomyces* · Parasiticide · qPCR · *Saprolegnia* · Saprolegniasis

1 Introduction

Walleye (*Stizostedion vitreum*) are large, cool-water sport fish popular among anglers throughout much of their range (Mitchill 1818; Quinn 1992; Nelson et al. 2016). These fish are common in the northern United States and Canada and inhabit a native range that extends southward, including the US states of Alabama, Mississippi, and Georgia (VanderKooy and Peterson 1998; Bednarski et al. 2010; Page and Burr 2011). Many state fisheries organizations, both in the core and on the fringe of the Walleye's range, operate Walleye propagation and stocking programs to support recreational fisheries. Most of these large Walleye hatching operations are faced with fungal-like oomycete infections of eggs in their hatching systems and must treat the eggs with parasiticides to prevent mortality (Colesante 1996).

Many fish species and their eggs are easily infected by organisms of the family Saprolegniaceae (Alderman and Polglase 1984; Gaikowski et al. 2003; Van Den Berg et al. 2013; Duan et al. 2018). Walleye eggs have only been reported to be infected by oomycetes of the genus *Saprolegnia*. These organisms, once considered to be fungi because of superficial similarities between their hyphal structures and those of true fungi, are ubiquitous in freshwaters throughout the world (Johnson Jr. et al. 2002; Magray et al. 2019). In natural environments, Saprolegniaceae species are an important aid to nutrient cycling in the ecosystem, where these organisms break down unfertilized eggs and other dead tissues (Van Den Berg et al. 2013; Magray et al. 2019). Under these circumstances, live eggs are not typically infected. However, when eggs are kept in high densities in hatcheries, the hyphal growth on dead embryos and unfertilized eggs can spread to live embryos and kill them (Smith et al. 1985).

There are existing methods for the treatment of saprolegniasis in hatcheries. The most straightforward of these is the removal of dead and infected eggs, though this is both labor intensive and not as effective as chemical treatments (Piper et al. 1982;

Barnes et al. 2003). Malachite green, an industrial dye, was once considered the standard for the chemical control of saprolegniasis (Van West 2006). However, malachite green was banned by the U.S. Food and Drug Administration (FDA) because the chemical has carcinogenic and teratogenic properties (Fitzpatrick et al. 1995). As of 2020, formaldehyde and hydrogen peroxide (hereinafter peroxide) are the only two chemicals approved by the FDA as parasiticides for Walleye eggs (United States Food and Drug Administration 2020).

The Georgia Department of Natural Resources (GADNR) treats their Walleye eggs with peroxide. Unlike most Walleye hatching operations, which incubate their eggs in flow-through systems, GADNR incubates eggs in a single, large recirculating system that supplies water to multiple hatching jars. Because peroxide quickly breaks down into nontoxic materials (water and oxygen), GADNR prefers it to formalin, which would persist in the recirculating system longer, possibly resulting in toxicity to the eggs. During incubation, Walleye eggs are exposed to peroxide on a regular basis as needed until the eggs begin to hatch (Clint Peacock, GADNR, personal communication). Jars cannot be treated individually because the system is composed of a single water source.

Some studies have evaluated the effects of various concentrations of peroxide on the hatching success of Walleye eggs (Rach et al. 1998; Gaikowski et al. 2003; Soupir and Barnes 2006). However, all of the studies were performed in flow-through systems. To our knowledge, the effects of parasiticidal peroxide treatments on the hatching success of Walleye eggs in recirculating systems have not been evaluated. Additionally, despite Walleye occupying a large range in North America, all relevant studies have been performed in the range core with none conducted anywhere on the fringe of the Walleye's range, such as Georgia (Rach et al. 1998; Gaikowski et al. 2003; Soupir and Barnes 2006). Such studies at the periphery of the Walleye's range are important because the suboptimal habitat or environmental conditions there may limit the production of high-quality gametes and cause intermittent recruitment failures. Hatcheries operating near the outer extent of the Walleye's range may face greater challenges in hatching Walleye eggs because of variable gamete quality and potentially increased susceptibility to pathogens. As a result, hatcheries on the fringe of the Walleye's range are very important for producing hatchery-reared individuals to supplement the production of wild-spawned Walleye and thereby maintain Walleye fisheries at range boundaries.

Further, whereas the direct purpose of peroxide treatment is to reduce the growth of Saprolegniaceae, few studies (of any species) have attempted to evaluate the growth of the pathogens (Barnes et al. 1998; Thoen et al. 2016; Straus et al. 2019). In the few studies that have tried to evaluate pathogen growth, outputs were qualitative or only semi-quantitative. Therefore, this work was undertaken to evaluate the effects of peroxide treatment concentration and frequency on both Walleye hatching success and pathogen growth in recirculating systems.

2 Methods

2.1 Facilities and System Design

All of our experiments were performed at the University of Georgia's Aquatic Biology and Ecotoxicology Lab. In 2018, 21 individual hatching systems (Fig. 1) were constructed based on a design that simulated the facilities of GADNR's Walleye hatching operation. Water in each system was cooled to approximately 12 °C by immersing the sumps in a chilled water bath. A Hach multiprobe® water quality meter was used to measure daily temperature, dissolved oxygen (DO), and pH in each system before the first treatment to ensure that these water quality variables were within an optimal range (mean $\text{temperature}_{2018} = 12.1$ °C ± 0.4 °C SD, mean $\text{temperature}_{2019} = 12.3$ °C ± 0.2 °C SD, mean $DO_{2018} = 10.2$ mg/

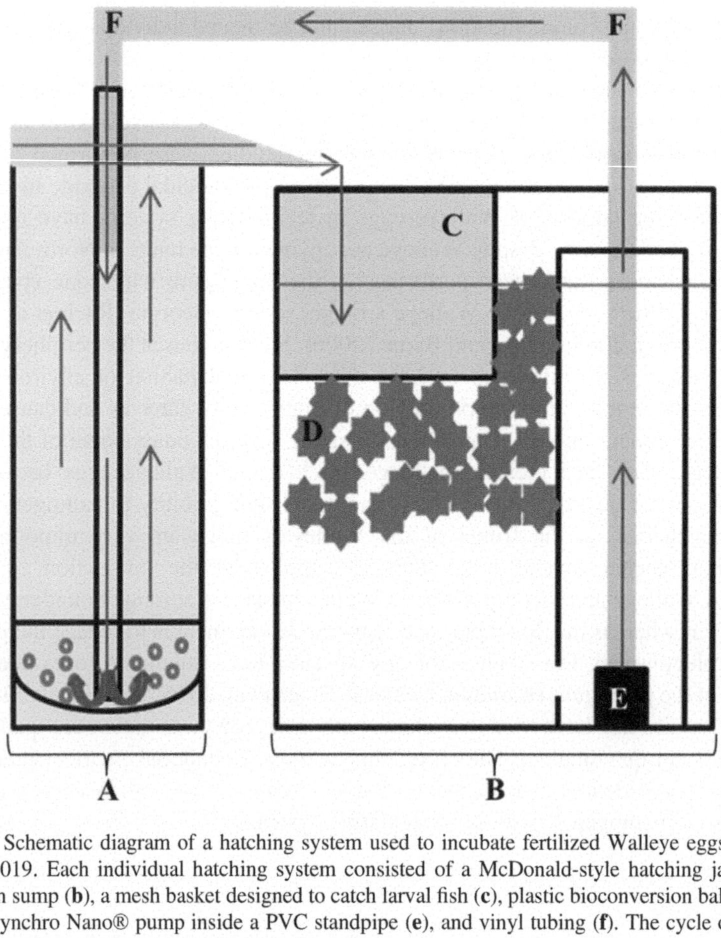

Fig. 1 Schematic diagram of a hatching system used to incubate fertilized Walleye eggs during 2018–2019. Each individual hatching system consisted of a McDonald-style hatching jar (**a**), a 5-gallon sump (**b**), a mesh basket designed to catch larval fish (**c**), plastic bioconversion balls (**d**), a Sicce Synchro Nano® pump inside a PVC standpipe (**e**), and vinyl tubing (**f**). The cycle of water circulation through the system is indicated by arrows

L ± 0.3 mg/L SD, mean DO_{2019} = 10.1 mg/L ± 0.2 mg/L SD, mean pH_{2018} = 7.87 ± 0.11 SD, mean pH_{2019} = 6.96 ± 0.38 SD). The systems were aerated by means of the spillway between the hatching jar and the sump. In 2019, hatching units were constructed in the same way, though there were 24 assembled to allow for an extra set of controls.

The sources of the pathogens for the experiments were different for both years of the study. In 2018, following the introduction of eggs to the hatching systems, each system was inoculated with five plugs (approximately 3 mm in diameter) of an oomycete culture grown on Sabouraud's dextrose agar. The isolate was cultured from a diseased Channel Catfish (*Ictalurus punctatus*) and possessed morphologic features consistent with *Saprolegnia* spp. These introductions were made to ensure that the pathogen was present in the system. However, in 2019, the systems were not inoculated with *Saprolegnia* at the onset of the experiment to better simulate conditions in a hatchery setting where the pathogen is not introduced intentionally. All systems were treated with peroxide from the beginning of the experiment, and none showed evidence of *Saprolegnia* growth at the experiment's onset.

2.2 Experimental Design

A 2 × 3 full-factorial experiment with a randomized complete block design was used to examine the effects of two treatment frequencies (once daily and twice daily) and three peroxide concentrations (100 mg/L, 250 mg/L, and 500 mg/L) on the hatching success of fertilized Walleye eggs. A daily 5-mL sham treatment of water was used as a control in both years. Separate graduated cylinders were used for treating with peroxide treatments and water-only controls to ensure that peroxide was not added to the control systems. With the exception of peroxide being replaced with water, the control systems underwent the same treatment procedures as the other experimental systems described below in Sect. 2.3. In 2019, an additional no-treatment control without any water transfer was added to account for the possible effects of the manipulation or disruption of eggs on the hatching success. These systems did not undergo the treatment procedures described in Sect. 2.3. There were three replicates of each treatment combination, as well as three replicates of the sham water control and three replicates of the no-treatment control in 2019. Each year, eggs from three Walleye were provided by the GADNR for the experiment and added to jars in a density of approximately 10,000 eggs per jar (estimated volumetrically based on standard practice density calculations from linear counts). Each hatching system used eggs from only one of the Walleye, and no replicates within a treatment group used eggs from the same fish. Thus, the female that produced the eggs represented a blocking factor. Each treatment combination was issued a unique number and then randomly assigned to a system with a randomized list of numbers generated in R (R Core Team 2017).

2.3 Peroxide Treatment and Hatching Success of Walleye Eggs

Our procedure for dosing eggs with peroxide changed after the initial exposure. Originally, our dosing procedure was modeled after that used by GADNR. Peroxide was added to the sump while the system cycled with the expectation that the peroxide would decay quickly, without adverse effects to the eggs, as is the case in GADNR's system. However, the peroxide did not decay as rapidly as expected. This problem was identified on the first day when bubbles began forming around the eggs, causing them to float out of the jars, particularly in systems treated with 500-mg/L peroxide. Therefore, continued treatment with this method was not possible. Each system underwent a complete water change after being exposed to peroxide for 8–18 h (10 h were needed to change the water in all 21 systems), eggs caught in mesh baskets were returned to their jars, and a new treatment procedure was initiated on day 2. Beginning on the second day of treatment in 2018, each jar was removed from its recirculating system and treated with the assigned concentration of peroxide for 15 min. After 15 min of exposure, the water/peroxide mixture in the hatching jar was flushed with fresh, dechlorinated water for 5 min. Finally, the hatching jar was reconnected to its sump, and flow was restored. In 2019, this new method was used for all dosing exposures during the experiment; the no-treatment control jars were never removed from their systems.

Treatments continued every 24 h for once-daily treatments and every 12 h for twice-daily treatments, until eggs hatched, after which the percent of viable eggs was evaluated. Following observations of hatching in a jar (usually from 9 to 11 days post fertilization), that system continued to cycle for 72 h without any peroxide treatment to allow for the hatching of eggs, after which the hatching system was taken down for evaluation. Though not all eggs had hatched after 72 h, this procedure was undertaken to prevent cannibalism among larval Walleye, which would negatively affect the evaluation of the hatching success (Chevalier 1973). If hatching was not observed, eggs were treated until day 21 because the 3-week mark was believed to be well past the hatching window based on personal communications with GADNR hatchery managers. The number of eyed eggs and hatched larvae for each experimental unit was systematically counted by hand following the conclusion of the experiment. Counting was considered complete if eyed eggs or larvae were not found for three consecutive passes. The hatching success was determined by calculating the percent viability for each system by dividing the sum of larvae and eyed eggs by the initial number of eggs placed in the system.

2.4 Evaluation of Pathogen Density

To quantify pathogen growth, we collected zoospores on filter membranes and extracted deoxyribonucleic acid (DNA) from those filters to measure the density of zoospores in the systems' water. During the hatching experiments, water samples were taken from each system on the morning of treatment days 3, 6, 9, and 12. Whatman® Nuclepore™ Track-Etched Membranes with 1-μm pores were used with a vacuum filtration system to filter three 100-mL subsamples from each initial sample for every system. Following filtration, each filter membrane was folded, sealed in a 2-mL polypropylene microvial, and frozen at −80 °C for storage. DNA was extracted from the filters by heating the filters in a Chelex® 100 solution based on a protocol described by Brewer and Milgroom (2010). Approximately 500 μL of supernatant, containing DNA, were then pipetted from the tube and deposited in a new microcentrifuge tube. Care was taken to ensure that no Chelex® 100 beads were transferred along with the supernatant. Samples were then stored at −20 °C.

Quantitative PCR polymerase chain reaction (qPCR) was used to evaluate the concentration of *Saprolegnia* DNA in samples generated from each filter. The primers used for the reactions were designed to be genus specific for *Saprolegnia* spp. by Rocchi et al. (2017). A hydrolysis probe based on that used in the aforementioned study, but with an additional quencher, was used to quantify amplification. Both the primers and the probe were manufactured by Integrated DNA Technologies, and the sequences for each are presented in Table 1. A Bio Rad CFX96 Touch Real-Time PCR Detection System was used to run reactions in 20-μL final volumes. Primer concentrations were 250 nM, and the concentration of the probe in each reaction was 200 nM. Five μL of DNA solution extracted from the filters were used for quantification. The remaining reaction volume consisted of 10-μL Bio Rad SsoAdvanced Universal Probes Supermix and DNA-free water. The amplification protocol used was also taken from Rocchi et al. (2017). Each run of samples included a standard dilution series repeated in triplicate, decreasing tenfold from 10^{-2}-ng DNA to 10^{-8}-ng DNA, as well as three no-template controls to determine the starting quantities of DNA in the unknown samples from their quantification cycles (Cq values). Premade filters with known concentrations of zoospores were used to relate the amount of DNA in solutions made from experimental filters to the numbers of zoospores that were on those filters. These zoospores were cultured in vitro and were added to the filters in known densities based on the numbers of zoospores counted using a hemocytometer and appropriate dilutions.

Table 1 Primers and probe used for the qPCR quantification of pathogen zoospore abundance in hatching systems being used to evaluate the effects of hydrogen peroxide concentrations and treatment frequencies on the hatching success of fertilized Walleye eggs in March 2018 and March 2019. The primers are based on those developed by Rocchi et al. (2017)

Primer/probe	Sequence
F primer	5'-GCATTCAAGTTTGTGGGAAC-3′
R primer	5'-CGGAAACCTTGTTACGACTTC-3′
Probe	5′/56-FAM/TCCTTAACC/ZEN/TCGCCATTTAGAGGAAGG/31ABkFQ/−3'

2.5 *Pathogen Identification*

A sample of hyphal tissue cultured on Sabouraud's dextrose agar from the 2019 experiment's naturally occurring infection was sequenced to identify the isolated organism. DNA was extracted from hyphae as described previously. Universal primers ITS1 and ITS4 were used along with an illustra™ PuReTaq™ Ready-To-Go™ PCR Bead tube to amplify an approximately 710 base-pair segment of the internal transcribed spacer region of the ribosomal ribonucleic acid (rRNA) gene through a standard PCR reaction. Sequencing was performed by Eurofins Genomics in Louisville, Kentucky. Sequence results were compared to known sequences published in GenBank® with a basic local alignment search tool (BLAST). While the *Saprolegnia* isolate used in 2018 was unavailable for sequencing, microscopic examination at the time of the experiment revealed morphologic features, terminal zoosporangia, typical of *Saprolegnia* spp. (Magray et al. 2019).

2.6 *Statistical Analyses*

A two-way analysis of variance (ANOVA) was used to evaluate differences in treatment means for percent viability among treatment combinations. A Tukey's Honestly Significant Differences posthoc test was used to identify which means differed from one another in cases where ANOVA indicated statistical significance. Two-way analyses of variance were also used to determine if the concentration of zoospores differed among treatments at each sampling day for both the 2018 and 2019 experiments. All statistical tests were conducted using R software, and an alpha value of 0.05 was used to evaluate significance in all cases (R Core Team 2017).

3 Results

3.1 *Hatching Success*

In 2018, the percent viability of Walleye eggs varied significantly among treatment concentrations but not between treatment frequencies. Hatching percentages for each concentration-frequency combination ranged from 0% in the 500-mg/L treatments to 45% in the 100-mg/L once-daily treatment (Table 2). Treatment concentration significantly affected the hatching success of the eggs ($P = 2.05 \times 10^{-6}$), but treatment frequency ($P = 0.240$) and the interaction between the two factors ($P = 0.286$) did not. Viability at treatments of 100 mg/L was significantly higher than other treatments at other concentrations, while none of the other treatment concentrations differed significantly from each other.

Table 2 Mean percent hatching success ± SE of Walleye eggs treated for 15 min with hydrogen peroxide at various concentrations and frequencies. In 2018, eggs treated with 100-mg/L peroxide, both once and twice daily, hatched at significantly higher percentages than all of the other treatments. None of the other treatments in 2018 differed from one another. In 2019, there was no effect of either treatment concentration or frequency

Year	Daily treatments	H_2O_2 conc. (mg/L)	Hatching success (%)
2018	1	000	9 ± 5
		100	45 ± 8
		250	7 ± 4
		500	0 ± 0
	2	000	¥
		100	33 ± 9
		250	7 ± 2
		500	0 ± 0
2019	0	000	4 ± 1
		Δ	Δ
		Δ	Δ
		Δ	Δ
	1	000	5 ± 5
		100	4 ± 4
		250	6 ± 3
		500	7 ± 4
	2	000	¥
		100	3 ± 3
		250	4 ± 2
		500	5 ± 3

¥ – Control systems that received sham water treatments were only treated once daily. Accordingly, no hatching success estimates exist for the twice-daily 0-mg/L treatment combination
Δ – Systems left untreated in 2019 necessarily did not receive hydrogen peroxide treatments at any concentration

The hatching success of Walleye eggs in 2019 was low across all treatment combinations, none of which were significantly different from each other. Hatching percentages for each combination ranged from 3% in 100-mg/L twice-daily treatments to 7% in 500-mg/L once-daily treatments (Table 2). Neither peroxide concentration ($P = 0.573$), dosing frequency ($P = 0.570$), nor the interaction ($P = 0.978$) between those two factors affected the percent viability of Walleye eggs.

3.2 *Pathogen Density and Growth*

There was great variability in estimated zoospore densities across treatments and days in both experiments. In 2018, the mean estimated zoospore density per 100 mL ranged from 138 in systems treated with 500-mg/L peroxide twice daily on day 6 to 25,800 zoospores in systems treated with 100-mg/L peroxide once daily on day

Table 3 The estimated number of zoospores ± (SE) per 100 mL of water sampled at four time points from hatching systems treated with hydrogen peroxide at varying concentrations and frequencies during a 2018 experiment. There was no significant effect of concentration or frequency on any day

Evaluation period	Treatments	Water control	100 mg/L	250 mg/L	500 mg/L
Day 3	Once daily	1.07×10^3 (755)	2.58×10^4 (2.41×10^4)	1.16×10^3 (701)	205 (89)
Day 3	Twice daily	–	2.82×10^3 (2.48×10^3)	2.02×10^3 (396)	4.47×10^3 (2.66×10^3)
Day 6	Once daily	982 (616)	1.04×10^3 (290)	2.08×10^3 (974)	1.84×10^3 (798)
Day 6	Twice daily	–	1.86×10^3 (1.00×10^3)	2.19×10^3 (2.18×10^3)	138 (125)
Day 9	Once daily	3.01×10^4 (2.86×10^4)	1.01×10^4 (4.68×10^3)	4.68×10^3 (2.90×10^3)	1.50×10^3 (810)
Day 9	Twice daily	–	5.95×10^3 (2.75×10^3)	7.21×10^3 (407)	1.05×10^4 (5.70×10^3)
Day 12	Once daily	1.43×10^3 (1.41×10^3)	6.27×10^3 (4.82×10^3)	1.31×10^4 (1.20×10^4)	1.99×10^3 (1.58×10^3)
Day 12	Twice daily	–	2.71×10^3 (2.46×10^3)	1.66×10^3 (806)	1.47×10^3 (1.15×10^3)

3 (Table 3). Results for 2019 were even more variable than in 2018. Estimated zoospore densities per 100 mL ranged from <1 in no treatment controls on sampling day 3 to 4.16×10^9 zoospores in systems treated with 100-mg/L peroxide twice daily on day 12 (Table 4). In most cases, the percent standard error for these estimates was greater than 50% of the estimate. In neither year did the peroxide concentration, treatment frequency, or interaction between the two factors affect the estimated density of zoospores in systems on any of the sampling days (all $P > 0.05$).

3.3 Pathogen Identification

The PCR used with the primers ITS1 and ITS4 produced a 700 base-pair product of predicted size from the oomycete isolated from diseased eggs in the 2019 experiment. A BLAST search of the sequenced code was most similar to species of the genus *Aphanomyces*, which, like *Saprolegnia* spp., belong to the family Saprolegniaceae and have similar life cycles. Specifically, the sequence aligned with over 99% similarity to sequences of multiple isolates of *A. laevis*, a pathogen known to infect fishes and result in a condition that resembles infections by *Saprolegnia* spp. (Table 5).

Table 4 The estimated number of zoospores (SE) per 100 mL of water sampled at four time points from hatching systems treated with hydrogen peroxide at varying concentrations and frequencies during a 2019 experiment. There was no significant effect of concentration or frequency on any day

Evaluation period	Treatments	No treatment	Water control	100 mg/L	250 mg/L	500 mg/L
Day 3	Once daily	–	12 (12)	77 (74)	113 (83)	43 (42)
Day 3	Twice daily	–	–	0.57 (0.49)	308 (302)	623 (617)
Day 3	No treatment	0.13 (0.08)	–	–	–	–
Day 6	Once daily	–	20 (14)	131 (73)	4.32×10^3 (3.22×10^3)	882 (437)
Day 6	Twice daily	–	–	6.07×10^5 (6.07×10^5)	1.4 (0.8)	288 (284)
Day 6	No treatment	1.23×10^3 (703)	–	–	–	–
Day 9	Once daily	–	6.12×10^3 (5.27×10^3)	6.79×10^3 (2.03×10^3)	86 (27)	146 (108)
Day 9	Twice daily	–	–	7.06×10^8 (7.06×10^8)	1.03×10^6 (8.86×10^5)	82 (80)
Day 9	No treatment	166 (96)	–	–	–	–
Day 12	Once daily	–	1.73×10^3 (1.19×10^3)	911 (859)	78 (37)	4.16×10^3 (4.10×10^3)
Day 12	Twice daily	–	–	4.16×10^9 (4.16×10^9)	2.84×10^5 (2.50×10^5)	664 (375)
Day 12	No treatment	283 (216)	–	–	–	–

Table 5 The top 10 GenBank® sequence comparisons based on total BLAST score for a DNA sequence amplified using universal ITS primers 1 and 4 and DNA extracted from a pathogen cultured from Walleye eggs that were used in an experiment evaluating the effectiveness of hydrogen peroxide concentration and dosing frequency on the hatching success of those fertilized eggs

Species	Percent similarity	Query cover	GenBank® accession ID
Aphanomyces laevis	99.86%	99%	AY310497.1
Aphanomyces laevis	99.70%	97%	AY683885.1
Aphanomyces laevis	99.85%	96%	AM947028.1
Aphanomyces sp.	99.85%	95%	HQ643123.1
Aphanomyces laevis	100.00%	92%	HQ111469.1
Aphanomyces laevis	98.01%	93%	FM999236.1
Aphanomyces cochlioides	95.84%	99%	AY647191.1
Aphanomyces sp.	96.65%	94%	AB533289.1
Aphanomyces laevis	96.76%	93%	KP006463.1
Aphanomyces sp.	100.00%	77%	GU014281.1

4 Discussion

4.1 Hatching Success and Egg Viability

At the onset of the experimentation, we hypothesized that an intermediate concentration of peroxide would result in the optimum hatching success of Walleye eggs because of the potential for toxicity of high peroxide concentrations to Walleye eggs. We also hypothesized that there would be a difference in hatching success between treating the eggs with peroxide once or twice daily. Peroxide concentration affected hatching success under some conditions, but exposure frequency did not. Specifically, in 2018, systems treated with 100-mg/L peroxide reached the eyed stage in percentages approximately four times greater than other concentrations. However, such differences were not seen in 2019. Although many of the experimental conditions were similar between experiments in 2018 and 2019, there were differences between the two isolates that might explain the observed results. Specifically, in 2018, a cultured *Saprolegnia* sp. was introduced to the hatching systems at the onset of experimentation. In 2019, a naturally occurring organism, later identified to be *Aphanomyces laevis*, infected the eggs in the systems.

There was a significant procedural change in our experimental methods in 2018 and 2019 that may have contributed to the different results between the 2 years. In 2018, our experimental systems experienced an unintended extended exposure to peroxide, which lasted 8–18 h on the first day of experimentation; this extended exposure did not occur in 2019. While the extended exposure appeared to have severe toxic effects on the Walleye eggs when treated at 500 mg/L, eggs treated at 100 mg/L had significantly higher hatching success than the other treatments and the control. These eggs also had better hatching success than the same concentration treated without the extended exposure in 2019. Perhaps exposure durations longer than 15 continuous minutes are needed to improve the hatching success of eggs when used at concentrations around 100 mg/L. The knowledge that peroxide in recirculating systems may persist longer than expected is valuable for agencies that use such systems to incubate their eggs but are unaware of the risks and/or existence of potentially longer exposures of eggs to parasiticides in recirculating systems.

Differences in baseline egg quality between experiments also could have contributed to the disparity in results observed between 2018 and 2019. Broodfish came from Lake Hartwell in 2018, whereas broodfish came from other reservoirs in 2019. Interannual variability in egg quality used between 2018 and 2019 may have been related to variable weather or other environmental conditions at the source reservoirs. The Walleye spawning migration in northern Georgia in 2018 occurred earlier and was much stronger than in 2019, possibly because of warmer water temperatures or greater river flows in 2018. The differences in water temperature between the 2 years also may have affected gamete development and therefore egg quality because gamete development in Walleye requires sufficiently cool (generally $<10\ ^\circ C$) water temperatures (Koenst and Smith Jr. 1976; Ney 1978). This explanation for potential differences in egg viability between years may be less likely, given that GADNR experienced similar hatching success in 2018 and 2019. Poor baseline egg quality in 2019 may have doomed those eggs to poor hatching success independent of any disinfectant treatments, which would explain the overall low hatching success observed during that year. The length and age of spawning females as well as mean egg diameter and mean egg mass can also affect baseline egg quality and viability (Johnston and Leggett 2002; Feiner et al. 2016). Evaluating these factors could be beneficial in providing guidance for procuring the best broodstock and gametes to increase the hatching success of hatchery-fertilized eggs.

We observed the maximum hatching success at a peroxide concentration of 100 mg/L in 2018 but not in 2019. This concentration was the lowest nonzero concentration tested in our study and is lower than any previously reported for the treatment of Walleye eggs. Lesser or equal concentrations have been effective in treating the eggs of Largemouth Bass (*Micropterus salmoides*) and Channel Catfish (Small and Wolters 2003; Matthews et al. 2012). The lowest previously reported peroxide treatment concentration used on Walleye eggs was a daily 15-min treatment of 200 mg/L, which was the only concentration used in that study and increased hatching success by 27–40% (Soupir and Barnes 2006). Although peroxide treatment concentrations of 500 mg/L (manufacturer recommendation) and greater have been successful in other species, these concentrations may be higher than optimal for Walleye. Although not optimal, treatments with the higher peroxide doses typically produce better hatching success than no treatment at all (Rach et al. 1998; Gaikowski et al. 2003). Results suggest that treating with lower peroxide concentrations for longer than 15 min could improve the hatching success of Walleye and reduce hatchery costs associated with parasiticides. Finally, although the percent viability of Walleye eggs at the end of our experiments was lower than those reported from these other northern studies, they fall within the range of hatching success typically found by GADNR (Clint Peacock, GADNR, personal communication). This lends credence to the idea that hatcheries on the fringe of the Walleye's range might experience greater challenges with hatching eggs than hatcheries in the core of the Walleye's range.

4.2 Pathogen Density and Growth

We hypothesized that pathogen density would be negatively affected by increasing peroxide concentrations and increasing exposure frequencies. Our experimental results did not support these hypotheses, possibly because our experimental design had limitations that we did not anticipate. Neither peroxide concentration nor exposure frequency seemed to affect the density of the pathogen zoospores. However, the expected increase in zoospores over time across all systems was not observed. In several instances, zoospore density decreased or fluctuated over the course of the experiments. This observation is inconsistent with the continuous hyphal growth observed while treating the systems throughout both experiments.

The possibility remains that the treatments may have affected pathogen growth or density, but our sampling schedule was insufficient to detect the effects.

There are several possibilities for the discordance between the observed hyphal growth and the lack of change in the number of zoospores. For example, the assumption that zoospore density is directly proportional to the amount of hyphal growth may be invalid. Even if the total number of zoospores produced was proportional to the amount of hyphal growth, the production may have come in bursts rather than continuously. If that were the case, detecting differences in zoospore density would be impossible unless each jar happened to be sampled following a mass production event. Additionally, methods used for collecting and evaluating zoospore density may not have worked properly. Water samples were collected at the spillway of the hatching jar where zoospores were assumed to have been washed out of the jar with the flow of the water. The presence of zoospores in the water was never confirmed with microscopy.

Even if the zoospores were sampled properly in a way that reflected the true densities in the systems, there are some possible reasons for the observed results. Though the hyphae in the hatching jars were always treated with peroxide, zoospores in the sumps were not. Consequently, an insufficient percentage of the zoospores may have been treated to see an effect of peroxide treatment. Unaffected zoospores flushed from the jars may also have served as a source of reinfection to the eggs. Further, PCR only detects copies of DNA and cannot discern between viable and nonviable zoospores. Additional variability may have been introduced by relying on natural infection to occur, even though each hatching jar theoretically received the same infective dose of spores. While the peroxide concentration and/or treatment frequency may have affected the number of viable zoospores, the quantification of DNA from live and dead zoospores may have obscured more promising treatment results.

The primers and probe used for this quantitative polymerase chain reaction (qPCR) analysis were designed to be genus specific for *Saprolegnia* spp. (Rocchi et al. 2017). However, our results demonstrate that the DNA of *A. laevis* can be amplified using the *Saprolegnia* primers. In designing the primers and probe, Rocchi et al. (2017) analyzed the sequences of six genera, including *Saprolegnia* and *Aphanomyces*, though they only considered *A. astaci* and not *A. laevis* from the latter. An experimental evaluation of the primers and probe only showed specificity for *Saprolegnia* when compared to *A. astaci*. In defense of the authors, a review of GenBank® reveals only nine sequences of the *A. laevis* 18S small subunit rRNA, all of which are partial sequences and do not include the corresponding sequence where the primers would anneal. Regardless, these primers were considered to be genus specific, but they evidently are not.

A method for properly measuring the growth of oomycete pathogens on fish eggs is needed to evaluate the effects of parasiticidal treatments on their density and growth. Although the amount of hyphae is typically the measurement of interest in the few studies that have attempted quantification, zoospore concentration may be a more appropriate measurement. Zoospores are the colonizing bodies of the pathogenic organisms and are the primary target of peroxide treatments because they are more sensitive to peroxide than hyphae. Additionally, although measuring hyphal growth in experimental systems where egg picking is not performed may be valuable, quantifying hyphal growth in real-world systems where infected eggs are commonly removed would be appropriate, though a reliable technique has not been developed to quantify hyphal growth. Thoen et al. (2010) developed a promising method for zoospore enumeration that employs the culture of viable zoospores from samples in microwell plates, though we are not aware of this technique being applied to a parasiticide treatment study. Although the qPCR methods described herein may not have been successful for evaluating pathogen growth, they might be refined to better assess growth in the future.

4.3 Pathogen Identity

The unexpected identification of *A. laevis* in association with Walleye eggs in 2019 suggests the possibility that the organisms causing the infections were different in the first and second experiments. If this were the case, the differences in hatching results between the 2018 and 2019 experiments could be related to the two pathogens' differing responses to peroxide. We would expect that if a difference in pathogens was responsible for the observed difference in results, a similar difference in results would be seen with respect to pathogen growth. Unfortunately, the methodologies used to quantify this growth provided inconclusive results. Published accounts of the effects of peroxide on eggs infected with *A. laevis* are lacking. Perhaps, this species has a higher resistance to peroxide than *Saprolegnia* spp., and resistance may be responsible for the observed differences between our experiments and the reported inconsistent results in hatching success on eggs incubated at hatcheries.

Studies of *A. laevis* are limited, and until now it has not been associated with disease in Walleye or their eggs. Although controlled experiments are needed to confirm the pathogenicity of *A. laevis* to incubating Walleye eggs, the organism has infected other species of fish, including Nile Tilapia (*Oreochromis niloticus*) and Blue Panchax (*Aplocheilus panchax*) (Mondal and De 2001; Ali et al. 2011). *A. laevis* belongs to the same genus as the highly parasitic *A. invadans* and *A. astaci*, which are responsible for epizootic ulcerative syndrome in fishes and the crayfish plague in crayfishes, respectively (Alderman et al. 1990; Iberahim et al. 2018). Although this finding was unexpected, it may prove to be pivotal for improving the efficacy of the treatment of Saprolegniaceae infections in Walleye hatching operations.

Most publications that have reported the effects of peroxide on the hatching success of fish eggs have assumed that the responsible pathogen was *S. parasitica*. Yet most studies fail to properly identify the pathogen. The results of our study indicate that assuming the pathogen responsible for saprolegniasis is *S. parasitica*, or other *Saprolegnia* spp., is inappropriate. Our results also suggest that different oomycetes may respond differently to peroxide treatments. If there are large differences in the treatment regimens required to effectively treat different Saprolegniaceae, hatcheries may want to tailor their treatment procedures to the organism or organisms infecting their eggs in a particular year. An experiment to determine if *A. laevis* reacts differently to chemical treatments than other commonly studied Saprolegniaceae could address this hypothesis and perhaps support the notion of targeting treatment by a pathogen.

4.4 Summary and Conclusions

The results of this study suggest that increasing the duration of peroxide treatment at lower treatment concentrations may improve hatching success. Correspondingly, recirculating systems that experience longer exposures to parasiticides should treat at lower doses. Prior to this study, the pathogen *A. laevis* had never been previously associated with any Walleye life stage. However, we identified this oomycete naturally colonizing Walleye eggs. This finding suggests that various Saprolegniaceae pathogens may be responsible for the losses of Walleye eggs at hatching facilities in different years. If that is the case, and if different pathogens respond differently to peroxide treatment, identifying the responsible organism and adjusting treatment based on this information may improve treatment efficiency.

Finally, our results support the Rocchi et al. (2017) findings that qPCR technology and the primers they designed can be used successfully to identify members of the Saprolegniaceae following the amplification and sequencing of a segment of the rDNA ITS region. Contrary to Rocchi et al. (2017), we found that primers were not genus specific for *Saprolegnia* spp. and identification of the oomycete is therefore not possible using these primers alone. Sequencing of the amplified ITS segment from the oomycete isolated in 2019 identified another Saprolegniaceae, *A. laevis*. Although a proper method of quantifying pathogen growth on fish eggs is still needed, refinement of qPCR techniques for this purpose may provide a solution to this problem in the future.

Acknowledgments We would like to thank GADNR for their technical consultation and for providing fertilized Walleye eggs and hatching jars for this project. We also thank John Leary for assisting with qPCR analyses and Gulcin Ekemen Canak for aiding in the identification of the pathogen. The Georgia Cooperative Fish and Wildlife Research Unit is sponsored jointly by the Georgia Department of Natural Resources, the University of Georgia, the US Fish and Wildlife Service, the US Geological Survey, and the Wildlife Management Institute. Any use of trade, products, or firm names is for descriptive purposes only and does not imply endorsement by the US government. This work was done under the auspices of the University of Georgia AUP No. A2018 01-022-Y2-A. Data associated with this study are available from the corresponding author upon reasonable request.

References

Alderman DJ, Holdich D, Reeve I (1990) Signal crayfish as vectors in crayfish plague in Britain. Aquaculture 86:3–6

Alderman DJ, Polglase JL (1984) A comparative investigation of the effects of fungicides on *Saprolegnia parasitica* and *Aphanomyces astaci*. Trans Br Mycol Soc 83(2):313–318

Ali EH, Hashem M, Al-Salahy MB (2011) Pathogenicity and oxidative stress in Nile tilapia caused by *Aphanomyces laevis* and *Phoma herbarum* isolated from farmed fish. Dis Aquat Org 94 (1):17–28

Barnes ME, Ewing DE, Cordes RJ, Young GL (1998) Observations on hydrogen peroxide control of *Saprolegnia* spp. during rainbow trout egg incubation. The Progressive Fish-Culturist 60 (1):67–70

Barnes ME, Stephenson H, Gabel M (2003) Use of hydrogen peroxide and formalin treatments during incubation of landlocked fall Chinook salmon eyed eggs. N Am J Aquac 65(2):151–154

Bednarski MS, Hendricks J, Higginbotham DL, Peterson DL (2010) Growth rates of stocked walleye in several Georgia reservoirs. Proceedings of the Annual Conference Southeastern Association of Fish and Wildlife Agencies 64:136–140

Brewer MT, Milgroom MG (2010) Phylogeography and population structure of the grape powdery mildew fungus, *Erysiphe necator*, from diverse *Vitis* species. BMC Evol Biol 10(268)

Chevalier JR (1973) Cannibalism as a factor in first year survival of walleye in Oneida Lake. Trans Am Fish Soc 102:739–744

Colesante RT (1996) Transportation and handling of walleye eggs, fry, fingerlings, and broodstock. In: Summerfelt RC (ed) Walleye culture manual, volume NCRAC culture series 101. North Central Regional Aquaculture Center Publications Office, Iowa State University, Ames, IA, pp 79–83

Duan J, Feng YH, Wang K, Geng Y, Deng Y, Ou Y, Chen D, Yang S (2018) A review: factors affecting outbreaks of Saprolegniosis on aquatic animals. Israeli JAquac:1–11

Feiner ZS, Wang H-Y, Einhouse DW, Jackson JR, Rutherford ES, Schelb C, Vandergoot CS, Zorn TG, Hook TO (2016) Thermal environment and maternal effects shape egg size in a freshwater fish. Ecosphere 7(5):1–16

Fitzpatrick MS, Schrek CB, Chitwood RL, Marking LL (1995) Evaluation of three candidate fungicides for treatment of adult spring Chinook salmon. The Progressive Fish-Culturist 57: 153–155

Gaikowski M, Rach J, Drobish M, Hamilton J, Harder T, Lee L, Moen C, Moore A (2003) Efficacy of hydrogen peroxide in controlling mortality associated with Saprolegniasis on walleye, white sucker, and paddlefish eggs. N Am J Aquac 65(4):349–355

Iberahim NA, Trusch F, Van West P (2018) *Aphanomyces invadans*, the causal agent of epizootic ulcerative syndrome, is a global threat to wild and farmed fish. Fungal Biol Rev 32(3):118–130

Johnson TW Jr, Seymore RL, Padgett DE (2002) Biology and systematics of the Saprolegniaceae. University of North Carolina at Wilmington. Department of Biological Sciences, Wilmington, NC

Johnston TA, Leggett WC (2002) Maternal and environmental gradients in the egg size of an Iteroparous fish. Ecology 83(7):1777–1791

Koenst WM, Smith LL Jr (1976) Thermal requirements of the early life history stages of walleye, *Stizostedion vitreum vitreum*, and sauger, *Stizostedion canadense*. J Fish Board Canada 33 (5):1130–1138

Magray AR, Lone SA, Ganai BA, Ahmad F, Dar GJ, Dar JS, Rehman S (2019) Comprehensive, classical and molecular characterization methods of *Saprolegnia* (Oomycota; Stramnipila), an important fungal pathogen of fish. Fungal Biol Rev 33(3–4):166–179

Matthews MD, Sakmar JC, Trippel N (2012) Evaluation of hydrogen peroxide and temperature to control mortality caused by saprolegniasis and to increase hatching success of largemouth bass eggs. N Am J Aquac 74:463–467

Mitchill SL (1818) Memoir on ichthyology. The fishes of New York, described and arranged. Pages 241-248 *in*, American monthly magazine and critical review
Mondal SK, De AB (2001) A fatal oomycotic disease of the freshwater fish *Aplocheilus panchax* in India caused by *Aphanomyces laevis*. Mycopathologia 154:21–24
Nelson JS, Grande TC, Wilson MVH (2016) Fishes of the world, 5th edn. Wiley, Hoboken, New Jersey
Ney JJ (1978) A synoptic review of yellow perch and walleye biology. Am Fish Soc Spec Pub 11: 1–12
Page LM, Burr BM (2011) Peterson field guide to freshwater fishes of North America north of Mexico, 2nd edn. Houghton Mifflin Harcourt, Boston
Piper RG, McElwain IB, Orme LE, McCraren JP, Fowler LG, Leonard JR (1982) Fish hatchery management, volume 3. United States Department of the Interior. Fish and Wildlife Service, Washington, D.C.
Quinn SP (1992) Angler perspectives on walleye management. N Am J Fish Manag 12:367–378
Rach J, Gaikowski M, Howe G, Schreier T (1998) Evaluation of the toxicity of hydrogen peroxide treatments on eggs of warm- and coolwater fishes. Aquaculture 165:11–25
Rocchi S, Tisserant M, Valot B, Laboissiere A, Frossard V, Reboux G (2017) Quantification of *Saprolegnia parasitica* in river water using real-time quantitative PCR: from massive fish mortality to tap drinking water. Int J Environ Res Public Health 27(1):1–10
Small B, Wolters W (2003) Hydrogen peroxide treatment during egg incubation improves channel catfish hatching success. N Am J Aquac 65(4):314–317
Smith SN, Armstrong RA, Springate J, Barker G (1985) Infection and colonization of trout eggs by Saprolegniaceae. Trans Br Mycol Soc 85(4):719–723
Soupir CA, Barnes ME (2006) Reduced formalin and hydrogen peroxide treatments during walleye egg incubation. N Am J Aquac 68(3):276–280
Straus DL, Ledbetter CK, Heikes DL (2019) Inhibiting fungus on largemouth bass eggs with copper sulfate and its toxicity to fry and juveniles. J World Aquacult Soc 51:214–223
Thoen E, Evensen O, Skaar I (2010) Microwell enumeration of viable Saprolegniaceae in water samples. Mycologia 102(2):478–485
Thoen E, Evensen O, Skaar I (2016) Factors influencing *Saprolegnia* spp. spore numbers in Norwegian salmon hatcheries. J Fish Dis 39:657–665
United States Food and Drug Administration (2020) Approved aquaculture drugs. U.S. Food and Drug Administration, Silver Spring, MD
Van Den Berg AH, McLaggan D, Dieguez-Uribeondo J, Van West P (2013) The impact of water moulds *Saprolegnia diclina* and *Saprolegnia parasitica* on natural ecosystems and the aquaculture industry. Fungal Biol Rev 27(2):33–42
Van West P (2006) *Saprolegnia parasitica*, an oomycete pathogen with a fishy appetite: new challenges for an old problem. Mycologist 20(3):99–104
Vander Kooy SJ, Peterson MS (1998) Critical current speeds for young gulf coast walleyes. Trans Am Fish Soc 127:137–140

Seasonal Movement Patterns and Distribution of Walleye in a Central Appalachian Hydropower Reservoir

Dustin M. Smith, Stuart A. Welsh, and Corbin D. Hilling

Abstract Understanding the ecology and spatial distribution of sport fishes is critical for fishery management. Recently, a Walleye (*Stizostedion vitreum*) population was reestablished in Cheat Lake, a 700-ha hydropower reservoir in northern West Virginia, USA; however, the movement patterns and seasonal distribution of this population were unknown. From 2012 to 2015, seasonal movements and distribution of telemetered Walleye in Cheat Lake were monitored using a stationary acoustic receiver array. Walleye locations were analyzed for seasonal changes, as assessed via seasonal distribution, home range, core range, and lake residency. Walleye movements and distributions varied seasonally and by sex. Overall, Cheat Lake Walleye used main lake habitats more heavily than riverine habitats. Seasonally, riverine habitats were most heavily used from March to August, with the greatest use occurring in March. In contrast, main lake habitats were most used from September to February, peaking in January. Additionally, male Walleye were more likely than females to occupy riverine habitats. Most Walleye demonstrated seasonal shifts in core range and linear home range. Additionally, male Walleye were more likely than females to have multiple core ranges. The number of monthly range shifts was higher than average from March to May and October to November. Also, male Walleye were more likely than females to emigrate from Cheat Lake into

D. M. Smith (✉)
West Virginia Division of Natural Resources, Wildlife Resources Section, Farmington, WV, USA
e-mail: Dustin.M.Smith@wv.gov

S. A. Welsh
U.S. Geological Survey, West Virginia Cooperative Fish and Wildlife Research Unit, Morgantown, WV, USA
e-mail: swelsh@wvu.edu

C. D. Hilling
U. S. Fish and Wildlife Service, Maine Field Office, East Orland, ME, USA

Present Address: U. S. Geological Survey, Great Lakes Science Center, Lake Erie Biological Station, Sandusky, OH, USA
e-mail: chilling@usgs.gov

J. C. Bruner, R. L. DeBruyne (eds.), *Yellow Perch, Walleye, and Sauger: Aspects of Ecology, Management, and Culture*, Fish & Fisheries Series 41,
https://doi.org/10.1007/978-3-030-80678-1_8

the Cheat River upstream. Overall, distribution and space use patterns indicated that Walleye distribution shifts were most likely in spring and fall months, associated with spawning activity and movement to overwintering habitats. Knowledge of these spatial patterns could help to inform management and monitoring efforts for this improving fishery.

Keywords Sex-specific habitat use · Home range · *Stizostedion vitreum* · Acoustic telemetry · Seasonal distribution · Spatial ecology

1 Introduction

Understanding the spatial ecology of fishes, including top predators, in aquatic ecosystems is a critical component to the conservation and management of fisheries (Landsman et al. 2011; Buckmeier et al. 2013; Cooke et al. 2016). Top predators such as the Walleye (*Stizostedion vitreum*) are important in structuring fish communities and are often popular sport fish (Craig 2000; Pothoven et al. 2016). Researchers are often interested in understanding the spatial ecology of such species, including home range and core range (Walsh et al. 2012). Home range is often defined as the area traveled by an animal for typical activities such as foraging, resting, and reproduction, whereas core range is often defined as an area of intensive use within an animal's home range (Crook 2004). Spatial distribution, home range, and core range often vary seasonally depending on habitat needs associated with spawning, foraging, and overwintering (DePhilip et al. 2005; Palmer et al. 2005; Bozek et al. 2011). Home and core ranges can also vary individually within a population or have sex-based differences (Palmer et al. 2005; Bozek et al. 2011; Hayden et al. 2014). Knowledge of the seasonal distribution and spatial ecology of top predators could benefit fishery management through the understanding of the spatial trophic structure and spatial vulnerability of populations to fishing pressure (Craig 2000; Pothoven et al. 2016; Bade et al. 2019; Matley et al. 2020).

Given the economic importance of recreational and commercial Walleye fisheries in North America (Schmalz et al. 2011), managers need information on the extensive movements and seasonal shifts in the distribution of Walleye in conjunction with spawning, foraging, and overwintering (Paragamian 1989; DePhillip et al. 2005; Bozek et al. 2011). Even though many tagging studies have examined the spatial ecology of Walleye (Paramagian 1989; DePhillip et al. 2005; Palmer et al. 2005; Hayden et al. 2014; Brooks et al. 2019; Matley et al. 2020), past studies in North America have focused mostly on Walleye movements and distribution in northern or midwestern states (Paramagian 1989; DePhillip et al. 2005; Brooks et al. 2019; Matley et al. 2020). Although much is known about Walleye life history, including spatial ecology in northern and midwestern states, as with many species, movement patterns and spatial ecology can have substantial variation between water bodies and regions (Bozek et al. 2011). Research on Walleye suggests that their movement can vary seasonally and with changing environmental conditions (Paragamian 1989;

Williams 2001; DePhillip et al. 2005; Palmer et al. 2005; Bozek et al. 2011). Most studies on Walleye distribution and habitat use have focused on activity during spawning with less focus on nonspawning periods (Paragamian 1989; Williams 2001; DePhillip et al. 2005; Palmer et al. 2005; Bozek et al. 2011). Although Walleye movement has been extensively studied in several regions of North America, such as the Great Lakes and the Midwest, few studies have been conducted in Appalachian reservoirs (Williams 2001; Palmer et al. 2005).

Cheat Lake is a central Appalachian hydropower reservoir located in northern West Virginia that historically supported Walleye but has been impacted by acidification. For over a century, Cheat Lake and the Cheat River watershed were severely impacted by acid mine drainage from abandoned mine lands and acid precipitation (Core 1959; Welsh and Perry 1997; Williams et al. 1999; Thorne and Pitzer 2004; Freund and Petty 2007). As a result, Walleye were reportedly extirpated from the reservoir in the late 1940s (Core 1959). The abatement of acid mine drainage pollution and acid precipitation, beginning in the 1980s, has led to improved water quality in the reservoir and throughout the watershed (Steelman and Carmin 2002; Thorne and Pitzer 2004; McClurg et al. 2007). In response to improved water quality, the West Virginia Division of Natural Resources (WVDNR) began stocking Walleye in 1999 and has continued on a biennial basis. Given the success of stockings and recent evidence of natural reproduction, more information on the life history of Walleye in Cheat Lake, including movements and distribution, could be beneficial to the future management of the population. The information gained on both spawning and nonspawning seasonal locations and movements would further enhance the management opportunities of the fishery. Therefore, research investigating seasonal distribution and space use patterns of Walleye within Cheat Lake, increasing knowledge of Walleye distribution, home range, and residency in Cheat Lake, could provide beneficial information to both managers and anglers.

The goal of this study was to determine the seasonal home and core ranges, lake residency, and seasonal distribution of Walleye in Cheat Lake, a West Virginia hydropower reservoir. Specifically, our objectives were to determine which reservoir areas were utilized, determine if Walleye distributions differed seasonally or between sexes, determine Walleye residency, assess emigration from the Cheat Lake system, and determine if there were temporal or sex-based differences in residency.

2 Methods

2.1 Study Area

Cheat Lake, formed in 1926, is a hydropower reservoir on the lower Cheat River, northern West Virginia, USA (Fig. 1). The reservoir has a surface area of 700 ha, is approximately 21 km in length, and has a maximum depth of 25 m near its dam (Hilling et al. 2018). The reservoir serves the needs of a hydroelectric generating

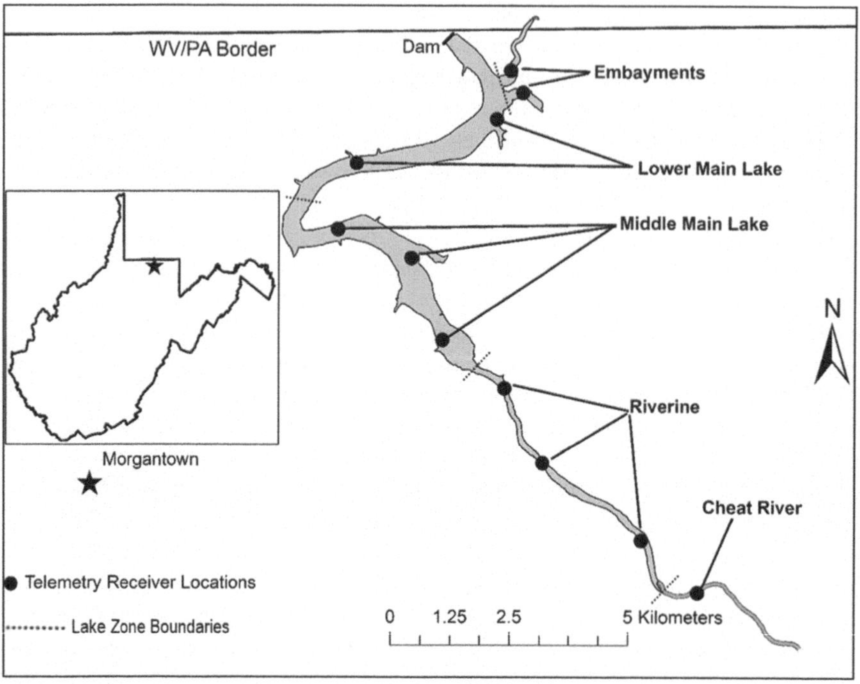

Fig. 1 Acoustic telemetry receiver locations and associated lake zones in Cheat Lake, WV. The city of Morgantown, WV, is indicated by the star symbol

facility, the Lake Lynn Hydropower Station, and experiences daily and seasonal water-level fluctuations due to hydropower operations. The minimum lake level is restricted to 0.6 m less than full pool (265.2 m above sea level) from May to October (Hilling et al. 2018). The largest permitted drawdowns occur during November–March when lake level can decrease 4.0 m less than full pool (Hilling et al. 2018). The minimum lake level is restricted to 2.1 m less than full pool during April in an attempt to protect the spawning habitat and activities of Walleye and Yellow Perch (*Perca flavescens*) (Hilling et al. 2018).

For this study, three spatial zones of Cheat Lake were designated for comparisons of Walleye movements and distribution: the riverine zone, middle main lake zone, and lower main lake zone (including embayments) (Fig. 1). Additionally, the Cheat River upstream of the reservoir was recognized as a separate zone (Fig. 1). The Cheat River was separated from Cheat Lake by the occurrence of the first visible riffle at the head of the lake. The separation of the lake zones was based on various factors, including reservoir morphology, bathymetry, water chemistry differences (Smith and Welsh 2015), and other longitudinal patterns common to reservoir systems (Miranda and Bettoli 2010). There are distinct morphological differences between the riverine zone, middle main lake zone, and lower main lake zone. The riverine zone is relatively narrow in cross-section, whereas the middle and lower main lake zones are typically 2.5–3 times the width of the riverine zone (Fig. 1). There is also a distinct difference in hydrological characteristics between the three zones. The riverine zone is heavily influenced by the incoming Cheat River in terms of the river current. In contrast, the middle and lower main lake zones are much more lacustrine in character as river current is spread out. This is apparent by the typical pattern of overwinter ice formation in the middle and lower main lake zones but the absence of ice in the riverine zone. Additionally, throughout most of the middle and lower main lake zones, average depths are greater than in the riverine zone. The lower main lake zone and middle main lake zone also differ in characteristics. Specifically, the middle main lake zone is more of a transitional area between the riverine habitat of the riverine zone and the lacustrine habitat of the lower main lake zone. The middle main lake zone typically has bathymetric and morphological characteristics intermediate of the riverine and lower main lake zones. Water chemistry differences also exist between the different zones. Specifically, during summer months, the riverine zone is typically cooler, and dissolved oxygen levels do not stratify compared to the middle main lake and lower main lake zones. The middle main lake zone is usually somewhat cooler with greater dissolved oxygen available in deeper water compared to the lower main lake zone. Measurements of water temperature and dissolved oxygen taken prior to this study (July and August 2011 and 2012) illustrate these differences (Table 1).

2.2 *Fish Collection and Tagging*

Fifty-two Walleye (30 males, 20 females, two undetermined sex, 432–708 mm TL) were collected and implanted with acoustic transmitters in the months of October–

Table 1 Summary of dissolved oxygen and water temperature measurements taken at Cheat Lake during July and August 2011–2012

Lake zone	Depth (m)	Dissolved oxygen (mg/L)	Dissolved oxygen (mg/L)	Water temperature (°C)	Water temperature (°C)
		mean (SE)	minimum	mean (SE)	maximum
Lower main lake	1	6.6 (0.2)	6	27.1 (0.3)	28.3
	5	5.3 (0.3)	4.7	25.5 (0.6)	27.2
	10	3.4 (0.4)	2.1	23.4 (0.3)	24.3
	15	1.6 (0.3)	0.5	20.8 (0.8)	23.3
Middle main lake	1	6.7 (0.2)	6.2	26.4 (0.3)	27.3
	5	6.1 (0.2)	5.1	24.5 (0.7)	26.6
	10	4.3 (0.4)	2.8	22.7 (0.6)	24.3
	15	3.4 (0.8)	1.9	20.5 (1.0)	23.4
Riverine	1	7.2 (0.1)	7.1	23.4 (0.4)	24.8
	5	7 (0.1)	6.6	22.5 (0.2)	23.5

February in 2011, 2012, and 2013. Walleye were collected using boat electrofishing and gill nets from all three reservoir zones. Prior to transmitter implantation, each Walleye was measured for total length (mm) and weighed (g). After anesthetization (MS-222, tricaine methanesulfonate, 100 mg/L), acoustic transmitters (Sonotronics CTT-83-3-I) were surgically implanted into the abdominal cavity of each Walleye (Hart and Summerfelt 1975). Acoustic transmitters were 62 mm in length and 16 mm in diameter, weighed 10 g in water, and had an estimated battery life of 3 years. Each fish was tagged with a numerically coded external T-bar anchor tag. Each anchor tag displayed contact information in the event fish were caught by anglers. Additional information was included on each tag recommending the release of the fish due to the 21-day hold time of MS-222. Fish were placed in a V-shaped trough during surgery, ventral side up, and the gills were continuously irrigated with water. Surgical instruments were sterilized prior to surgery, and betadine was applied to the incision site as an antiseptic. To insert the transmitter, an incision of approximately 20–30 mm was made, and 3–4 sutures of nonabsorbable monofilament were used to close the incision. Surgical procedures always lasted less than 7 min. After surgery, fish were placed in a live well to recuperate and were monitored until they were swimming upright and behaving normally (usually a period of 5--10 minutes). To reduce tag-induced behavioral changes, transmitter weight was never more than 2% of the fish's weight (Winter 1996). A recovery period of 4 weeks was included prior to data collection to monitor abnormal behavior associated with gear-induced and post-surgery stress or injury (Gilroy et al. 2010). When possible, Walleye were sexed by the examination of the gonads through a surgical incision or by the expulsion of milt for males. Some Walleye that were initially difficult to sex were later recaptured via fish surveys or anglers, and then sex was verified.

2.3 Telemetry

The movements and locations of tagged Walleye were monitored from January 2012 to April 2015. The tagged Walleye were monitored year-round using an array of stationary receivers (Sonotronics omnidirectional submersible receivers) deployed throughout Cheat Lake (Fig. 1). Receiver position was relatively equidistant from each other to maximize effective coverage. Receivers were either attached to buoys or tethered to the shoreline via root systems using a 9.5-mm steel cable and anchored by using two 20- × 20-cm cinder blocks. Receivers attached to the shoreline were dropped approximately 20–30 m away from and perpendicular to the shoreline. At most, ten acoustic receivers were active within the reservoir, with an additional receiver placed approximately 1 km upstream of Cheat Lake (upstream of first riffle/run complex). The two receivers located within the large embayments near the dam were lost in December 2013. The mean distance between each receiver was approximately 2.4 river kilometers (rkm). The tag detection range of acoustic receivers can be influenced by thermal stratification, acoustic disturbances (bridges), and sinuosity

(Shroyer and Logsdon 2009). The tag detection range varied seasonally in Cheat Lake due to thermal stratification. Specifically, thermal stratification reduces the effective range of receivers (Shroyer and Logsdon 2009). Range detection tests determined that the average detection range of acoustic receivers during periods of thermal stratification was 200–500 m. During periods without stratification, the range of receivers was between 400 and 900 m.

2.4 Data Analysis

Telemetry data were retrieved from stationary receivers and processed using Sonotronics SURsoftDPC© software (Sonotronics, Inc.). Afterward, data were exported to Microsoft Excel for further data processing and analysis. Acoustic telemetry data can produce false detections due to background noise (sonar units, other disturbances) and multiple tagged fish close to a receiver at once (Pincock 2011). Possible false detections were eliminated from the dataset by omitting single detections from individual fish within a 24-hour period (Harasti et al. 2015). Additionally, records of individual fish occurring in multiple locations simultaneously (<0.01% of detections) were eliminated from the dataset. We used manual tracking to help determine if fish had died or dropped their transmitter. If a transmitter became stationary for long periods, we used manual tracking to check for movement. If manual tracking concluded no movement and the transmitter remained stationary thereafter, the fish was considered dead or the transmitter dropped from the initial point it became stationary. Due to the large number of detections per individual fish that often include hours or days of continuous detections at the same receiver, data were transformed into a manageable format for analysis. Specifically, data were summarized to reflect arrival and departure dates/times and the direction of travel for individual fish for each receiver (Rosenblatt and Heithaus 2011).

Overall and seasonal distribution and range of tagged Walleye was summarized from processed telemetry data. Due to the potential bias of using the number of detections at a receiver from unequal detectability, fish locations were instead summarized by the amount of time each fish spent at a receiver or in a lake zone (Walsh et al. 2012; Ramsden et al. 2017). Specifically, the percentage of at-large time each fish spent near each receiver or in a lake zone was determined. The calculation of percent time spent near a receiver allowed for the determination of the proportional use of lake zones by individual fish and also the proportional use of lake zones by month for all fish. When referencing distribution seasonally, seasons were defined as the following: winter (December–February), spring (March–May), summer (June–August), and fall (September–November). The proportional use of lake zones was examined for differences across months. Comparisons were also made to determine if there were differences in the proportional use of lake zones by fish sex.

For the analyses of core range, range expansion, and residency, considerations of normality and autocorrelation helped determine the appropriate statistical methods to use. Normality of data was assessed using residual and QQ probability plots, as well as tests such as the Shapiro-Wilks tests (Brown and Guy 2007; Rogers and White 2007). If there were significant concerns with normality, nonparametric tests (e.g., Kruskal-Wallis test, etc.) were used to analyze data (Brown and Guy 2007). Additionally, repeated measure data collected on individual fish may lead to issues with autocorrelation (Rogers and White 2007). To account for a potential autocorrelation, mixed effects models that incorporate individual fish as random effects were utilized where necessary (Rogers and White 2007).

Due to the linear setup of our array system and the coarse detail of locations due to relocation data from receivers, traditional home range calculation techniques were not utilized (Vokoun 2003; Walsh et al. 2012). Instead, the approach used by Walsh et al. (2012) was adopted by calculating probability intervals using Pareto cumulative frequency density plots. This method calculates a utilization distribution that is based on the probability of an individual fish using a particular area (Vokoun 2003; Walsh et al. 2012). As described by Walsh et al. (2012), receiver area boundaries were designated as mid-points between receivers. Sections that encompassed 50% of the receiver areas used by a fish were considered the core-use area (Walsh et al. 2012). Similarly, sections that encompassed 95% of the receiver areas used by a fish were considered the home range of the fish (Walsh et al. 2012). The home range length for individual fish was described as the distance between the furthest downstream and the furthest upstream areas encompassed in the home range of a fish (Walsh et al. 2012).

The spatial distribution of Walleye was examined in several ways utilizing core range calculations. The overall core range was calculated for each individual fish. The number of overall core use areas was calculated and analyzed using a Kruskal-Wallis test to determine if there were sex-based differences ($\alpha = 0.05$). A Kruskal-Wallis test was also used to determine if there were sex-based differences in the inclusion of riverine habitat in the overall core use areas ($\alpha = 0.05$). Specifically, the Walleye core range was labeled as either including the riverine zone and/or Cheat River or not including these zones. Monthly core range calculations for each Walleye were used to determine the frequency that receiver areas were included in core use areas across months. Using monthly core range calculations, the lake zone encompassed by the core use area was determined for each month. Additionally, changes in core range shifts were calculated. During months when tagged Walleye shifted lake zones in core range (e.g., core range shift from the middle main lake to riverine zone), a "1" was assigned for that month. If no shift occurred, then a "0" was assigned. This binary setup allowed us to determine what months had the highest frequency of core range shifts among tagged Walleye. Repeated-measures binomial logistic regression (package "lme4" in program R, Bates et al. 2015) was used to determine if there was a significant effect ($\alpha = 0.05$) of sex and/or month on core range shifts.

Changes in monthly Walleye movement were evaluated by comparing linear range expansion and contraction. To accomplish this, the number of receivers by which individual fish were detected each month was extracted and a yearly mean calculated for the number of receivers visited (Topping et al. 2006). To get an estimate of monthly range deviations, the yearly mean for each fish was subtracted from the number of receivers that fish visited each month. Positive deviations from the mean number of receivers visited indicated range expansion, while negative deviations indicated range contraction (Topping et al. 2006). A linear mixed effects model (package "lme4" in program R, Bates et al. 2015) was used to test for significant effects of month and/or sex ($\alpha = 0.05$) in monthly range deviations, where fixed effects were month and sex and random effects were the individual fish.

Lake residency of tagged fish in our study was affected by both emigration downstream of the lake (via dam spillway or turbine passage) or emigration upstream of the lake into the Cheat River. Downstream movement via the dam spillway or turbine passage resulted in permanent emigration from the lake (and possibly mortality in some instances), whereas upstream movement into the Cheat River allowed for later immigration back into the lake. Potential environmental factors (river discharge, lake elevation, water temperature) associated with permanent emigration over or through the dam were evaluated. Due to the relatively small number of fish that escaped via the dam, no statistical analyses were conducted on these movements. Dam escapements were simply described with associated environmental conditions (water temperature (°C), lake elevation, and river discharge) and temporal patterns through simple summary statistics (e.g., mean, standard errors, etc.) or with a graphical approach. A residency index was calculated as the number of days fish were present within the lake divided by the total number of monitoring days for a given fish. The calculation of this index provided an indication of what proportion of time fish remained in the lake boundaries versus the time spent upstream in the Cheat River. An overall residency index (including the entire tagged life of a fish) and a monthly residency index were calculated. Sex-based differences in the overall residency of tagged Walleye was assessed using a Kruskal-Wallis test ($\alpha = 0.05$). The effects of month and/or sex on differences in residency were assessed using a linear mixed effects model ($\alpha = 0.05$), where fixed effects were month and sex and random effects were individual fish.

Environmental data were acquired from the US Geological Survey (USGS) Water Watch website (http://water.usgs.gov/waterwatch). Water temperature and river discharge data were obtained from the USGS streamgage on the Cheat River at Albright, West Virginia (USGS streamgage 03070260, US Geological Survey 2019). Lake elevation data were obtained from a monitoring gage at the Cheat Lake hydrostation near Stewartstown, West Virginia (USGS streamgage 03071590, US Geological Survey 2019).

3 Results

From January 2012 to April 2015, a total of 40 Walleye (19 males, 19 females, two undetermined sex) provided data on seasonal distribution and range (Tables 2 and 3). The number of fish monitored per year included six individuals in 2012, 31 individuals in 2013, 20 individuals in 2014, and 15 individuals in 2015. Twelve of the fish

Table 2 Summary of individual Cheat Lake Walleye telemetry histories

ID	Sex	Total length (mm)	Tag detections	Days monitored	Days detected	RI	No. core areas	Core zone	Home Range Zone
33	F	466	126,026	876	765	1	1	M	M
41	F	516	181,523	888	792	0.99	1	M	M
52	F	708	113,147	880	679	0.98	1	M	M-R
53	F	600	109,467	372	333	1	1	M	L-M-R
55	F	485	91,428	827	649	1	1	M	L-M-R
57	F	459	52,193	496	401	1	1	M	M
59	F	559	114,804	876	692	1	1	M	M-R
60	F	476	79,394	524	484	1	1	M	L-M
79	F	542	59,752	606	439	1	1	M	L-M
83	F	499	16,500	494	131	0.57	2	M-C	M-R-C
84	F	480	123,248	867	585	1	1	L	L-M-R
85	F	570	40,697	334	275	1	1	M	M
87	F	449	65,013	473	370	1	1	M	L-M
88	F	465	156,416	871	688	0.93	1	M	L-M-R-C
96	F	518	38,921	863	411	0.81	2	L-M	L-M-R-C
157	F	617	4168	176	62	0.86	1	M	M-R-C
179	F	652	24,320	195	142	0.91	1	M	L-M-R-C
185	F	568	98,932	541	464	1	1	M	M
190	F	580	67,026	492	368	1	2	L-M	L-M-R
35	M	446	55,040	828	479	1	1	R	M-R
38	M	432	15,713	50	44	1	1	M	M-R
39	M	437	62,214	746	432	0.8	2	M-R	M-R-C
40	M	459	73,191	918	397	0.8	2	M-R	M-R-C
42	M	443	23,450	68	62	1	2	M-R	M-R
43	M	467	114,870	746	546	0.84	2	M-R	M-R-C
50	M	475	86,944	800	334	0.56	2	M-C	M-R-C
51	M	435	28,906	227	183	1	1	M	M-R
58	M	430	71,744	397	331	0.63	2	M-C	M-R-C
80	M	495	188,272	871	626	0.96	1	M	M-R-C
82	M	505	70,594	620	417	0.99	1	M	M-R
86	M	452	132,350	876	729	0.92	1	M	L-M-R-C
89	M	430	35,368	667	449	0.79	2	M-R	M-R-C
90	M	440	23,483	328	220	1	1	M	L-M-R
93	M	450	11,072	259	130	0.98	2	L-M	L-M-R
94	M	500	39,589	620	394	0.96	2	M-R	M-R-C
98	M	487	50,550	608	239	0.62	2	M-C	M-R-C
189	M	556	41,557	113	93	1	1	M	M-R
193	M	487	16,890	452	141	0.59	2	M-C	L-M-R-C
44	U	476	21,991	823	198	0.73	2	R-C	M-R-C
151	U	490	14,132	46	46	1	2	L-L	L-M

RI residency index; zone abbreviations: *L* lower main lake, *M* middle main lake, *R* riverine zone, *C* Cheat River

Table 3 Average overall proportional use based on daily values (with standard errors in parentheses) of lake zones by tagged Walleye from 2012 to 2015. U = unknown sex

ID	Sex	Monitoring Period	Lower Main	Middle Main	Riverine	Cheat River
33	F	2013–2015	0.0% (0)	99.0% (9.8)	1.0% (0.1)	0.0% (0)
41	F	2012–2014	0.1% (0.1)	95.4% (7.4)	3.6% (0.5)	0.9% (1.0)
52	F	2012–2014	0.6% (0.3)	69.3% (7.1)	28.2% (2.2)	1.9% (1.4)
53	F	2013	6.5% (1.2)	76.2% (9.0)	17.4% (2.8)	0.0% (0)
55	F	2013–2015	4.8% (0.9)	88.0% (7.1)	7.3% (1.4)	0.4% (0.2)
57	F	2013–2014	0.0% (0)	100.0% (8.9)	0.0% (0)	0.0% (0)
59	F	2013–2015	0.4% (0.2)	87.0% (9.4)	12.1% (1.7)	0.4% (0.3)
60	F	2013	20.4% (2.3)	79.6% (7.2)	0.0% (0)	0.0% (0)
79	F	2013–2014	16.1% (3.6)	81.4% (8.3)	2.6% (0.6)	0.0% (0)
83	F	2013–2014	0.0% (0)	40.7% (6.6)	16.3% (2.1)	43.0% (13.8)
84	F	2013–2015	90.4% (7.7)	4.0% (0.9)	5.6% (0.8)	0.0% (0)
85	F	2013	0.0% (0)	95.0% (9.3)	5.0% (0.7)	0.0% (0)
87	F	2013	49.6% (7.3)	50.4% (6.8)	0.0% (0)	0.0% (0)
88	F	2013–2015	9.4% (2.4)	65.6% (6.8)	18.5% (2.2)	6.5% (3.6)
96	F	2013–2015	28.5% (5.4)	36.1% (6.0)	16.8% (1.9)	18.7% (9.4)
157	F	2014–2015	2.3% (1.1)	65.7% (13.6)	18.3% (3.5)	13.7% (14.4)
179	F	2014–2015	11.2% (2.4)	65.8% (9.9)	14.3% (2.3)	8.7% (6.6)
185	F	2014–2015	0.0% (0)	98.5% (8.8)	1.5% (0.2)	0.0% (0)
190	F	2014–2015	31.2% (4.5)	62.7% (6.0)	6.1% (1.0)	0.0% (0)
35	M	2012–2014	0.0% (0)	41.7% (6.2)	58.3% (5.7)	0.0% (0)
38	M	2012	0.0% (0)	92.2% (20.6)	7.8% (2.3)	0.0% (0)
39	M	2013–2014	0.0% (0)	34.0% (5.6)	45.6% (3.8)	20.5% (8.2)
40	M	2012–2014	0.7% (0.3)	27.1% (6.2)	52.5% (3.7)	19.8% (7.9)
42	M	2012	0.0% (0)	46.4% (19.1)	53.6% (10.2)	0.0% (0)
43	M	2012–2013	1.6% (0.5)	25.7% (4.9)	57.0% (3.6)	15.7% (6.5)
50	M	2012–2014	0.0% (0)	39.0% (7.2)	17.2% (2.3)	43.8% (13.2)
51	M	2013	0.4% (0.2)	81.8% (8.3)	17.8% (3.7)	0.0% (13.2)
58	M	2013	0.0% (0)	46.7% (7.3)	16.4% (3.4)	36.9% (14.5)
80	M	2013–2015	0.2% (0.1)	82.0% (9.8)	13.6% (2.1)	4.2% (3.1)
82	M	2013–2014	1.7% (0.5)	54.4% (7.0)	42.9% (3.3)	1.1% (0.8)
86	M	2013–2015	7.5% (1.0)	77.2% (8.6)	7.1% (1.2)	8.3% (4.9)
89	M	2013–2014	0.0% (0)	26.7% (5.6)	52.8% (3.0)	20.5% (7.4)
90	M	2013	17.3% (4.0)	72.6% (9.2)	10.1% (2.2)	0.0% (0)
93	M	2013	44.9% (9.2)	34.9% (7.7)	18.6% (3.5)	1.7% (1.1)
94	M	2013–2014	1.4% (0.4)	37.9% (7.1)	56.5% (4.0)	4.2% (1.6)
98	M	2013–2014	0.0% (0)	33.5% (6.3)	29.0% (3.5)	37.6% (12.9)
189	M	2014–2015	0.0% (0)	78.9% (15.6)	21.1% (5.1)	0.0% (0)
193	M	2014–2015	21.1% (3.4)	31.5% (5.4)	6.2% (1.0)	41.2% (14.0)
44	U	2013–2015	0.0% (0)	21.3% (3.4)	51.5% (3.5)	27.2% (6.9)
151	U	2014	82.6% (9.0)	17.4% (6.7)	0.0% (0)	0.0% (0)

originally tagged did not provide sufficient data on seasonal distributions due to either mortality, emigration downstream of the dam within 30 days of tagging, or transmitter failure. Of these twelve fish, two fish were suspected to have died as a result of surgery, while six fish were suspected of dam passage within days after surgery. Four of the fish disappeared without explanation, possibly due to transmitter failure or mortality/dam passage, which could not be confirmed. Fish that were tagged in the winter of 2014 did not provide summer-winter movement data as acoustic receivers were removed from the reservoir the following spring.

A total of 2,769,936 detections were recorded for 40 acoustically tagged Walleye (Table 2). The most detections for an individual fish were 188,272 (fish #80). The mean number of detections (all fish) was 69,248 (SE = 7461). The study period encompassed 1216 days of total monitoring. Individual fish were monitored for a mean of 568 days (SE = 43.5), and the most days monitored for an individual fish were 918 days for fish #40. The fish with the most days detected on at least one receiver was fish #41 (792 days). However, the days detected were mainly influenced by whether fish spent long periods of time in areas without receivers such as the Cheat River or embayment areas. The temporal distribution of Walleye showed substantial individual variation, but proportional use of lake zones was similar for many individuals (Table 3). Distribution often varied by month or season for individuals. Although Walleye used all lake zones, the middle main lake zone was used most frequently overall by both male and female Walleye (females: mean = 71.6%, SE = 5.8%; males: mean = 50.7%, SE = 5.1%; Figs. 2 and 3). The lower main lake zone was the overall least used area for males (overall mean = 5.1%, SE = 2.6%; Fig. 2), whereas the Cheat River was the overall least used area for females (overall mean = 5.0%, SE = 2.4%; Fig. 2). The riverine zone and Cheat River were used substantially by males and accounted for an overall average of 30.7% (SE = 4.6%) and 13.4% (SE = 3.7%) of time and for a combined overall mean of 44.2% (SE = 5.6%; Fig. 2). For males, the riverine zone and Cheat River were used most heavily during spring and summer (March–September, monthly mean = 72.0%), and use decreased substantially in fall and winter (October–February, monthly mean = 14.5%) (Fig. 2). For females, the riverine zone and Cheat River were primarily used in spring (March–April, monthly mean = 35.2%,), and use was substantially lower in other months (May–February, monthly mean = 9.0%) (Fig. 3). In contrast to the riverine zone and Cheat River, males primarily used the middle main lake zone in fall and winter (October–February, monthly mean = 80.6%,), while the use of this zone decreased from March to September (monthly mean = 21.8%) (Fig. 2). Females utilized the middle main lake zone heavily during all months, but use was particularly high from May to February (monthly mean = 73.9%) and was lower in March and April (monthly mean = 56.8%) (Fig. 3). The use of the lower main lake zone was comparatively low for both males and females during all months, but females did utilize this zone more frequently than the riverine zone and Cheat River from June to February (Figs. 2 and 3).

The core, home, and total linear ranges of tagged Walleye varied across individuals, although similarities in space use patterns were apparent in different groups of

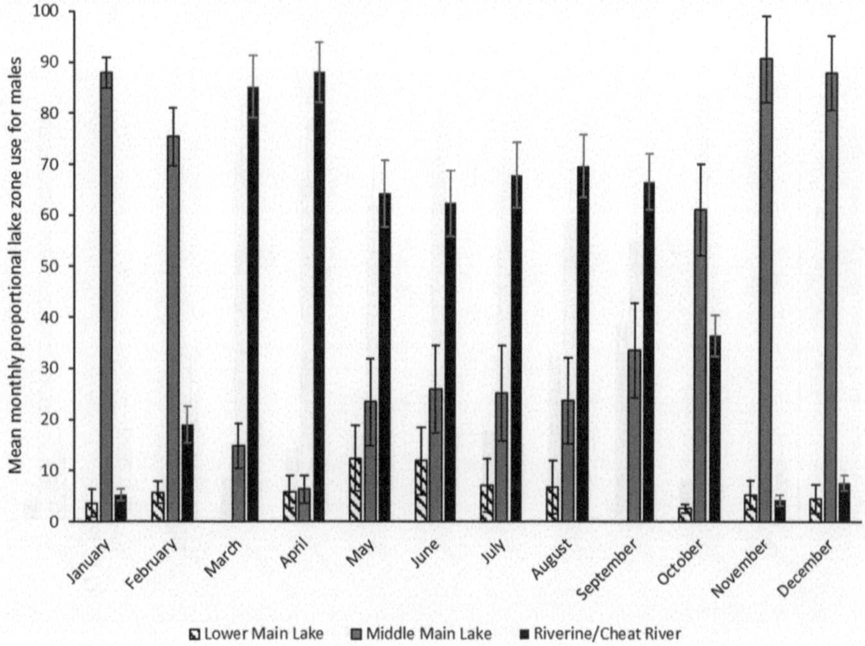

Fig. 2 Mean proportional monthly lake zone use by male Walleye in Cheat Lake, WV. Error bars are ± standard error

tagged fish. Walleye could be grouped by the number of overall core areas, including those individuals that occupied one core use area and those that occupied two separate core use areas (Table 2). Specifically, 60% of tagged Walleye occupied one overall core use area, whereas 40% occupied two overall separate core use areas (Table 2). There was a significant difference in the number of core use areas between male and female Walleye (Kruskal-Wallis: df $= 1$, $H = 7.05$, p value $= 0.008$). Specifically, most females (84.2%) only had one core use area, whereas most males (57.9%) had two core use areas (Table 2). Most Walleye had a core range encompassing or including the middle main lake zone (90% of Walleye; Table 2). The riverine zone and Cheat River were included in the core use areas by fewer Walleye (20% and 15%, respectively; Table 2). The lower main lake was included in the core range by the fewest proportion of fish (10%; Table 2). Additionally, the inclusion of the riverine and/or Cheat River zones in Walleye core use areas significantly differed between sexes (Kruskal-Wallis: df $= 1$, $H = 11.86$, p value <0.001). Specifically, only one female (5.3% of females) utilized riverine habitats as part of its core use area, whereas, 57.9% of males utilized riverine habitats as part of their core range. The home range of tagged Walleye also varied individually. Some Walleye utilized nearly the entire lake as part of their overall home range (16.4 km) while one fish had the smallest overall home range that only included two receivers (2.1 km). The total linear range of Walleye likewise varied individually. The largest

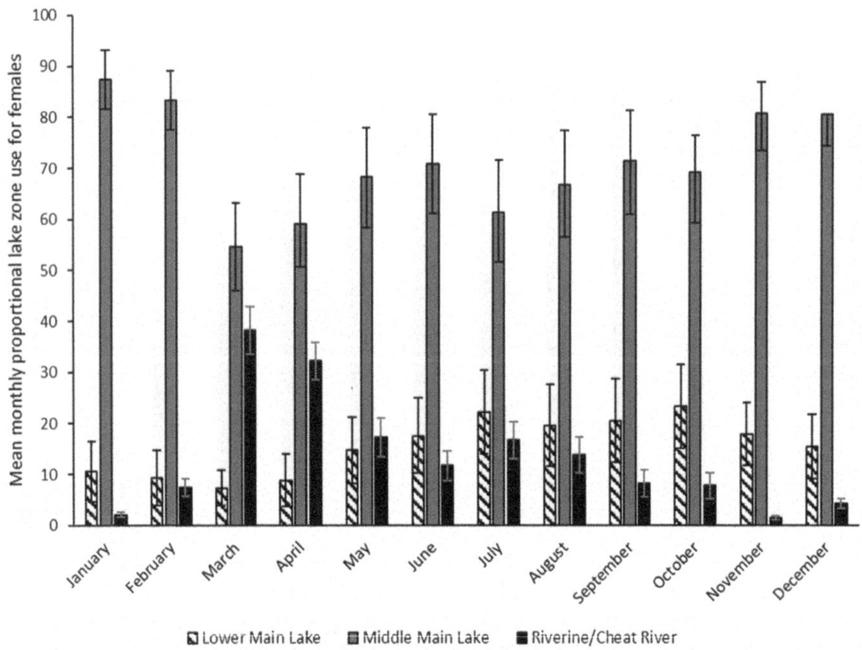

Fig. 3 Mean proportional monthly lake zone use by female Walleye in Cheat Lake, WV. Error bars are ± standard error

linear range encompassed nearly the entire reservoir and the monitored portion of the Cheat River upstream of the lake (19.6 km), whereas the smallest linear range only included three receivers (3.7 km). The mean total linear range for Walleye was approximately 14.3 km (SE = 0.75).

Residency of tagged Walleye varied individually, with significant sex-based differences. Overall residency of tagged Walleye also varied, with some tagged fish never leaving the reservoir, while others temporarily exited the reservoir by swimming upstream into the Cheat River. Overall, 52.5% of tracked Walleye at some point exited the reservoir via the Cheat River resulting in a residency index of less than 1 (Table 2). There was a significant difference in residency between male and female Walleye (Kruskal Wallis: df = 1, $H = 4.48$, p value = 0.03; Fig. 4). Male Walleye were more likely to leave the lake for the Cheat River (mean residency index = 0.81, SE = 0.05) and have a lower residency than female Walleye (mean residency index = 0.95, SE = 0.01; Fig. 4). Specifically, male Walleye spent an average of 58 days per year (SE = 14.72, range = 0–160 days per year) in the Cheat River, whereas females spent an average of 15 days per year (SE = 9.75, range = 0–157 days per year) in the river.

In addition to the overall space use patterns, the examination of the monthly space use patterns provided insight into Walleye movements and distribution. The examination of monthly core range shifts, residence time, and linear range change all revealed similar patterns in Walleye distribution and space use in Cheat Lake.

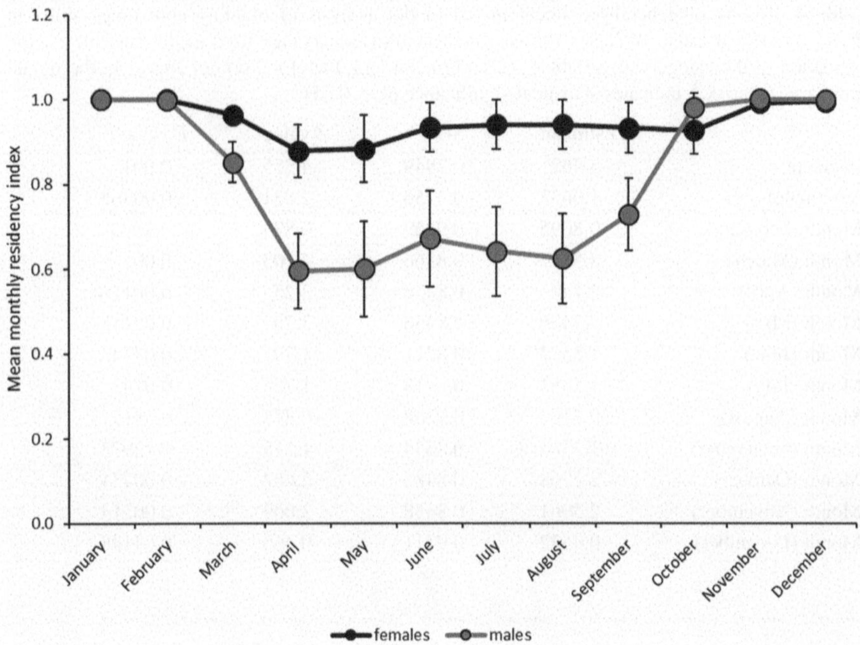

Fig. 4 Mean monthly deviation in residency index by male and female Walleye in Cheat Lake, WV. Error bars are ± standard error

Logistic regression results suggested core range shifts differed significantly across months and by sex (Table 4, Fig. 5). Specifically, logistic regression results suggested Walleye range shifts were significantly different in the months of March, April, May, October, and November and male range shifts were significantly different than female range shifts ($p < 0.05$; Table 4). Male Walleye, on average, experienced more core range shifts (monthly mean = 5.2, SE = 1.1) than females (monthly mean = 3.5, SE = 0.71). The largest peak in core range shifts occurred in spring (March–May) and fall (October–November) (Fig. 5). The monthly mean number of core range shifts by lake zone was 8.7 (SE = 1.6), which was greater than the mean from March to May and October to November (Fig. 5). The highest number of core range shifts was in March (19 individuals with core range shifts; Fig. 5). The middle main lake zone, on average, occurred most frequently in the monthly core ranges of tagged Walleye (Tables 5 and 6). However, in March and April, the riverine zone occurred most frequently, on average, in the core ranges of tagged Walleye (Tables 5 and 6). Additionally, when combining the use of the riverine zone and Cheat River, there was a small peak in the use of these areas in the month of July (Table 5). In July, the riverine zone and Cheat River combined occurred more frequently than the middle main lake zone in the core ranges of tagged Walleye (Table 5).

The monthly residency index also revealed patterns in Walleye distribution. Based on the linear mixed model analysis, the residency index significantly differed

Table 4 Results of generalized linear mixed model analysis of monthly core range shifts in Walleye in Cheat Lake, WV. Sex (female) and month (January) are used as the baseline for the estimation of the categorical variables sex and month and, therefore, do not appear in the model summary. Asterisk * indicates statistical significance ($\alpha < 0.05$)

	Estimate	SE	*Z* value	*P* value	
Intercept	−3.462	0.7949	−4.355	<0.001	*
Sex (male)	1.0631	0.3756	2.831	0.00465	*
Month (February)	0.8095	0.9269	0.873	0.3825	
Month (March)	3.5147	0.8566	4.103	<0.001	*
Month (April)	2.748	0.8456	3.25	0.00116	*
Month (May)	2.7486	0.8456	3.25	0.00115	*
Month (June)	1.5592	0.8711	1.79	0.07348	
Month (July)	1.5551	0.8713	1.785	0.0743	
Month (August)	0.8106	0.9268	0.875	0.38177	
Month (September)	1.3384	0.8834	1.515	0.12975	
Month (October)	2.2768	0.8488	2.682	0.00731	*
Month (November)	2.5961	0.8458	3.069	0.00214	*
Month (December)	0.4522	0.9711	0.466	0.64148	

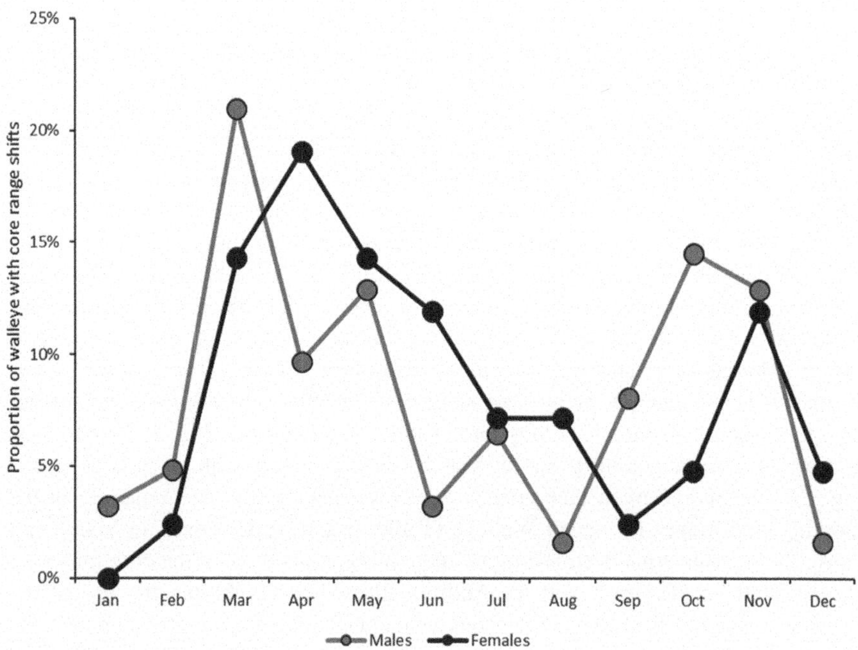

Fig. 5 Mean proportion of monthly core range shifts by male and female Walleye in Cheat Lake, WV

Table 5 Mean proportional occurrence of lake zones in monthly core use areas occupied by Walleye in Cheat Lake

Zone	Jan	Feb	Mar	Apr	May	Jun	Jul	Aug	Sep	Oct	Nov	Dec
Lower main lake	4.9	2.2	2.3	7.0	18.0	13.2	11.9	11.1	8.1	15.0	12.5	9.8
Middle main lake	92.7	80.00	29.6	30.2	43.6	44.7	40.5	47.2	48.7	60.0	85.0	87.8
Riverine	2.4	17.8	59.1	34.9	15.4	23.7	31.0	22.2	24.3	20.0	2.5	2.4
Cheat River	0.0	0.0	9.1	27.9	23.1	18.4	16.7	19.4	18.9	5.0	0.0	0.0

across months ($F = 7.57$; df = 11,330; $p < 0.001$) and between males and females ($F = 5.77$, df = 1.29, $p = 0.02$). Specifically, males were more likely to leave Cheat Lake (and have a lower residency index) than females (Fig. 4). April had the lowest mean residency index of all months (mean for both sexes combined = 0.75, male mean = 0.60, female mean = 0.88) due to more Walleye leaving the lake for the Cheat River (Fig. 4). The monthly residency indices from April to September were significantly low compared to other months due to the increased use of the Cheat River during this time period (Fig. 4). January had the highest residency index (mean RI = 1.0) as no Walleye utilized the Cheat River during this month (Fig. 4).

The monthly change in receiver use (i.e., linear range) of tagged Walleye was consistent with those of the monthly core range and residency index. The linear mixed model analysis suggested that linear range change significantly differed across months ($F = 2.83$; df = 11,330; $p < 0.01$) but was not significantly different between males and females ($F = 0.43$, df = 1.29, $p = 0.52$). The only months with evidence of mean linear range expansion were February–April and October (Fig. 6). The mean deviation in the linear range for these months was >0, indicating an expansion of the linear range and an increased movement for tagged Walleye during these months. However, linear range expansion was significantly different only during the month of March ($p < 0.001$).

A total of 12 individuals (23.1% of tagged Walleye) passed over or through the dam during the study. Most dam passage events occurred in November or December (75%). Four individuals were caught by anglers in the tailwater pool shortly after passing over the dam. Two individuals (March and December) likely passed through the dam turbines as lake elevation was decreasing and hydropower generation was occurring. These fish were considered as likely deceased from the passage event as the transmitters were continuously detected near the turbine outflow for several months without movement. No other tagged fish that exited via the dam were continuously detected in the tailwater, potentially indicating survival. The ten fish that potentially survived the passage of the dam (including the four caught by anglers) passed during high water events (mean lake elevation 265.1 m above sea level $\pm$ 0.04 standard error, river discharge 344.6 m^3s^{-1} $\pm$ 61.9) when lake elevation was increasing (mean daily lake elevation increase 0.5 m $\pm$ 0.2 standard error) and

Table 6 Monthly core use area lake zones occupied by Walleye in Cheat Lake (L = lower main lake zone, M = middle main lake zone, R = riverine zone, and C = Cheat River). Walleye with core use areas overlapping multiple zones are indicated by hyphenated abbreviations. U = unknown sex

ID	Sex	Jan.	Feb.	Mar.	Apr.	May	Jun.	Jul.	Aug.	Sep.	Oct.	Nov.	Dec.
33	**F**	M	M	M	M	M	M	M	M	M	M	M	M
41	**F**	M	M	M	M	M	M	M	M	M	M	M	M
52	**F**	M	M	R	M-R	M	M	R	M	M	M	M	M
53	**F**	M	M	R	M	M	M	M-R	R	M	M	M	M
55	**F**	M	M	M	M	M-R	M	M	M	M	M	M	M
57	**F**	M	M	M	M	M	M	M	M	M	M	M	M
59	**F**	M	M	R	M	M	M	M	M	M	M	M	M
60	**F**	M	M	M	M	L-M	M	L-M	M	M	M	M	M
79	**F**	M	M	M	M	M	M	M	M	M	L	M	M
83	**F**	M	M	R	C	C	C	C	C	C	C	M	M
84	**F**	L	L	L	L-R	L	L	L	L	L	L	L	L
85	**F**	M	M	M	M	M	M	M	M	M	M	M	M
87	**F**	M	M	M	M	L-M	L	L	L	L	L	L-M	M
88	**F**	M	M	R	R-C	M	M	M	M	M	L-M	M	M
96	**F**	M	M	R	C	C	L	L	L	L	L	M	M
157	**F**	M	M	R	C	.	.	.	.	.	.	.	M
179	**F**	M	M	M-R	R-C	.	.	.	.	.	M	M	M
185	**F**	M	M	M	M	M	M	M	M	M	M	M	M
190	**F**	L-M	M	M	L	L	M	M	M	M	M	M	L-M
35	**M**	M	M	R	R	R	R	R	R	M-R	M	M	M
38	**M**	M	M	M	.	.	.	.	.	.	.	.	.
39	**M**	M	R	R	R	R	R	C	C	C	M-R	M	M
40	**M**	M	M	C	C	R	R	R	R	R	M-R	M	M
42	**M**	M	M	R	.	.	.	.	.	.	.	.	.
43	**M**	M	M-R	R	R	C	R	R	R	R-C	M-R	M	M-R

50	**M**	M	M	R-C	C	C	C	C	C	R	M	M	M
51	**M**	M	M	R	R	M	M	M	.	.	.	M	M
58	**M**	M	M	M-R	R	C	C	C	C	C	R	M	M
80	**M**	M	M	R	M-C	M	M	M	M	M	M	M	M
82	**M**	M	M	R	R	M	M	R	M-R	R	M-R	M	M
86	**M**	M	M	R-C	C	M	M	M	M	M	M	M	M
89	**M**	M	M-R	R	C	C	R-C	R	R	R	R	M	M
90	**M**	M	M	R	R	L	L	M	M	M	M	M	M
93	**M**	M	M-R	R	L-R	L	L	L	L	.	.	M	M
94	**M**	M	M	R	R	R	R	R	R	M	M	M	M
98	**M**	M	M	R	R	C	C	C	C	C	R	M	M
189	**M**	M	M-R	.	.	.	.	.	.	.	.	.	M
193	**M**	M	M	R-C	C	C	C	C	C	C	M	L-M	L-M
44	**U**	M-R	R	R	R-C	R-C	R-C	R-C	C	R-C	C	M-R	M
151	**U**	.	.	.	.	.	.	.	.	.	.	L	L

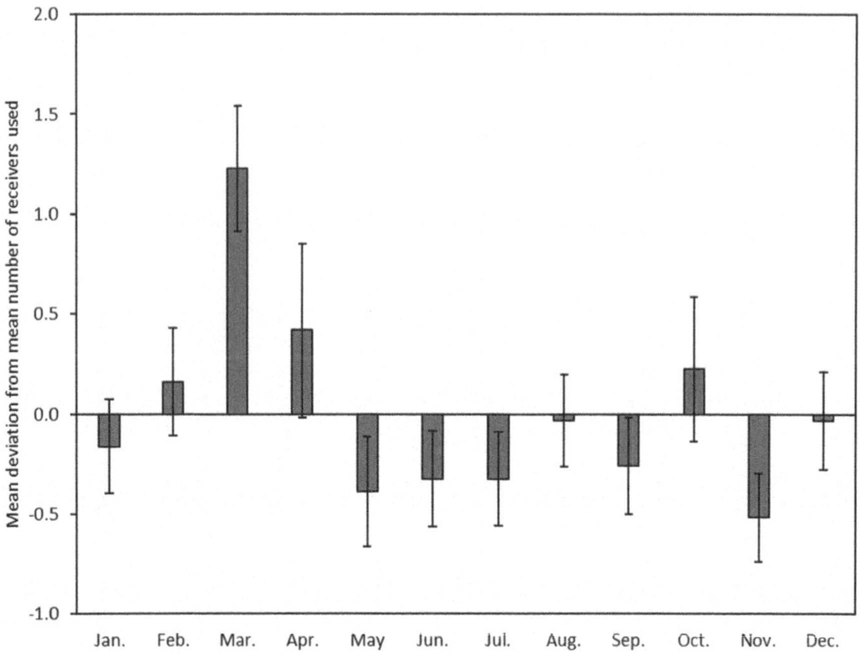

Fig. 6 Mean monthly deviation in receiver use (linear range) by Walleye in Cheat Lake, WV. Error bars are ± standard error

approaching full pool (265.2 m above sea level). In contrast, the two fish that likely died during passage passed during comparatively lower water periods (263.0 m above sea level ± 0.7, river discharge 75.8 m^3s^{-1} ± 0.1) when lake elevation was decreasing from hydropower operations (mean daily lake elevation decrease −0.5 m ± 0.4 standard error).

4 Discussion

This study found that Walleye often demonstrated range shifts and changing movement patterns during periods associated with spawning, postspawn, and overwintering. Most Walleye (80%) made upstream movements and range shifts from lake to riverine environments in conjunction with the spawning period. After spawning, a portion of the tagged individuals (58% overall), including 85% of females, migrated back to main lake areas within a few weeks, while many males (61%) remained in riverine habitats for several months. Some individuals (10%) displayed shifts in range toward riverine habitats during peak summer, possibly in relation to increasing water temperatures and declining oxygen conditions in the main lake. By fall, most individuals (98%) remaining in riverine habitats made return trips and range shifts to the main lake where overwintering occurred.

Examining overall Walleye distribution patterns, it appeared tracked fish largely favored middle main lake habitats where depths and water quality characteristics (dissolved oxygen, water temperature, flow, etc.) were intermediate compared to upstream and downstream habitats. The middle main lake is shallower than the lower main lake but deeper than the riverine zone. Likewise, the middle main lake often offers slightly cooler water temperatures and can provide higher dissolved oxygen levels at cooler depths than the lower main lake but is warmer with lower dissolved oxygen compared to the riverine zone (Table 1). On average, male Cheat Lake Walleye spent over 50% of their time and females spent over 71% of their time in the middle main lake zone. Other studies have reported Walleye primarily utilizing lacustrine reservoir habitats during nonspawning periods (Palmer et al. 2005; Williams 2001). The middle main lake zone has shallow flats juxtaposed next to deep water areas, two large coves, and the most abundant and diverse forage of all areas of the lake. Walleye have been reported to select shallow flats and coves to forage on at night (Fitz and Holbrook 1978; Haxton et al. 2015)—the same habitat that is most prevalent in the middle main lake zone. Although the middle main lake zone does stratify, stratification can be weaker compared to the lower main lake zone. For instance, during the summers of 2011 and 2012, dissolved oxygen levels at cooler depths (10–15 m) were often less than 3 mg/L (Table 1). During this same period, dissolved oxygen levels in the middle main lake zone also decreased at these depths but were typically above 3 mg/L (Table 1). Although every lake zone generally supports an abundance of fishes common to Walleye diets (e.g., Yellow Perch, Gizzard Shad (*Dorosoma cepedianum*), Emerald Shiner (*Notropis atherinoides*), and *Lepomis* species), past surveys suggest that during summer and fall, main lake areas tend to support a greater abundance and diversity of forage opportunities compared to riverine areas (Smith and Welsh 2015). Additionally, the lower main lake zone, although providing deep water with cool summer temperatures, tends to strongly stratify. It could be difficult for Walleye to locate preferred water temperatures (~ 22 °C; as summarized in Bozek et al. 2011) with suitable dissolved oxygen (> 5 mg/L preferred, > 3 mg/L necessary for survival; as summarized in Bozek et al. 2011). The lower main lake also has sharply sloped banks, leading to limited littoral zone areas on which to forage. Therefore, it is possible the heavy use of the middle main lake zone is tied to foraging opportunities, habitat, or a combination thereof, which has been suggested in previous studies on Walleye distribution (Wang et al. 2007; Raby et al. 2018; Brooks et al. 2019; Matley et al. 2020). However, further data collection and investigation would be necessary to confirm these explanations.

The core and home range size of tracked fish were similar throughout the year, although seasonal shifts in areas used by Walleye were documented. There was individual variation in the size of home and core ranges, with some individuals occupying relatively small areas (<5 km) and other individuals occupying the entire reservoir (>19 km), similar to other studies that have noted a wide range in individual range variation (Williams 2001; Palmer et al. 2005; Kirby et al. 2017). The total ranges of Walleye in Cheat Lake were small compared to what has been reported in some other studies when considering the distance traveled (e.g., river kilometers) (DePhilip et al. 2005; Palmer et al. 2005). However, this may be due to the small size of Cheat Lake (20.9 rkm) and the limited monitoring area of the Cheat River compared to other water bodies where telemetry studies have taken place. Several Walleye occupied the entirety of Cheat Lake and also utilized some of the upstream Cheat River. These fish had total linear ranges of at least 19.6 rkm based on the distance between the Cheat River receiver and the most downstream receiver in Cheat Lake. However, due to a lack of receiver coverage, there is uncertainty surrounding how far upstream fish traveled into the Cheat River, so it is possible these fish had much larger linear ranges than realized. When considering the proportion of lake area occupied, Walleye in Cheat Lake had similar total ranges to those reported in other studies (Williams 2001; DePhillip et al. 2005; Palmer et al. 2005; Foust and Haynes 2007; Kirby et al. 2017), with some Walleye only occupying a very small percentage of the reservoir and other Walleye utilizing the entire reservoir. As previously alluded to, the middle main lake zone was most frequently included in core use areas of tracked fish. A total of 90% of tracked fish utilized the middle main lake zone as part of their overall core range, reinforcing the importance of this area to Cheat Lake Walleye. Walleye differed in that individual fish either occupied one or two overall core use zones. This suggests fish with only one core use zone had a more overall restricted high use range, while fish with two core use zones exhibited more plasticity or temporal variations in areas of high use. Other studies have only alluded to the multiple core use areas of Walleye via a description of the temporal changes in range and distribution (DePhilip et al. 2005; Palmer et al. 2005; Raby et al. 2018; Brooks et al. 2019) but have not specifically quantified them, so comparisons with other populations are difficult.

Walleye exhibited seasonal variations in core ranges and lake zone use. In spring months and to a lesser extent in mid-summer, fish shifted core use areas from the middle main lake to the riverine zone and Cheat River. Spring range shifts were most likely a factor of prespawning and spawning activities. These spawning-related range shifts are typical of what occurs in other systems (Palmer et al. 2005; Pritt et al. 2013; Bade et al. 2019; Brooks et al. 2019; Matley et al. 2020). The small peak in range shifts evident in mid-summer (July) could potentially be related to challenging physicochemical conditions within the main lake areas during summer months. Specifically, water temperatures were highest in July (maximum of 28.3 °C during the study period), and dissolved oxygen concentrations were also noticeably stratified during this time period (Table 1). This could result in an oxygen-temperature squeeze (Coutant 1985; Williams 2001; Clark-Kolaks 2008; Bozek et al. 2011), forcing some Walleye to make range shifts in search of cooler, more oxygenated water, which is most likely to be found near the inflow of the Cheat River. Movements in search of optimum water temperature conditions (~ 22 °C; as summarized in Bozek et al. 2011) have been suggested in other studies (Wang et al. 2007; Hayden et al. 2014; Raby et al. 2018; Brooks et al. 2019; Matley et al. 2020). However, only a small number of Walleye made this mid-summer habitat shift, indicating that other main lake residents chose to remain in stratified main lake habitats. Cheat Lake typically experiences fall turnover in September (Smith and

Welsh 2015), which is also the time when tagged fish began to increase the use of the middle lake zone. Once water temperatures cooled and fall months arrived, nearly all Walleye shifted core use areas again to occupy primarily the middle main lake zone. Use of the middle lake zone peaked during late fall/early winter. Walleye may retreat to the main lake zone during this period to locate deeper water or concentrated prey (Paragamian 1989; DePhillip et al. 2005). The convergence of Walleye into deeper waters in the fall has been commonly reported in other studies (Paragamian 1989; Williams 2001; DePhilip et al. 2005).

In general, there were two groups of Walleye in our study: lake resident fish (60%), which spent most of their time in main lake habitats, and riverine resident fish (40%), which spent a substantial portion of their time in riverine habitats in addition to overwintering in main lake habitats. Lake resident fish typically occupied main lake core ranges during all time periods except for months associated with spawning. Riverine resident fish occupied riverine core ranges during all time periods except fall and winter, in which most of these fish switched to occupying main lake habitats. Riverine resident fish also were more likely to emigrate from Cheat Lake via the Cheat River upstream, resulting in lower residency indices for these fish. Although both males and females were often lake resident fish, riverine resident fish were much more likely to be males (58% of males, 5% of females). Other Walleye movement studies have reported similar segregation of Walleye occupying the lacustrine or riverine environments (DePhilip et al. 2005; Palmer et al. 2005; Wang et al. 2007; Hayden et al. 2014). In most of these studies, differences in overall distribution were often tied to variations in postspawning movements between males and females or genetically induced behavior (DePhilip et al. 2005; Palmer et al. 2005; Hayden et al. 2014). However, many studies have not quantified core and home ranges for individuals but instead have qualitatively described seasonal movements. Additionally, in most of these studies, fish had left spawning areas by late spring (Palmer et al. 2005; Hayden et al. 2014; Raby et al. 2018). In our study, it was not uncommon for fish to remain near spawning areas until fall. DePhilip et al. (2005) did have similar results in Au Sable River, Michigan, USA. In their study, 25% of tagged Walleye did not migrate out from spawning areas to the downstream reservoir until fall. The authors concluded that some Walleye delayed their return to the reservoir to take advantage of optimal foraging conditions. Palmer et al. (2005) found lake resident fish spawned in riverine habitats but subsequently returned to main lake habitats, whereas river resident fish spawned and remained in riverine habitats. Differences in their study were thought to be the result of genetic differentiation (Palmer et al. 2005). However, genetic testing of Cheat Lake Walleye indicated movement differences were not genetically based (WVDNR unpublished data, Dustin Smith WVDNR). In the present study, nearly all individuals occupied the main lake at some point during a given year (almost always to overwinter), but differences existed in ranges occupied in postspawn and summer periods.

Sex-based differences in distribution and movement patterns were apparent in our study. Other studies have noted the apparent link between Walleye sex and seasonal distribution (Wang et al. 2007; Hayden et al. 2014; Raby et al. 2018; Bade et al. 2019; Matley et al. 2020). Specifically, other studies have reported a dichotomy in postspawn distributions between males and females (DePhilip et al. 2005; Wang et al. 2007; Hayden et al. 2014; Raby et al. 2018). Several theories have been posited as to why males and females segregate. Other researchers have suggested theories related to maximizing spawning success for males, occupation of preferred water temperatures, and optimal foraging theory (DePhilip et al. 2005; Wang et al. 2007; Hayden et al. 2014; Raby et al. 2018; Matley et al. 2020).

It is possible that the sex-based differences in distribution in our study were related to differences in the spawning behavior between males and females. Some authors have suggested that male Walleye extend the time spent on spawning grounds to maximize their potential spawning attempts with as many females as possible (Hayden et al. 2014; Raby et al. 2018). Males often spawn with multiple females, whereas females typically deposit all of their eggs in a short time frame (Colby et al. 1979). By extending the time spent on spawning grounds, males may increase their interaction with females and increase their spawning attempts (Hayden et al. 2014; Raby et al. 2018). However, this theory does not explain the residence of males past the month of April in riverine habitats.

Another possible explanation for sex-based differences in distribution is related to variations in habitat needs between males and females. Other researchers have suggested females are more likely to search out optimal habitat (e.g., water temperature) conditions compared to males after spawning to maximize their energy intake (Wang et al. 2007; Raby et al. 2018; Matley et al. 2020). In many waters, as surface water temperatures warm, deep waters in main lake areas could provide thermal refuge for female Walleye, potentially optimizing growth potential and body condition. In Cheat Lake, deeper main lake areas do offer cool water temperatures compared to the uniform temperatures found throughout the riverine zone during summer months. However, in Cheat Lake, the riverine zone still consistently offers Walleye water temperatures near their reported optimal water temperature of 22 °C throughout the summer months (Table 1). Additionally, during summer, the stratification of main lake areas may limit the ability to find cool waters as oxygen levels at cooler depths often decrease below 5 mg/L for weeks at a time. In summer, some females (16%) displayed a propensity to make forays from the main lake back into the riverine zone. Potentially, these fish were searching for cooler, more oxygenated water, which the incoming Cheat River provides during summer periods. However, only a small number of females displayed this behavior, suggesting other main lake residents were able to find suitable habitats without making movements into the riverine zone.

Researchers have also suggested that some Walleye (especially females) migrate to areas after spawning that offer optimal foraging opportunities in terms of preferred prey (Bowlby and Hoyle 2011; Hayden et al. 2014; Raby et al. 2018; Brooks et al. 2019; Matley et al. 2020). Although the riverine zone supports prey fish for Walleye, during summer and fall, a higher proportion of these fish tend to be smaller shiner species (e.g., Mimic Shiner (*Notropis volucellus*), Emerald Shiner, etc.), Logperch (*Percina caprodes*), and riverine centrarchids (e.g., Smallmouth Bass (*Micropterus dolomieu*), Rock Bass (*Ambloplites rupestris*)). In contrast, during summer and fall, the main lake offers a greater diversity of prey fish and a greater size spectrum of potential prey. Specifically, the main lake supports a strong population of Yellow Perch and Gizzard Shad, especially during summer and fall months (Smith and Welsh 2015; Hilling et al. 2018), and Cheat Lake Walleye have demonstrated a strong propensity to prey on Yellow Perch and Gizzard Shad (Smith and Welsh 2015). Yellow Perch were over three times as abundant in main lake areas compared to riverine areas during fall electrofishing surveys conducted from 2011 to 2015 (Smith and Welsh 2015). Similarly, Gizzard Shad were over 15 times as abundant in main lake areas compared to riverine areas during fall electrofishing surveys from 2011 to 2015 (Smith and Welsh 2015). Although foraging-based movements may be a contributing factor, further data collection and investigation would be required to support this explanation.

Most Walleye (80%) increased their use of upstream riverine habitats and the incoming Cheat River during spring. Range shifts during spring months provided evidence that Walleye used the headwaters of Cheat Lake and the Cheat River for spawning. Given that Cheat Lake is an ecosystem recovering from decades of acidification, the identification of available spawning habitat for a once extirpated species such as the Walleye is important. Cheat Lake experiences seasonal changes in lake-level fluctuations, which can impact Walleye spawning (Johnson 1961; Chevalier 1977). The lake area utilized for spawning was a relatively small area (approximately 1 rkm) just downstream of the incoming Cheat River. This limited spawning area creates an inherent risk of disruption to spawning activity. Poorly timed lake-level decreases in spring could lead to spawning failure for fish utilizing this area. Some Walleye were evidently utilizing the Cheat River to spawn and as a summer habitat. These Walleye are protected from lake-level fluctuations, but if larval Walleye subsequently drift downstream to the main lake, they would still be susceptible to changing water levels given their poor swimming ability (Walburg 1971). Additionally, should acidification issues arise again in the future, this area would be the first to receive acidic water from upstream prior to dilution in the larger body of the main lake. Acidic conditions are not conducive to successful Walleye reproduction (Hulsman et al. 1983; Rahel and Magnuson 1983), so the protection of suitable water quality, especially around spawning habitat, is critical. Evidence of spawning being restricted to the upper portion of the lake should improve the ability of researchers to monitor the impacts of lake-level fluctuations on spawning activity in future years. However, this seasonal clustering of adult Walleye also potentially increases their susceptibility to angling (Palmer et al. 2005; Bade et al. 2019). The use of the Cheat River for spawning and summer habitat lends some evidence that a portion of the population may be protected from lake-level fluctuations and angling pressure due to reduced boat access.

Residency indices of tagged Walleye provided information on the frequency of fish temporarily leaving the reservoir for the upstream Cheat River. Overall, over 50% of tagged Walleye at some point temporarily exited the reservoir for the river upstream. When examining the residency of tagged Walleye monthly, clear patterns of temporal emigration from the reservoir dependent on the time of year are evident. The heaviest use of the Cheat River occurred in April. This is likely due to Walleye leaving the reservoir to spawn in the Cheat River upstream. Although female

Walleye typically returned to utilizing primarily main lake habitats in the summer with occasional forays into upstream riverine habitats, a large proportion of males continued to utilize upstream riverine habitats throughout the summer until fall. Some Walleye (primarily males) continued to periodically utilize the Cheat River upstream, while others continued to occupy the Cheat River until fall. It is unknown why some fish choose to remain in the Cheat River or utilize it frequently compared to others. Walleye remaining in the river may simply be choosing to limit postspawn movement and instead focus on immediately foraging upon available prey in the river (DePhilip et al. 2005).

The dichotomy in habitat use between males and females could affect management strategies and angling pressure (Palmer et al. 2005; Wang et al. 2007; Pritt et al. 2013; Hayden et al. 2014; Bade et al. 2019). It could benefit fisheries managers to account for unequal fishing pressure between males and females. Specifically, female Walleye in Cheat Lake may experience higher susceptibility to angling, given their closer proximity to the angler access areas in the main lake. Weekly observations during telemetry work suggest most angling occurs in Cheat Lake from May to October, when the reservoir fluctuations are restricted for recreational activity. This coincides with the time period female Walleye largely utilize main lake habitats. A recent discussion with approximately ten avid Cheat Lake Walleye anglers has suggested main lake areas are preferred for techniques such as trolling from May–October, a technique in which a large area can be covered by anglers targeting Walleye. Interestingly, Bade et al. (2019) found that the use of trolling techniques caught more female Walleye, whereas the use of jigging techniques caught more male Walleye. An angler creel survey, extensive tagging efforts, and research into the effort and harvest habits of Cheat Lake anglers would be beneficial for future management.

This study demonstrated seasonal patterns and sex-based differences in the Walleye distribution in Cheat Lake. Understanding seasonal movements and distributions can improve the management of Walleye populations (Williams 2001; Palmer et al. 2005; Pritt et al. 2013; Brooks et al. 2019). Specifically, this information would be useful from a management perspective as knowing when and where congregations of Walleye will occur seasonally could help direct survey efforts and the management of angler harvest (Williams 2001; Palmer et al. 2005; Pritt et al. 2013; Brooks et al. 2019). Future research aimed at monitoring angling impacts on seasonal congregations could help prevent overharvest. Creel survey and angler effort research on Cheat Lake could also help determine the potential impacts of these seasonal, sex-based distributions. Additionally, future stock assessment surveys could consider the sex-based segregation of Walleye within Cheat Lake to improve the interpretation of results. Overall, results from this study may provide fisheries managers with valuable information that could be beneficial in the future management of this reestablished fishery. The information gained will help guide future monitoring and research and could aid in directing future management actions to maintain and potentially improve this fishery.

Acknowledgments This work was supported by Cube Hydro Partners, LLC, and the West Virginia Division of Natural Resources. We would like to thank Joni Aldinger, Melissa Braham, Brian Crabill, Kyle Hartman, Wendell Hofer, Frank Jernejcic, Pat Mazik, Scott Morrison, Amanda Renick, Austin Rizzo, David Smith, Nate Taylor, Bri Tierney, Patricia Thompson, Zack Vucurevich, and David Wellman. This study was performed under the auspices of West Virginia University IACUC protocol 11-0403. Any use of trade, firm, or product names is for descriptive purposes only and does not imply endorsement by the US government.

References

Bade AP, Binder TR, Faust MD, Vandergoot CS, Hartman TJ, Kraus RT, Krueger CC, Ludsin SA (2019) Sex-based differences in spawning behavior account for male-biased harvest in Lake Erie walleye (*Sander vitreus*). Can J Fish Aquat Sci 76:2003–2012

Bates D, Maechler M, Bolker B, Walker S (2015) Fitting linear mixed-effects models using lme4. J Stat Softw 67:1–48

Bowlby JN, Hoyle JA (2011) Distribution and movement of bay of Quinte walleye in relation to temperature, prey availability, and Dreissenid colonization. Aquat Ecosyst Health 14:56–65

Bozek MA, Haxton TJ, Raabe JK (2011) Walleye and Sauger habitat. In: Barton BA (ed) Biology, management and culture of walleye and Sauger. American Fisheries Society, Bethesda, Maryland, pp 133, 601pp–197

Brooks JL, Midwood JD, Gutowsky LFG, Boston CM, Doka SE, Hoyle JA, Cooke SJ (2019) Spatial ecology of reintroduced walleye (*Sander vitreus*) in Hamilton harbour of Lake Ontario. J of Great Lakes Res 45:167–175

Brown ML, Guy CS (2007) Science and statistics in fisheries research. In: Guy CS, Brown ML (eds) Analysis and interpretation of freshwater fisheries data. American Fisheries Society, Bethesda, Maryland, pp 625–676

Buckmeier DL, Smith NG, Daugherty DJ (2013) Alligator gar movement and macrohabitat use in the lower Trinity River, Texas. Trans Am Fish Soc 142:1025–1035

Chevalier JR (1977) Changes in walleye (*Stizostedion vitreum vitreum*) population in rainy Lake and factors in abundance. J Fish Res Board Can 34:1696–1702

Clark-Kolaks S (2008) Distribution and movement of walleye (*Sander vitreus*) in Monroe Reservoir, Indiana 2008 and 2009. Fish Research Final Report, Indiana Fisheries section, Indiana Department of Natural Resources, Division of Fish and Wildlife

Colby PJ, McNicol RE, Ryder RA (1979) Synopsis of biological data on the walleye *Stizostedion v. vitreum*. FAO Fisheries Synopsis 119, Rome

Cooke SJ, Martins EG, Struthers DP, Gutowsky LFG, Power M, Doka SE, Dettmers JM, Crook DA, Lucas MC, Holbrook CM, Krueger CC (2016) A moving target—incorporating knowledge of the spatial ecology of fish into the assessment and management of freshwater fish populations. Env Mon Assess 188:239

Core EL (1959) Biological investigations of cheat Lake. West Virginia University, Morgantown, WV

Coutant CC (1985) Striped bass, temperature, and dissolved oxygen: a speculative hypothesis for environmental risk. Trans Am Fish Soc 114:31–61

Craig JF (2000) Percid fishes: systematics, ecology and exploitation. Blackwell Science, Oxford, UK

Crook DA (2004) Is the home range concept compatible with the movements of two species of lowland river fish? J Anim Ecol 73:353–366

DePhilip MM, Diana JS, Smith D (2005) Movement of walleye in an impounded reach of the au Sable River, Michigan, USA. Environ Biol Fish 72:455–463

Fitz RB, Holbrook II, JA (1978) Sauger and walleye in Norris Reservoir, Tennessee. In: Kendall RL (ed). Selected coolwater fishes of North America. Bethesda, American Fisheries Society, Special Publication Number 11, pp. 82–88

Foust JC, Haynes JM (2007) Failure of walleye recruitment in a lake with little suitable spawning habitat is probably exacerbated by restricted home ranges. J Freshw Ecol 22:297–309

Freund JG, Petty JT (2007) Response of fish and macroinvertebrate bioassessment indices to water chemistry in a mined Appalachian watershed. Environ Manag 39:707–720

Gilroy DJ, Jensen OP, Allen BC, Chandra S, Ganzorig B, Hogan Z, Maxted JT, Vander Zanden MJ (2010) Home range and seasonal movement of taimen, *Hucho taimen*, in Mongolia. Ecol Freshw Fish 19:545–554

Harasti D, Lee KA, Gallen C, Hughes JM, Stewart J (2015) Movements, home range and site fidelity of snapper (*Chrysophrys auratus*) within a temperate marine protected area. PLoS One 10(11):e0142454. https://doi.org/10.1371/journal.pone.0142454

Hart LG, Summerfelt RC (1975) Surgical procedures for implanting ultrasonic transmitters into flathead catfish (*Pylodictis olivaris*). Trans Am Fish Soc 104:56–59

Haxton T (2015) Characteristics and spatial segregation of sympatric saugers and walleyes in the Ottawa River, Canada. N Am J Fish Manage 35:418–430

Hayden TA, Holbrook CM, Fielder DG, Vandergoot CS, Bergstedt RA, Dettmers JM, Krueger CC, Cooke SJ (2014) Acoustic telemetry reveals large-scale migration patterns of walleye in Lake Huron. PLoS One 9(12):e114833. https://doi.org/10.1371/journal.pone.0114833

Hilling CD, Taylor ND, Welsh SA, Smith DM (2018) Population characteristics of yellow perch in a central Appalachian hydropower reservoir. J Fish Wildl Manag 9:486–495

Hulsman PF, Powles PM, Gunn JM (1983) Mortality of walleye eggs and rainbow trout yolk sac larvae in low pH waters of the La Cloche Mountain area, Ontario. Trans Am Fish Soc 112:680–688

Johnson FH (1961) Walleye egg survival during incubation on several types of bottom in Lake Winibigoshish, Minnesota, and connecting waters. Trans Am Fish Soc 90:312–322

Kirby LJ, Johnson SL, Ringler NH (2017) Diel movement and home range estimation of walleye (*Sander vitreus*) within a no-take urban fishery. J Freshw Ecol 32:49–64

Landsman SJ, Nguyen VM, Gutowsky LFG, Gobin J, Cook KV, Binder TR, Lower N, McLaughlin RL, Cooke SJ (2011) Fish movement and migration studies in the Laurentian Great Lakes: research trends and knowledge gaps. J Great Lakes Res 37:365–379

Matley JK, Faust MD, Raby GD, Zhao Y, Robinson J, MacDougall T, Hayden TA, Fisk AT, Vandergoot CS, Krueger CC (2020) Seasonal habitat-use differences among Lake Erie's walleye stocks. J Great Lakes Res 46:609–621

McClurg SE, Petty JT, Mazik PM, Clayton JL (2007) Stream ecosystem response to limestone treatment in acid impacted watersheds of the Allegheny plateau. Ecol Appl 17:1087–1104

Miranda LE, Bettoli PW (2010) Large Reservoirs. In: Hubert WA, Quist MC (eds) Inland fisheries management in North America, 3rd edn. American Fisheries Society, Bethesda, Maryland, pp 545–586

Palmer GC, Murphy BR, Hallerman EM (2005) Movements of walleyes in Claytor Lake and the upper New River, Virginia, indicate distinct lake and river populations. N Am J Fish Manage 25:1448–1455

Paragamian VL (1989) Seasonal habitat use by walleye in a warmwater river system, as determined by radiotelemetry. N Am J Fish Manage 9:392–401

Pincock DG (2011) False detections: what are they and how to remove them from detection data. VEMCO whitepaper document DOC-004691, v03. Amrix systems, Inc. Halifax, NS, Canada

Pothoven SA, Madenjian CP, Hook TO (2016) Feeding ecology of the walleye (Percidae, *Sander vitreus*), a resurgent piscivore in Lake Huron (Laurentian Great Lakes) after shifts in the prey community. Ecol Freshw Fish 26:676–685

Pritt JJ, DuFour MR, Mayer CM, Kocovsky PM, Tyson JT, Weimer EJ, Vandergoot CS (2013) Including independent estimates and uncertainty to quantify total abundance of fish migrating in a large river system: walleye (*Sander vitreus*) in the Maumee River, Ohio. Can J Fish Aquat Sci 70:803–814

Raby GD, Vandergoot CS, Hayden TA, Faust MD, Kraus RT, Dettmers JM, Cooke SJ, Zhao Y, Fisk AT, Krueger CC (2018) Does behavioural thermoregulation underlie seasonal movements in Lake Erie walleye? Can J Fish Aquat Sci 75:488–496

Rahel FJ, Magnuson JJ (1983) Low pH and the absence of fish species in naturally acidic Wisconsin lakes: inferences for cultural acidification. Can J Fish Aquat Sci 40:3–9

Ramsden S, Cotton CC, Curran MC (2017) Using acoustic telemetry to assess patterns in the seasonal residency of the Atlantic stingray *Dasyatis sabina*. Environ Biol Fish 100:89–98

Rogers KB, White GC (2007) Analysis of movement and habitat use from telemetry data. In: Guy CS, Brown ML (eds) Analysis and interpretation of freshwater fisheries data. American Fisheries Society, Bethesda, Maryland, pp 625–676

Rosenblatt AE, Heithaus MR (2011) Does variation in movement tactics and trophic interactions among American alligators create habitat linkage? J Anim Ecol 80:786–798

Schmalz PJ, Fayram AH, Isermann DA, Newman SP, Edwards CJ (2011) Harvest and exploitation. In: Barton BA (ed) Biology, management, and culture of walleye and sauger. American Fisheries Society, Bethesda, Maryland, pp 375–402

Shroyer SM, Logsdon DE (2009) Detection distances of selected radio and acoustic tags in Minnesota Lakes and Rivers. N Am J Fish Manage 29:876–884

Smith DM, Welsh SA (2015) Biological monitoring of aquatic communities of Cheat Lake, and Cheat River downstream of the Lake Lynn hydro station, 2011–2015. Final Project Report. FERC Project Number 2459

Steelman TA, Carmin J (2002) Community based watershed remediation: connecting organizational resources to social and substantive outcomes. In: Rahm D (ed) Toxic waste and environmental policy in the 21st century United States. McFarland Publishers, Jefferson, North Carolina, pp 145–178

Thorne B, Pitzer K (2004) An update on the Cheat River restoration. Proc Am Soc of Mining Recl 2004:1894–1905

Topping DT, Lowe CG, Caselle JE (2006) Site fidelity and seasonal movement patterns of adult California sheephead *Semicossyphus pulcher* (Labridae): an acoustic monitoring study. Mar Ecol Prog Ser 326:257–267

U.S. Geological Survey (2019) USGS 03070260 Cheat River at Albright, WV, in USGS water data for the nation: U.S. Geological Survey National Water System database, accessed 15 August 2019 at URL doi:https://doi.org/10.5066/F7P55KJN

Vokoun JC (2003) Kernel density estimates of linear home ranges for stream fishes: advantages and data requirements. N Am J Fish Manage 23:1020–1029

Walburg CH (1971) Loss of young fish in reservoir discharge and year-class survival in Lewis and Clark Lake, Missouri River. In: Hall GE (ed) Reservoir fisheries and limnology, American fisheries society, special publication 8. Bethesda, Maryland, pp 441–448

Walsh CT, Reinfelds IV, Gray CA, West RJ, van der Muelen DE, Craig JR (2012) Seasonal residency and movement patterns of two co-occurring catadromous percichthyids within a south-eastern Australian river. Ecol Freshw Fish 21:145–159

Wang HY, Rutherford ES, Cook HA, Einhouse DW, Haas RC, Johnson TB, Kenyon R, Locke B, Turner MW (2007) Movement of walleyes in lakes Erie and St. Clair inferred from tag return and fisheries data. Trans Am Fish Soc 136:539–551

Welsh SA, Perry SA (1997) Acidification and fish occurrence in the upper Cheat River drainage, West Virginia. J Am Water Resour As 33:423–429

Williams JD (2001) Walleye movement, distribution, and habitat use in Laurel River Lake, Kentucky. Proc Annu Conf Southeast Assoc Fish Wildl Agencies 55:257–269

Williams DR, Clark ME, Brown JF (1999) Stream water quality in coal mined areas of the lower Cheat River basin, West Virginia and Pennsylvania, during low-flow conditions, July 1997. Water Resources Investigation Report 98–4258, U.S. Geological Survey

Winter J (1996) Advances in underwater biotelemetry. In: Murphy BR, Willis DW (eds) Fisheries techniques, 2nd edn. American Fisheries Society, Bethesda, Maryland, pp 555–585

Managing Tribal Fisheries and Employees on the Reservation

Carl A. Klimah

Abstract Tribal natural resource management agencies continue to provide employment opportunities within the fishery field, and while much is known about the western ways of managing fisheries, tribal culture, history, and views on fishery management are largely unknown to non-Natives. Thus, fishery management on tribal reservations can present a set of unique challenges for fishery managers who are unfamiliar with the tribal aspect of natural resource management, as well as the sovereign status of tribes and their ability to set their own regulations. In this chapter, we give guidance on how to prepare to work for a tribe, effectively manage employees, create fishery regulations, and maintain open communication within the tribal community. In addition, we also discuss tribal history, including treaty rights, sovereignty, and the different important government acts that impacted Natives. Working for tribes and managing their fisheries can be a life-changing and rewarding experience. Therefore, it is beneficial to discuss the phases that many tribal employees experience (optimism, frustration, burnout, acceptance) and how to mitigate the frustration and burnout phases. This chapter aims to inform those who are interested in managing tribal fisheries so that they're better prepared to effectively manage fish populations and make a positive impact within tribal communities and workplaces.

Biindaajimowin Odanokiitoon Anishinaabe bagwaji-aabajichiganan ogimaakaan nitaawigi'amegwewin. Gigikendaamin Gichi-mookomaani-nitaawigi'amegwewin. Gigikendaanziimin Gichi-mookomaanag Anishinaabe izhitwaawin gaye Anishinaabe ganawenjigewin gaye dash Anishinaabe nitaawigi'amegwewin. Oshki-gwiinawi-inendam nitaawigi'amegwewi-mizhinaweg Anishinaabewi-Ishkoniganing giishpin gikendanzigwaa Anishinaabe bagwaji-aabajichiganan gaye Anishinaabe inakaaneziwin gaye dash Anishinaabe onashowewin. O'omaa, ningikinoowidoomin gaa-ozhitaa-anokiig, miinawaaj deboomangid anokiimaaganag minawaaj nitaawigi'amegwewi-onashowewinikeyaang miinawaaj dash wiindamawangid biindigeyi'iing Anishinaabe

C. A. Klimah (✉)
Department of Natural Resources, Mille Lacs Band of Ojibwe, Onamia, MN, USA
e-mail: Carl.Klimah@millelacsband.com

J. C. Bruner, R. L. DeBruyne (eds.), *Yellow Perch, Walleye, and Sauger: Aspects of Ecology, Management, and Culture*, Fish & Fisheries Series 41,
https://doi.org/10.1007/978-3-030-80678-1_9

oodenaang. Bekish, nindazhindaamin Anishinaabe-ganawenjigewin gaye gichi-ashodamaagewining dibinawewiziwinan gaye inakaaneziwin gaye onashowewin. E'-anokiitawaag Anishinaabeg miinawaaj deboondang nitaawigi'amegwewin jiikenda-agwadoon. Nindawaa dash onizhishin ji-dazhindang azhiganeg dash baatayiinowag Anishinaabe-anokiimaaganag inendang (apenimowin gaye mamiidaawendamowin gaye zhagadendamowin gaye dash go debwetamowin) miinawaa dash wiido-okamawaawag iniw mamiidaawendamowin gaye dash zhagadendamowin. Gaa-maane-babaamendang Anishinaabe nitaawigi'amegwewin gikinoo'amawaawag. Owii'-bami'aan giigoonyan gaye dash netaawichigeng biindigeyi'iing Anishinaabe oodenawang miinawaa endazhi-anokiing. [Translated into Ojibwe by: Chares Lippert, Mille Lacs Band of Ojibwe, Department of Natural Resources, 43408 Oodena Drive, Onamia, Minnesota 56359].

Keywords Native American · Treaty rights · Tribe · Fishery management · Ojibwe

1 Introduction of the First Nations

> We have been called the Indians, We have been called Native American, We have been called hostile, We have been called Pagan, We have been called militant, We have been called many names.... The callers of names cannot see us, but we can see them.
> –Song: We Are the Halluci Nation, Artist: A Tribe Called Red

Simply put, indigenous peoples are people who descended from ancestors who were already living in North America prior to Columbus's arrival in 1492. The people of the First Nations have been labeled and called many names since European colonization, and many result from a misunderstanding and lack of knowledge of historical events and the effect on present-day indigenous peoples. How the first peoples got to North America is a subject of great debate (Bernardo et al. 2017), which will not be discussed further due to its cultural sensitivity (Goodsky Jr., 2020, personal communication). What we do know for sure is that the people of the First Nations adapted, thrived, and evolved on the North American landscape for more than 15,000 years (Adovasio et al. 1978) and live among us to this day in every region of the United States.

Prior to 1492, North America was unrecognizable compared to how it looks today. Trees that had lived for hundreds of years grew thick dying of old age, fish and wildlife were abundant and plentiful, prairies were burned for thousands of years to the point in which plants evolved, and people were free to move and live as desired as dictated by existing tribes and civilizations (McCann 1999). At this time, North America was populated, with potentially more people living in North America than Europe. Estimates range from study to study, but many scholars now estimate the pre-Columbian population in North America to be between 75–100 million (Thornton 1987). Cahokia is a good example of an ancient city that was located across the Mississippi River from modern-day St. Louis, Missouri. At Cahokia's peak from 1050–1200 A.D., the city had between 10,000 and 20,000 residents, making it bigger than London at the time (Jarus 2018). Cahokia also had expansive trade routes, which reached to Wisconsin, the Gulf of Mexico, Illinois, Oklahoma, as well as other areas in the current southern United States (Emerson and Lewis 1991).

In 1492, the Columbian exchange introduced new diseases to North America, many of which spread from tribe to tribe before many indigenous people even met a European (Treuer 2013). The exchange of diseases, as well as acts of genocide by Columbus and other representatives of colonizing governments (Bigelow 1992), precipitated more than a 95% decline of the indigenous population within 100 years (Treuer 2013). Despite the decimation, the indigenous population persisted and started to grow again by the early 1900s (Thornton 1987). Currently, there are an estimated 6.7 million indigenous peoples living in the United States alone (US Census 2017). Tribes today have invested in modern economies, political systems, and agricultural practices and are spread throughout urban and rural areas across the country.

2 History of the First Nations

> When Indian incomes are level with yours, when our schools are as good as yours, our houses as warm, our kids as safe and our woods and streams clean as yours, when our babies first open their eyes to as bright a future as yours, then we we'll talk about level playing fields. Whether out of racism or out of ignorance, there are always some who will go after Indian self-determination and economic development in ways as old as Columbus, as bold as Custer and as devious as any federal land grabber.
>
> —Biidaabanookwe (Arrival of Speech Woman), Marge Anderson, Mille Lacs Band of Ojibwe Ogimaa (Chief Executive) from 1991 to 2000 and from 2008 to 2012

Prior to American colonization and the arrival of Christopher Columbus, there were over 500 Native languages and 300 unique indigenous nations found in North America (Treuer 2013), while today only 169 Native languages exist, with only 20 having more than 2000 speakers (US Census Bureau 2010). During this time period, modern conceptions of race did not yet exist and people discerned friend from foe by language, culture, and geographic area. After 1492, the British, Russians, French, and Spanish increasingly arrived seeking furs, minerals, timber, and people to enslave (Pevar 2012). Natives, meeting with European traders, usually were eager to trade goods (Venables 2004). Metals, guns, axes, and pots were highly sought after, as they greatly reduced labor efforts that came with living a subsistence lifestyle, and were acquired through the trade of furs by the first peoples (Thomas et al. 1993). The desire to have western goods brought indigenous peoples into closer contact with European immigrants and traders (Kay 1984). As a result, the French, British, Spanish, Russians, and later Americans would see the resources held and owned by the indigenous peoples and want them in their entirety (Fig. 1). The demand for natural resources would bring about profound changes on the landscape, which is still seen to this day (Treuer 2015). In 1874, the mining of the South Dakota

"First, walk into the waters of the lake and die; second, march aimlessly into the wilderness never to return, and third, fight for the amelioration of these wrongs"

Fig. 1 Chief Mee-Gee-Zee of Mille Lacs (1901) describing the choices his people had when all of their lands were stolen and they were "trespassing" on their own land. Photo courtesy of the Minnesota Historical Society

Black Hills began, despite resident Lakota not agreeing to opening their lands to mining and having agreed upon treaties (Ostler 2010). In 1877, the Black Hills Act opened their lands up to mining, and thousands of gold prospectors came from all over the country (Lazarus 1991). Today, hundreds of abandoned mines have been found to have been leaking arsenic into the water supply and valley where most hunting, fishing, and gathering are done. As a result, tribal members who are closer to arsenic sources in the Black Hills are suspected to have higher rates of autoimmune disease (Ong et al. 2014). Nationwide, tribal members living near abandoned mines have an increased likelihood of kidney disease, hypertension, diabetes, and more (Lewis et al. 2017).

During the 1600s and early 1700s, the current United States was sparsely populated by European immigrants (Haines and Steckl 2000). Most of these immigrants at first immigrated to North America to make money off the fur trade (Dolin 2010). Eventually, other sources of income and promises of land would draw even more European immigrants to the "new world" (Dinnerstein and Reimers 2009). By the time of the American Revolution in 1776, Native land loss through theft and violence had already begun (Treuer 2013). By the early 1800s, mass European immigration to the Midwest and West Coast began putting even more demand on indigenous lands, confining Native peoples to an even smaller landmass (Wedll 1985). Settlers arriving in places like the Midwest found themselves in need of farmable land, even though much of it already belonged to other immigrants or tribal peoples. Such was the demand for farmable land that many wetlands were trenched in an attempt to drain them (Treuer 2015). The attempts were usually not successful, and in Minnesota and Wisconsin Ceded Territories (Fig. 2), over 77% of white homesteaders/farmers failed and went into tax forfeiture (Treuer 2015). Timber, fish, mineral, and fur companies would eventually use all their resources to steal Native land to gain profit from the natural resources on and within native lands. Treaty fraudulence, forced removal by law enforcement agencies and timber companies, arson of Native villages, and an exclusion of representation in the United States government chipped away at Native lands, but not as fast as the US government wanted by the early and mid-1800s (Wedll 1985; Routel 2013).

Beginning in 1830, the Indian Removal Act was passed, which ordered the forced relocation of tribes from the southeastern United States to lands west of the Mississippi River (Stewart 2007). The Trail of Tears is the most commonly known example. On the Trail of Tears, the Cherokee were forced to the west, far away from their homelands. Of the Cherokee that left, 25–35% would perish (Sturgis 2007) from starvation, disease, and harsh travel conditions. Other tribes, such as the Creek, Chickasaw, and Choctaw, would also be removed and subjected to similar treatment (Cave 2003). The survivors of the removals would settle in Oklahoma and other western states, where their reservations are found in the modern day (Stewart 2007). In 1851, the "Indians Appropriations Act" would set aside land for Natives to live on and would create the modern-day reservation (Fig. 2; Kladky 2013). The goal of reservations was to, first, take the natural resources from tribal people and, second, destroy tribal cultural and societal norms in the process.

Once concentrated on reservations, tribal leadership was replaced by US "Indian agents," who were generally corrupt and undermined tribal chiefs' authority and traditional governance structures (Talbot 2006). One such governance structure that was severely impacted for many tribes was the clan system (Treuer 2015). Within Ojibwe and many tribal societies, clans determined leadership roles/eligibility, tasks and jobs to do, who to marry, and more (Gagnon 2012). The clan system stabilized indigenous communities as well as helped grow the culture. At the time, though, many officials in the United States federal government were determined to abolish it and Native culture altogether (Talbot 2006). In 1882, the Bureau of Indian Affairs (BIA) introduced a list of "Indian Offenses" that would criminalize Native cultural practices (Philp 1973). Following the creation of reservations, Native children were

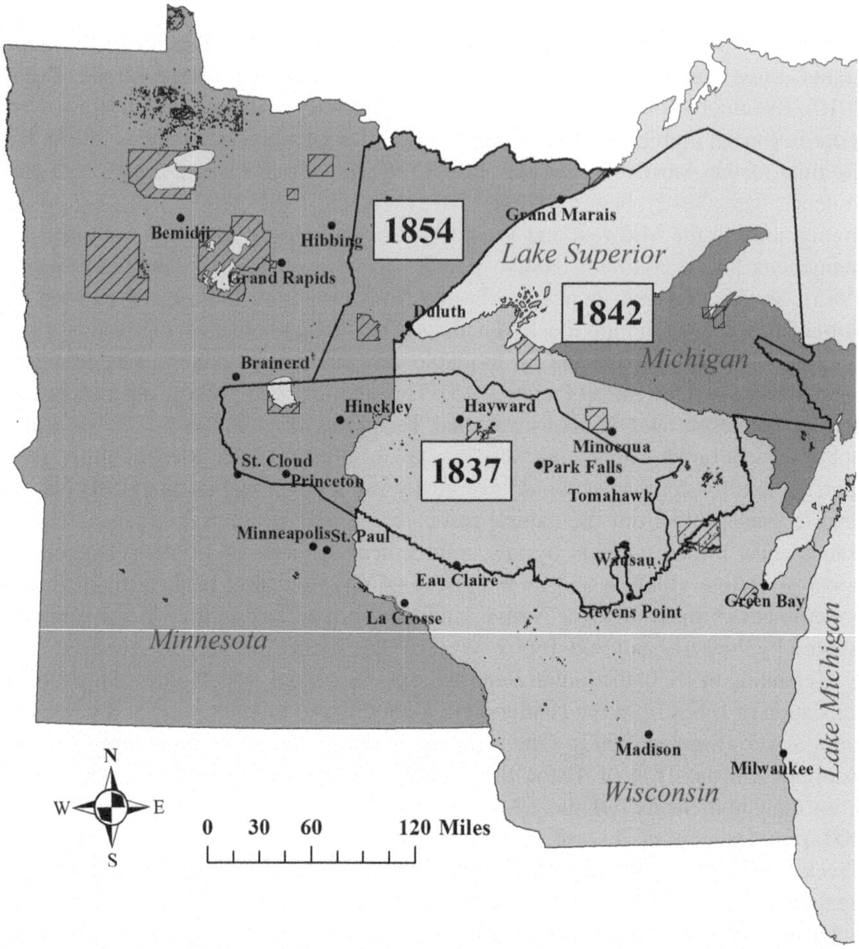

Fig. 2 Map of Minnesota, Wisconsin, and Michigan 1837, 1842, and 1854 Ceded Territories. Ceded Territories are represented by double black lines and contain a box with the year in which a treaty was made. Current reservation land is represented by slanted black lines. Much of the dots and darker clusters are small reservation land plots

abducted to be sent to boarding schools, which were strife with physical, emotional, and sexual abuse. This would be called the residential boarding school era and would last from 1860 to 1978. In Minnesota, 35% of Anishinaabe children were separated or stolen from their families and placed into foster care (Treuer 2010). To this day, the trauma seen by those who went to residential boarding schools continues to cause substance abuse, mental health issues, crime, and cultural loss (Smith 2004; Bombay et al. 2014). No reparations have been considered on behalf of the United States today despite it negatively impacting generations of Natives who never experienced the residential boarding schools firsthand (Pember 2019). Educational limitations were also placed on Natives to prevent them from attaining higher education, Christian missionaries tried to eliminate traditional Native religious and spiritual beliefs, and on some reservations, the USA government tried and succeeded at starving people to death by eliminating big game, taking more land, and reducing/eliminating annuity payments (Tinker 1993; Irwin and Roll 1995; Wishart 2007).

To makes matters worse for the indigenous peoples, a new congressional act would solidify the American government's desire to eliminate the tribes. The Dawes Act of 1887, passed without tribal consent (Treuer 2015), privatized, illegally taxed, and effectively disposed of almost all the land held in tribal possession (Fig. 1; McLaughlin 1996). It would not be until 1934, with the Indian Reorganization Act (IRA), that Native land received any sort of protection and recognition (Kelly 1975). The IRA would aim to reverse the goal of assimilation for Native peoples by protecting tribal customs and culture as well as encouraging tribal sovereignty (Stricker et al. 2020). Due to the IRA, tribes could draft their own constitutions as well as organize under a common government, even though many of the newly made tribal constitutions were foreign (American) in concept (Deloria and Lyttle 1983). Many smaller tribes, families, and clans were also organized under a common government, which was not a traditional concept (Taylor 1980). As a result, there were and still are Natives who criticized and still criticize the IRA, with tension still existing due to this issue within some tribes (COIA 2010). Many historians, though, still argue that without the IRA, tribes would not exist today in the United States (Washburn 1984).

After 1934, however, attempts by the United States government to reduce Native land, rights, and protections continued in earnest. In 1954, the United States began the "policy of termination," which provided for the ability to eliminate the legal recognition of tribes through executive administrative procedures (as opposed to congressional action) (Treuer 2010). The main goal of the policy was assimilation and to eliminate the unmet fiscal obligations the federal government had to the tribes (Wilkinson and Biggs 1977). While many tribes avoided the policy before it was discontinued, several tribes were severely affected and were no longer recognized by the United States. While some have since reobtained their recognition, some tribes have not, with the fight continuing to this day (Treuer 2012a). Other major reforms would not begin again until the 1960s and 1970s. Despite Natives receiving citizenship in 1924, it would not be until 1962 that all Native Americans in the United States could vote (Roche 2019). Several congressional acts would also be passed that protected Native children from adoption into non-Native families as well as Native burial grounds (George 1997).

In today's modern times, many tribal members have mixed feelings about reservations. The passage of the Indian Gaming Act in 1988 brought wealth to many tribes over the next decade and transformed their reservations (Evans and Topoleski 2002). Still, for many tribes, the opening of casinos has not alleviated all the problems, with only around 40–50% of tribes profiting from casinos (Treuer 2010). There are tax breaks that Natives are eligible for, but Natives usually still need to pay federal and state income tax as the tax exemptions are for revenues only earned from the reservation (BIA 2021). Given that an estimated 22% of Natives live on the reservation, the tax benefits are usually nonexistent (HHS 2021). Poverty, cultural loss, land loss, and lack of employment opportunities remain a problem to this day on many reservations despite major progress being made over the last few decades. This progress includes new community centers, wastewater treatment facilities, government centers, tribal language schools, tribal colleges, and a wide variety of nongaming businesses (NCAI 2020). For some tribal members, the reservation brings up hurtful memories and feelings, while other tribal members view the reservation as a place of pride and would never want to live anywhere else (Treuer 2012b).

3 Sovereignty of First Nations

Tribal sovereignty remains misunderstood by mainstream American society and is a subject of confusion among many nontribal governmental institutions. Most are unaware that geographic regions of the United States belonged to other nations and the US had to purchase and negotiate for land (Kukla 2003). Tribes considered themselves part of their own nations as they had existed before the creation of the United States. In the 1830s, a series of cases were ruled in tribal peoples' favor and later became known as the Marshall Trilogy, named after Supreme Court Justice John Marshall. In the decision, John Marshall ruled that tribes are sovereign nations that are not subject to state or county laws and have control over their internal affairs (Prygoski 1995). In addition, these decisions and answers became known as the Federal Indian Trust Responsibility. The Federal Indian Trust Responsibility is a legal obligation under which the United States agreed to the protection of Native lands, assets, resources, treatys, and other recognized rights (Jewell 2014).

Despite the Marshall Decision, many laws and acts have been passed in an attempt to eliminate and weaken tribal sovereignty. Some examples are the Major Crimes Act of 1885 and Public Law 280 (Trafzer 2009), with both allowing for non-Native law enforcement officers to prosecute tribal citizens on tribal lands. In fact, Public Law 280 allows for non-Native law enforcement agencies to make arrests within reservation boundaries and has led to lowered income levels and increased crime in tribal communities (Dimitrova-Grajzl et al. 2014). Today, tribal sovereignty has improved, with many tribes having greater jurisdiction over land/water within their boundaries. Many tribal nations are able to set and enforce laws/ordinances, establish a court system, as well as utilize, protect, develop, and manage their natural resources. Despite these efforts and improvements, tribal sovereignty can be limited or under threat by corporations, laws, and permits and at times by local, state, and federal governmental entities (Pevar 2015; Lhamon 2018; Marohn 2018). One of the biggest challenges that remain is for nontribal people and agencies to understand tribal sovereignty, accept it, and recognize it. For the Ojibwe in Minnesota, Wisconsin, Michigan, Montana, and other states, though, they still view sovereignty as meaning tribes have complete control over internal affairs and the right to use their lands to hunt, fish, and gather and that sovereignty exists independent of other governments (Isham and Zorn 2015).

4 Treaties with the First Nations

Treaties are the supreme law of the land
—US Constitution Article 6

The image of the indigenous person has changed from time to time. Two stereotypes emerged over the centuries: the peaceful native and the opposite of peaceful. In truth, the indigenous peoples are a mixture of both, but both peace and violence did not occur without reason and were often strategic (Treuer 2015). The indigenous peoples historically exercised their own military power. They occupied and survived in North America by having extensive knowledge of the landscape, control over the river trade routes, superior fighting skills, Tribal Ecological Knowledge (TEK; knowledge of the ecosystem acquired over generations), and advanced agricultural practices (Turner 1897; McCann 1999). Even into the 1800s, indigenous peoples defending their homelands were a strong force to be contended with. For example, in 1889, the Battle of Sugar Point on the Leech Lake Reservation in Northern Minnesota was won by less than 20 Ojibwe warriors when over 100 US soldiers came to arrest Chief Bug-O-Nay-Ge-Shig (Hole-In-The-Day). There were no Ojibwe casualties, but seven US soldiers were killed and 16 were wounded. The US soldiers eventually retreated to Walker, MN, and Chief Bug-O-Nay-Ge-Shig escaped (Duoos 2020).

Despite being a powerful military force, most tribes chose not to fight Americans and other colonizers. The complex web of different ethnicities and political alliances that existed meant that many tribes had alliances with the Americans, French, and British, while many other tribes did not (O'Brien 1989). The common identification and cohesion of "Native Americans" had not yet taken form, with fighting still occurring at times between tribes. Still, tribal peoples fought in every war to assist the United States, including the American Revolution, World War I, and World War II. In World War II, Navajo US soldiers known as "Code Talkers" passed messages to one another in their Native language, which the enemy could not decode, to help defeat the axis forces (Aaseng 1992; Treuer 2012a). For most European immigrants, reliance on tribal people was the only way to survive the harsh North American landscape (Wedll 1985). Many tribal and nontribal people went about their ways without ever having conflict. For many European immigrants, their experience with tribal people was the opposite, as it is portrayed in American propaganda/history (Wedll 1985).

Still, there were attacks on European settlers from time to time (Nunnally 2007). Most of these attacks were not unprovoked and were exaggerated and used as propaganda by politicians for purposes of fearmongering (Treuer 2015). To ensure that settlers would be able to homestead safely and to ensure American companies exploited and took indigenous lands, the United States government entered into a treaty period with tribes, which lasted from 1778 to 1871 (Venables 2004). Under these treaties, tribes would negotiate in exchange for their land, reservations, payments, government benefits, and the most well-known treaty-reserved rights.

Treaty rights are known as usufructuary rights in the American legal system (Spangler 1997). In the United States, when you sell a parcel of land, you can put into the sales agreement that you and your family can hunt and fish forever on that parcel you sell. This is essentially what many tribes did during the treaty rights period, though on a much larger geographic scale. Many of these reserved rights differed by area, as well as if they were explicitly stated in the treaties. The Midwest Ojibwe tribes signing treaties in 1836, 1837, 1842, and 1854 had their right to hunt, fish, and gather on Ceded Lands, clearly stated in the terms of the treaties. For example, Article 5 of the 1837 Treaty of St. Peters states that “the privilege of hunting, fishing, and gathering the wild rice, upon the lands, the rivers and lakes included in the territory ceded, is guarantied (sic) to the Indians, during the pleasure of the President of the United States” (Kappler 1904). For many other tribes that signed treaties, these reserved rights were not explicitly mentioned, and therefore they still do not have their treaty rights recognized (Enger 2016).

After 1871, the recognition of treaty rights faded, were forgotten by newly established states and territories, and were inevitably not recognized. Tribal members were prosecuted by local and state authorities for exercising their right to hunt, fish, and gather (Wedll 1985). Starting in the early 1960s, tribal members in the Pacific Northwest and upper Midwest began using the legal system in an attempt to make federal courts affirm their reserved rights. In 1964, Robert Satiacum and Billy Frank Jr. organized tribal “fish-ins,” in which tribal members defied state law to fish with the intent of getting arrested (Wilkinson 2000). The arrests from these fish-ins would lead to the 1970 case *US v. Washington* to determine if the tribal members arrested did in fact have reserved rights (Brown 1994). In 1974, United States District Court Judge George Boldt concluded that reserved hunting, fishing, and gathering rights did exist for many Northwest tribes (Ferguson 1998). The Boldt Decision would pave the way for other tribes to have their treaty rights affirmed, especially for those in the Midwest.

Around the same time the “fish-ins” on the Northwest Coast were occurring in the early 1960s, Ojibwe in Michigan and Wisconsin were also violating fish and game laws to have their treaty rights recognized (Wilkinson 2000). Using similar tactics as the Northwest tribes, Ojibwe in Michigan got arrested, fought in court, and eventually had their treaty rights affirmed in 1979, which is known as the Fox Decision (McGruther 1999). In 1974, a Lac Courte Oreilles member named Fred Tibble violated Wisconsin fish and game laws and took the case to a Federal Circuit Court of Appeals (Loew 1997). After 9 years of litigation, the courts upheld his claim, ruling that his treaty rights existed. Since then, many tribes have fought successfully and unsuccessfully to have their treaty rights affirmed. It would not be until 1999 with *Mille Lacs vs Minnesota* that the US Supreme Court would rule that many Midwestern tribes did in fact have treaty rights due to treaties signed in 1837, 1842, and 1854 (Fig. 2; GLIFWC 2020).

To this day, tribal members exercising their treaty rights remain controversial within the white community. In Wisconsin, shortly after fishing treaty rights were first exercised in 1984, centuries of hatred, racism, and ignorance were unleashed. Hate groups including Stop Treaty Abuse, Equal Rights for Everyone, and Proper Economic Resource Management (PERM) began to harass and terrorize tribal members that were exercising their treaty rights (Wilkinson 1991). Slurs such as "Timber N-words," "Save two Walleyes spear a pregnant Squaw," and many others were chanted and yelled at tribal members trying to exercise their rights. Many times the situation would turn violent with Ojibwe being murdered, physically assaulted, or shot at or having their nets cut and destroyed (Nesper 2002). After treaty rights were affirmed for tribes in the Pacific Northwest (after 1974), similar incidents were common as well (Sherman 2020).

In 1994, the antitreaty right protest movement in Wisconsin, Minnesota, and Michigan was diminished by federal Judge Barbara Crabb, who ruled in a suit brought by the American Civil Liberties Union that the harassment of tribal harvesters was illegal and racially motivated (Nesper 2002). The 1994 ruling did help curtail the masses of racists at the boat landings protesting Walleye (*Stizostedion vitreum*) harvest, but it did not stop harassment and violence entirely. To this day, gunfire, harassment, and property damage still happen (LDF 2019; Lafond 2019) across the upper Midwest Ceded Territories as well as in the Pacific Northwest (Mulier 2006). In Elk River, Minnesota, PERM, the last remaining treaty rights hate group still has members who commit criminal offenses, some of which go unreported or are never investigated further (Koskinen 2017).

Despite the controversy and negativity, for those tribes that did get their treaty rights affirmed, it opened the door to what is now known as "Co-Management." Many court orders mandated that tribal and state governments work together to preserve shared natural resources (Pinkerton 1989) both on and off many reservations. Today, this includes harvest-level setting, fishery management plans, data sharing, and collaborative research between state and tribal governments. Due to the immense workload associated with comanagement, many tribes have now created their own natural resource departments.

5 Modern-Day Tribal Fisheries Management

Over the years, tribal governance structures have changed dramatically, and many tribes have invested resources in fisheries management practices. In 2018, over 15 million fish were raised and stocked by Midwestern tribes (GLIFWC 2019). Tribes on the West Coast also have many hatcheries and raised over 30 million fish in 2017 alone (NTT 2017). There are hundreds of additional fisheries management activities undertaken by tribes as well, and these include acoustic telemetry studies, fish habitat remediation, fish passage projects, improved fisheries regulations, conducting fisheries surveys, and more (USFWS 2013). Many of these projects are already having positive impacts on shared user fisheries, which include tribal restocking to restore crashed fish stocks, reducing fishing pressures on Salmon fisheries, removing dams, being able to develop better management plans through more frequent fish recruitment surveys, and gaining a better understanding of fish behavior and habitat preferences through fish tagging studies (Bougher 2006; Conder et al. 2008; Gilbert et al. 2019; Nairn 2020).

In addition to having fisheries departments, tribes that have reserved rights also have a complex and large amount of regulations for their tribal members to abide by. For example, the 1837 Treaty of St. Peters signatory tribes have to abide by a set of regulations called the 1837 Ceded Territory Conservation Code. This code requires permits for tribal fish harvesters, mandated harvest reporting, equipment regulations, season dates, as well as specific regulations for certain water bodies (GLIFWC 2016). Ojibwe harvesting fish at Mille Lacs Lake with gill net and spear have every single fish counted and weighed so harvest data is accurate (Erickson 2003). There are also requirements between the states and the 1837 signatory tribes, many of which are outlined in Federal Protocols that were determined after the Supreme Court ruled in favor of the tribes (Krogseng 2000). For the Ojibwe tribes in Wisconsin and Minnesota that harvest fish within the 1837 Ceded Territory, this includes mandated reporting to state fisheries departments when gill nets or spears are used, state/nontribal monitors at boat landings where tribal harvest is taking place, mandatory notification and permits when stocking on and off reservation land, and restrictions on species that tribal people can take (MNDNR 2020). In Michigan, five Anishinaabe tribes, within the 1836 Ceded Territory of Michigan, entered into the 2007 Inland Consent Decree in order to avoid going to trial with the state over the disagreement on whether or not tribal members could fish in Lake Superior and other inland lakes according to their own tribal regulations. This decree placed prohibitions on the tribal harvesting of fish with gill nets and snagging. It also imposed some state season restrictions and mandatory harvest reporting, despite language to the contrary in the 1836 Treaty of Washington. Commonly, tribes have complex processes for changing rules and regulations, which may include altering agreed-upon state and tribal management plans, having agreements at interagency task forces, presenting proposed regulations to state agencies, and sometimes federal agency approval. People need to realize that many of the state compacts/agreements with tribes have violated tribal sovereignty, and future professionals must work to correct that. Many decrees, protocols, and agreements can be modified with the use of memorandaum of understanding, committee consensus, and other legal means.

The Northwest tribes that retained treaty rights after the Boldt Decision found themselves in a similar situation as the Ojibwe tribes in the Midwest but with a different fish species: Salmon (Clark 1985). Northwest treaty tribes today also have regulations, which are set by individual tribes, but are also coordinated by the Pacific Fisheries Management Council (PFMC), which is a combination of tribal, state, and federal personnel that work together to set fisheries regulations and harvest limits (NWIFC 2020). After the Boldt Decision in 1974, the Northwest Indian Fisheries Commission (NWIFC) was created to assist tribes in biological surveys, legal matters, interagency collaboration, and more (NWIFC 2020). In the Midwest, the Great Lakes Indian Fish and Wildlife Commission (GLIFWC) was created in 1984 by 11 Ojibwe tribes, which functions/objectives are the same as the NWIFC (Oberly 2014). In addition, these Ojibwe tribes belong to the 1837 Ceded Territory Fisheries Technical Committee (FTC), which is also a collaboration between state and tribal agencies to set regulations and Walleye harvest throughout the 1837 Ceded Territory in Minnesota (FDL 2020). Today, both GLIFWC and NWIFC commonly coordinate, work, and communicate with one another (GLIFWC 2013). Both agencies, in addition to providing excellent biological services, also provide employment opportunities to tribal members, engage in community outreach programs, publish cultural-educational documents, host youth events, as well as fund ceremonies (Neumeyer 2015; Chase 2016; Boyd 2017). The future is bright for tribal natural resource management, and as more tribes develop their individual fisheries departments, the potential for jobs in the fisheries field will expand, giving many Natives and non-Natives experience in an already competitive field.

6 Contrasting Views on Natural Resources

The earth and its resources belong of right to its people.

The western natural resource view is very distinct, but yet it has become the status quo. For most professionals, the western view is so engrained in their outlook that they are not aware of it. When contrasted with Native natural resource views, the differences become apparent. The two views are not competing with one another but rather contrast and can be complimentary. It is this contrast that has fueled disagreements between tribes and outside entities at times. By understanding both points of view, not only will it prepare future fisheries professionals with working for tribes, but it can also assist those who already work with tribes.

The present-day western view found in America has deep roots in European history dating back to the middle/medieval ages. European feudalism existed between the tenth and thirteenth centuries (Cartwright 2018). During this time, land was given to nobles by the monarch or ruler. Many peasants worked these farms for very low or no pay and rarely ever traveled more than 25 miles from their home manor (Tuchman 1978). It would not be until the Black Death (Bubonic Plague) that the feudal system would collapse (Hilton 1976). Still, though, the lands remained in the possession of the wealthy, while the peasant population was restricted from owning land as well as hunting game.

The lack of land ownership in the Middle Ages had a long-lasting impact on the European people. By the time European immigrants started to arrive in North America, many of Europe's natural resources had been eliminated (Hoffmann 2005). Generations of European immigrants had worked as peasants, and now their descendants wanted their own land, free to own and farm. Upon arrival in America, many immigrants saw something that had been denied to them for all their lives: land (Treuer 2015). These settlers were especially influenced by concepts of Manifest Destiny, which was the belief that settlers were to claim all the North

Fig. 3 Gifford Pinchot, the first head of the US Division of Forestry, considered the father of the western conservation movement

"The earth and its resources belong of right to its people"

American land under the name of god and capitalism (Mountjoy 2009). The quest was on to settle and stake claims for European immigrants even though the land they were staking claim to was not theirs. Over the next century, fences would come up and land would be privatized (Henig 1989). Eventually, the American landscape would look similar to medieval Europe's, with the new immigrants creating a quasi-feudal system, a system that they came to America to escape.

Another factor that has played a role in the western view is the influence of Christianity. Christianity is believed to have played a role in the western viewpoint as it does contain concepts of human superiority to animals and natural resource domination (White 1967). Both of these factors rely on a common opinion: that Earth's natural resources are for man's purpose and use. The western natural resource point of view is that humans are unique, special, and different from other animals (Fig. 3; Coates 2013). In regard to fisheries, professionals and other people holding western views believe that fisheries must be managed for humans. They also believe that left alone, fisheries cannot manage themselves.

7 Native Natural Resource View

> Everything in the world talks, just as we are now- the trees, rocks, everything. But we cannot understand them, just as white people do not understand Indians.
>
> —Nomlaki tribal member, California, date unknown

When thinking about the Native natural resource point of view, it is first important to remember that there are many different tribes across North America. Many tribes' views have evolved over the ages and can differ greatly between one another. For this next section, views on the natural resources are talked about generally, with some focused on Ojibwe views. Exceptions to these views will be found in almost every tribe, and beliefs can vary greatly from tribal member to tribal member.

A key contrast between the western and Native natural resource viewpoints is man's role in the environment. Most tribal people believe that human beings are a part of nature and not separate from it (Thomas et al. 1993). For example, the Ojibwe believe that mankind cannot survive without the other elements put in place, such as fire, water, air, plants, medicines, and animals. But there is humility needed as all these other elements can survive without mankind (Isham and Zorn 2015; Applegate 2020). In fact, almost all tribes have ceremonial processes when an animal needs to be harvested. Examples of these processes include praying before and after a kill, conducting specific ceremonies before harvesting, and following preestablished intratribal norms on how to harvest animals (Treuer 2012a).

For most tribes, the harvest of an animal still is a spiritual process (Treuer 2012a). The act of taking an animal is a spiritual activity that can be compared in religious magnitude to taking communion in Christian faiths. Many tribal people believe that animals can in fact communicate with them as they also have spirits (Pomedli 2014). According to Ojibwe beliefs, the animal will offer itself in a sacrificial way out of pity for man so that he may eat. This is the way designed by the creator (Applegate 2020). In Ojibwe belief, many objects that western society considers not to be living are in fact alive and have a spirit. For example, rocks, trees, birds, and plants all have spirits and therefore are referred to as animate in the Ojibwe language (Goose 2020). For this reason, many tribal people pray to animal spirits, have ceremonies, as well as have animal clan names and artwork. All is to honor the animal spirits that provide food and also help teach Native people how to survive (Applegate 2020).

Another key aspect is how many Natives view land. Land also can have a spirit and is commonly prayed to. While many tribes fought over land-use rights, for the most part, land ownership was a foreign concept prior to 1492 in North America (Isakson and Sproles 2008). Mille Lacs Band of Ojibwe elder Terry Kemper stated that you cannot own land, trees, or other human beings. When fences came up after land privatization, many Ojibwe, including Terry Kemper, were horrified. The Native natural resource belief to this day is that land cannot be owned to the point where complete exclusion of others occurs (Wedll 1985). In the modern era, there usually is an exception for non-Natives on tribal land. The taking of Native land has left many deep scars, and therefore many tribes are protective of who utilizes their land in the modern era (Brown-Rice 2013). At tribes like the Mille Lacs Band of Ojibwe, tribal members cannot own any land but lease rights to land owned by the tribal government. For non-Natives who get permission to access the tribal land, a permitting system has been devised. Many other tribes have adopted similar systems to regulate modern-day tribal land usage.

Finally, many tribes believe that the land, fisheries, and ecosystems can be left alone to manage themselves. Much of this stems from the fact that tribes did live sustainably on the landscape of North America for thousands of years (Treuer 2013). To many tribes, the systems that they had in place worked, and as long as nature was undisturbed and the overharvesting of resources did not occur, the earth would keep providing (Smith 2009). Elder Natalie Weyaus stated in a Mille Lacs Band Department of Natural Resources meeting that once humans leave the earth alone and stop trying to manage it, it would heal itself. To this day, biodiversity of plants and animals remain higher on tribal lands compared to nontribal lands (Schuster et al. 2019).

8 ABCs of a Tribal Employee

Generally, this next section is written for nontribal members who are interested in working for a tribe, although the techniques outlined here can be applied to any constituency, culture, or people. On many reservations, there is a tight-knit community and culture. Much of this culture can be withheld from outsiders, as it has been exploited in the past, as well as in the present, and was almost destroyed by outside influences. Because of this and historical trauma, nontribal members need to be extra patient, open-minded, and willing to learn and grow.

Learning the specific tribe's culture is one of the most important things to do as a nontribal employee. Culture defines people's beliefs, rituals, as well as day-to-day practices (Keesing 1974). Not understanding cultural differences can lead to many problems when first starting employment. The best thing to do is to make an attempt to learn the tribe's culture you work for in its entirety. Reading books, talking to tribal members, attending ceremonies (if invited), and going to community activities open to the public, such as the Pow Wow, are great ways to learn. The learning process can take years, and may never stop, but when you give learning an effort, tribal members will know. Learning a specific tribe's language is also another great thing to do. It shows respect to the tribe and willingness to learn. In addition, much of tribal culture and history that was lost due to colonization is still preserved within tribal languages. For example, the sacred Dakota area called "Bdote," meaning where the two rivers meet, is a reference to the joining of the Mississippi and Minnesota rivers in Minnesota (DeCarlo 2017). Being able to teach this culture and history from the language as well as communicate to people in their language can go a long way.

Humility and respect are two important things to have when working for a tribe. When employed by a tribe, there will be many times that you make mistakes with work. Being able to be corrected and not dictate tribal culture will make more tribal members interested in teaching you. Respecting those around you even if they do not show respect for you initially is key. By showing respect, respect will be reciprocated. Lastly, many tribal employees may need to have thick skin. Many tribal members are dealing with many personal and work-related stresses. Discomfort, poor performance, and anger directed at you often do not have anything to do with you specifically. Showing respect and taking little offense are usually some of the best actions tribal employees can take should this type of situation arise. Many tribes also have teachings that have been passed down from elders and have been incorporated into many tribal human resource departments. For Ojibwe people, these are called the seven grandfather teachings, which include Gwayakwaadiziwin (Honesty), Debwewin (Truth), Inendizowin (Humility), Zaagi'idiwin (Love), Nibwaakaawin (Wisdom), Zoongide'iwin (Courage), and Manaaji'idiwin (Respect).

9 Managing Tribal Employees

When managing tribal employees, it is important to remember that it is not much different than managing nontribal employees off the reservation. Many tribal governments are set up similar to the US federal government, and therefore many of the techniques can be cross-applied. Despite similarities, there are several considerations that may be unique to your tribal employees that managers should be aware of.

Cultural considerations are some of the most important considerations there are. For the Ojibwe, putting out tobacco ("Asemaa" in Ojibwe) as an offering to the Manidoo (creator) before electroshocking fish or setting nets must be completed prior to undergoing those activities. Many tribes also have certain times of the year for different ceremonies, and managers will most likely have to accommodate employees attending cultural functions at some point. These cultural functions could include Pow Wows, wild rice camps, fish camps or harvest events, first-kill ceremonies, and more. Understanding and accommodating cultural considerations can motivate employees and reduce workplace tension. Accommodating cultural functions also helps teach the culture to employees who may not be as knowledgeable. Cultural training workshops are also another great way to implement tribal culture into the workplace.

Economic as well as personal considerations may have to be made as well for tribal employees. For tribal members who are employed with tribal governments, stresses associated with living on or near a reservation do not end once the workday is over. Many reservations across the United States live in economically depauperate areas, and so even with steady pay, many tribal member employees may be supporting other family members or have other unknown needed expenditures. Transportation remains one of the top barriers inhibiting many tribal members from getting out of poverty, and you may have to work with employees to get them to work. Be prepared for unique situations and to accommodate tribal employees when they have economic and personal obstacles to overcome. If training can alleviate some of the problems, then take the time to train your employees. Consider sending employees back to school, to related job-mentoring positions, or to technical training sessions to boost their confidence and skill set.

10 Phases of a Tribal Employee

Working on the reservation can be rewarding, fun, and a learning experience, but it does not come without its challenges. Below we will discuss the four phases of a tribal employee. These phases were identified by the author by observing and talking to tribal employees, friends, and coworkers for years, as well as based on significant input from Kelly Applegate, Director of Resource Management, at Mille Lacs Band of Ojibwe, who has worked on the Mille Lacs Reservation since 2002. By understanding these phases, managers can improve employee retention and prepare employees for working on a reservation.

10.1 Optimism Phase

The first phase that almost all tribal employees will go through is optimism. Many individuals who do work for a tribal government do so because it is more than a job for them. Many tribal members decide to work for the tribal government to make a change and help their people (Nayquonabe 2019). Many non-Native allies will also try their best in tribal governments to make positive changes. Whether the employee is Native or not, they all will enter this first phase.

In the optimism phase, the employee who has just recently been hired has a lot of energy and eagerness to work. The employee usually has lots of ideas, goals, and plans on how they are going to make a positive impact. Idealism is a sign that an employee is in this stage, and many times the idealism is unrealistic unbeknownst to them. Part of the problem is that even if an employee may know their culture and the people in the community, they do not have experience in government work or in really working on the reservation for the most part. They do not have the connections or technical skills or are unfamiliar with day-to-day government work. At this stage, managers need to be supportive of the employee and channel their goals and ideas into concrete work projects. Managers should work to keep the employee upbeat while getting them to think in more realistic terms.

10.2 Frustration Phase

The frustration phase is a phase almost every tribal employee will go through. The frustration phase follows the optimism phase usually after a series of failures. These include unsuccessful projects and/or failed attempts at making a change. In this phase, the tribal employee's beliefs and goals are being challenged. Many tribal employees start to clearly see the obstacles to being successful on reservations, as well as the flaws in many of their ideas. While the employee is gaining experience, they still need more experience and to be more flexible to achieve the success they desire. Manager support is critical at this stage. If managers start to see employees undertake projects that are unrealistic or unsuccessful, then they need to step in to reprioritize their employees' activities. Managers should also help with projects so that not every undertaking fails. A few small successes for tribal employees and managers can go a long way in keeping workplace morale up and preventing frustration from escalating further.

10.3 Burnout Phase

If the frustration phase is not mitigated, then the employee will enter the burnout phase, which is the riskiest position for an employee to be in. When this phase occurs, it may or may not be apparent at first. Typically, employees will react in one of two ways in this phase. The first way is doubling down on work. The series of failures (and sometimes successes) will build to the point where an employee undergoes lots of work stress. Instead of taking a break, the employee will work twice as hard and twice as long. Sometimes this does work in producing success but usually will come at the price of the employee's mental health. Many employees will actually take on more work during this phase as they are determined to be successful, working overtime and outside of work hours. Unfortunately, this will reduce work performance over time and create a situation where the employee may quit from the stress.

The other response is the "do-nothing" response. Instead of doubling down, the employee essentially gives up trying to be successful. In both of these cases, pessimism runs high. Many employees in this position have realized that success and change on the reservation are difficult with many unforeseen barriers impeding them. At times it can seem for these employees that nothing can be changed without experiencing great difficulty. They are right to some degree, but the stress will usually not allow them to see the solutions.

Once an employee has entered the burnout phase, quitting is almost inevitable, and it is hard to bring people back to perform well once they have burned out. The only way to prevent burnout is through close managerial oversight to make sure the employee is not overwhelmed. Managers should force the employee to take a vacation when they enter this phase. Support for the employee and positivity work, and if managers provide structure, employee loss can be prevented.

10.4 Acceptance Phase

The acceptance phase is the phase of success. If an employee manages to make it past the previous phases, then most likely they have changed something in their work routine. Reprioritization goes a long way at this phase. Success cannot come overnight on the reservation, and change can be slow and frustrating. Therefore, by making more realistic goals, and reprioritizing what needs to be done, the tribal employee will eventually achieve success.

By this phase, almost all of the employee's ideas and workload have changed. They still may have the goals they originally had when first starting to work for tribes, but they realize it may take longer than expected or may require a different approach. Employees will start to get more experience in the tribal government by this point. Procedures, rules, and lessons learned from the past will translate into some undertakings being successful. Most who enter this acceptance phase will come to rely on support from other tribal employees, which will form valuable connections, especially in departments and agencies other than the one they are in. These employees have the potential to be long-term employees. They have the potential to eventually make positive change and be successful like they wanted to. Managers at this point need to provide support, but only if they need the support. Managers should focus on making these tribal employees happy at the workplace as their happiness will eventually manifest itself into a positive change on the reservation.

11 Fishery Regulation Proposal

Proposing fisheries regulations in tribal communities can be tricky due to several factors. First, western fisheries regulations and science were not a part of decision-making for most of the indigenous people's history. For many tribes, members regulated their own harvest using firsthand experiences and TEK (and still do to this day). Even if a specific lake has a bag limit of ten, many tribal members will not take the full amount (Kalk 2019). "Take only what you need" is a mantra that is commonly found among different tribes. For this fact, managers need to consider whether or not a regulation is often even necessary. Often, regulations will make little or no impact as there are already unspoken regulations that tribal members have enacted.

Situations will arise, though, where there is a need for regulation. Some tribal members may be dissatisfied with a certain fish population in a lake. This usually happens when extreme change occurs to the system, such as a major shift in size structure, lack of certain species of fish, or aquatic plant habitat destruction. If the lake or river is not in balance both spiritually and biologically, then you will probably have to propose a fishery regulation. This can be contentious if not done correctly, with sensitivity and with great respect for individual tribal members' rights. New regulations should have a demonstrated ability to affect the desired change, and other management strategies that minimize or eliminate the need for regulation should be considered.

There cannot be too much communication with the tribal members over new regulations. A potential first step to proposing a fishery regulation is holding fishery community meetings. Community meetings will allow tribal members to address the problem and suggest regulations. If there is a particular regulation you want to impose, then make it known to tribal members well in advance of these meetings. Posting any fishery regulatory changes in tribal newspapers and on social media sites will help inform people who cannot attend community meetings. A PowerPoint presentation can also help convey the biological need and can be done using Zoom© and other video-sharing platforms if in-person attendance is not possible. If there is confusion over the biological need, make sure and spend time to explain. Always clearly define the biological need for the regulation and explain simply.

When proposing a fishery regulation for a tribe, make sure to never make assumptions. A mistake made by the author was to assume that Mille Lacs Band of Ojibwe tribal members wanted a ten Black Crappie (*Pomoxis nigromaculatus*) a day bag limit and a three Walleye a day bag limit on a newly restored impoundment that was opened to fishing 3 years after being restocked. Instead of this proposal, many tribal members wanted a limit of zero Walleye a day until spawning was successful, and some members wanted the Black Crappie bag limit reduced. Managers must realize that sometimes proposed fishery regulations on a tribal lake may have to encompass other tribal beliefs. For most tribal members, the health of a lake is more important than the harvest of fish.

Any proposed regulations from tribal members need to be taken with the utmost seriousness. If any proposed regulations go against cultural beliefs or are not accepted by the tribal community, then a "no regulation" option must be considered. Sometimes spiritual beliefs, cultural customs, and TEK, the knowledge that tribal members acquired through generations of observations of animals and the natural world (Schmidt and Stricker 2010), will not always be conveyed to you. For example, scientific surveys in 1977 indicated that the Bowhead whale population was under 1000. But Inuit whale hunters said that the numbers were underestimating the population due to their in-depth knowledge of Bowhead whale behavior. After scientists took TEK into consideration, new survey methods were developed and found the population to be closer to 8000 (Schmidt and Stricker 2010). It is equally important as the intended biological impacts of regulation propositions to consider TEK, spiritual beliefs, and cultural customs. The treaty rights to hunt, fish, and gather belong to the tribe you work for, and their decisions need to be implemented based on their will.

12 Fishery Regulation Implementation

Once the community meetings have concluded and there is a general consensus that a regulation does need to be implemented, there are a couple of important steps managers should take. The first step is to write a draft plan and send it to coworkers and tribal leadership for peer review. As more people review the plan, comment, and edit, the risk of the community not accepting the plan will go substantially down. If you are non-Native, then many times tribal coworkers will notice something that could be misconstrued a certain way or find cultural problems. The extra set of eyes from a coworker can mean the difference between a successful and unsuccessful fishery regulation proposal. Another effective method to reduce the chances of tribal community members being dissatisfied with your fisheries regulations and decisions is to conduct a survey to quantify tribal community opinions. When quantifying community opinions, make sure to use a variety of methods to survey tribal members. These methods include sending out paper surveys, creating online surveys, taking opinion polls at community meetings, and talking to tribal elders. By quantifying community opinions on proposed regulations, consensus can be met and justification for your decisions will be defendable.

When the time comes to implement fishery regulations, realize that many tribes have the legal authority to set their own regulations and each tribe may have a specific set of legislative representatives, review boards, commissions, and/or administrative paths required for policy change. The pathways to approve policy change will differ between tribes, but understanding the internal specific governance structure of the tribe you work for will be critical to implementing any fishery regulation. Many times tribal managers do not understand the sovereignty that the tribe they work for has. If possible, try and set the regulations internally within the tribe. At times, though, this may not work as there may be a need to coordinate with other state agencies. Lakes that are shared by both state and tribal users will most likely require communication and coordination with state agencies. For some tribes, when the treaties were adjudicated, the courts mandated regulation coordination with non-Native natural resource agencies. Other tribes also volunteer to participate in advisory committees that are typically made up of tribal and nontribal citizens, scientists, and fishing guides/resort owners (USFWS 2021). One such group, called Northern Inland Lakes Citizens Fishery Advisory Committee (NILCFAC), reviews fishery surveys, recommends management goals, as well as provides policy/management inputs from tribal members who are on the committee (MIDNR 2021). Tribal involvement on these committees can help reduce community tension and misunderstanding between tribal and state users and help promote a more guided management approach. When in doubt, check with tribal leadership and other tribal natural resource employees/managers on how to proceed.

Coordination with non-Native agencies and other tribes will often be required not just for implementing tribal fisheries regulations but also for undertaking other fisheries projects and activities. For example, in 2005, a diet study of Double-crested Cormorants (*Phalacrocorax auritus*) on Leech Lake was initiated by the Leech Lake Band of Ojibwe and completed with funds from the United States Fish and Wildlife Service (USFWS). The project had many non-Native collaborators and partners, including the Minnesota DNR, University of Minnesota, USFWS, US Department of Agriculture, and Leech Lake Association (LLBO 2004). Having a diverse number of partners can be beneficial as the whole community and other professionals can work together to make a difference and make projects more impactful. Many grantors like to see this collaboration, so project proposals to external funders typically are more likely to be chosen for funding, which have non-Native agencies collaborating. The public relations benefits to tribes that choose to collaborate with nontribal entities are high and can transform the non-Native public opinion as it allows them an inside view on how tribes study and manage their natural resources. Posting successes on social media sites such as Facebook and Instagram only helps to increase transparency and highlight project successes.

While coordination with other agencies has its benefits, it can also have its risks. In Minnesota, the construction of the Sandpiper Pipeline, now known as Line 3, required the Minnesota Public Utilities Commission (MNPUC) to consult with impacted tribes in order to draft an Environmental Impact Statement (Bishop 2020). Despite the tribes providing a wealth of data and information, most was left out of the final document and in the decision-making process. In addition, by the MNPUC consulting with the tribes, the MNPUC could say that their tribal consultation was finished and there was nothing the tribes could do about the proposed pipeline (TCA 2018). Other times, grant or federal money can come with stipulations. The Natural Resources Conservation Service and many other federal agencies require that projects are maintained up to 10 years after completion, and if they are not, they have the ability to cancel contracts and demand payment to recoup losses (NRCS 2018). Tribal managers must evaluate collaboration by individual agencies and make sure all of the details of any agreement or project are clear and known beforehand. Success is more probable with improved communication with external funders before, during, and after the funding application process.

There are also times where collaboration with other agencies or projects funded with federal dollars may be extremely difficult to obtain. Many federal agencies require state and federal permits before funding is awarded. As a result, one agency denying a permit can prohibit many other federal agencies from working with you. Many grantors also require matching funds to initiate a project, which may be prohibitive to tribes that restrict match funds or do not have them, to begin with. Working with the federal government and other agencies can also be time-consuming and painstaking. Forms to apply for federal funds can contain hundreds of pages, and many tribal governments simply do not have the administrative staff or time to be able to apply to multiple grants. Lastly, many times grant funding for tribes is competitive, so there may not be enough funds for all the tribes applying. All of these are just some of the barriers to collaboration with outside agencies. Tribal managers will have to find unique ways around these barriers or find different sources of funds/collaboration.

Just because there are risks and work associated with collaborating with non-Native agencies does not mean tribal managers and employees should avoid them entirely. As tribal managers, one of our first tasks is to protect the tribe against any outside influence or policies that are detrimental to tribes. Tribal managers need to pay extra attention to details when applying for external funds or working with outside agencies. Having budget tables, scopes of work, and detailed proposals will help keep collaborators on track and prevent confusion on projects. Taking detailed notes, recording meetings, and having agreements in writing all help to prevent other agencies from using information against you or from speaking for your tribe. State and federal policy will have to be reviewed to make sure it does not negatively impact the tribe or its sovereignty. Throughout their careers, tribal managers will have to review these policies, some of which may have negative impacts on the tribe, which may not be apparent at first. Great communication should be had with tribal leaders when questions about specific outside agency policy intentions are questioned.

13 Tribal Community Meetings

Tribal community meetings on fisheries issues should be held regularly to keep tribal members informed on your department's progress. They can also provide valuable direction on what projects managers should undertake. When wanting to hold a community meeting, make sure to advertise in all the ways that are possible. Hang up flyers in common areas, make social media posts, personally invite elders and other tribal members, and send out mass emails/texts. Advertise weeks in advance, on the week of the meeting, and on the day of the meeting.

Incentivizing meetings is also another great way to increase attendance. Free traditional food is a great way to turn people out to your meetings, especially elders. At the end of the meetings, raffles and prizes are also a good idea as they will keep people listening and attending. Finally, games for kids before or after the meeting can also increase turnout and provide youth an experience to learn and have fun. Having a presentation or a video can also help tribal members get interested and participating in a discussion.

When giving a presentation at a tribal meeting, make sure to keep things "short and sweet" and have plenty of pictures and maybe even a video. All people, regardless of culture, will retain less information the longer the lecture/presentation is (Savoy et al. 2009). Too much information can drag the meeting on and spawn distracting side discussions. The shorter the presentation is and the longer the discussion with tribal members is, the better. As a tribal employee holding one of these meetings, you are a moderator. You are there to moderate discussions to accomplish objectives. Always let everyone speak and do not speak for the tribal members present. If things get tense or people seem bored, do not be afraid to take breaks and resume the meeting after.

14 Managerial Tips for Working on a Reservation

When employees first start out as a manager, it can be important to remember to start small when it comes to projects. It is important that new managers achieve some small successes before taking on much bigger projects. Fisheries project ideas that are inexpensive and attainable and can make an impact include building fishing docks, holding kids and community fishing days, and holding youth fisheries workshops (Fig. 4). All ideas, big or small, need to have one thing in common: they are for the good of the community. By making a positive impact on the community, your department will get more support and more tribal members will benefit.

Fig. 4 *Gii-pagida'waawag*—"They set a net." Mille Lacs Band of Ojibwe fisheries technician Cameron Weous showing tribal students how to set a fyke net. The fyke net contained Walleye, which were stripped of eggs then fertilized and placed in a hatchery. This was a successful youth involvement idea

Funding for full-time staff positions can remain hard to come by, depending on which reservation people may find themselves working on. If staff is needed, there are several outlets that tribes can pursue. The United States Department of Labor has a program for Natives in which it will fund them to work for you as well as provide financial assistance for employees, equipment for departments, and assistance to employees in obtaining their driver's licenses. Grant writing skills can help you achieve goals that would be financially untenable by the tribe alone. Grant funding may be available from state, federal, nonprofit, or private sources, and creativity with a particular project can unlock those resources. Another great way to get help on a budget is to hire tribal youth interns. Tribal youth interns are a great source of help and offer tribal youth the opportunity to obtain in-field experience. It provides them structure and is a great way to get youth career oriented.

15 Staying Positive

There is no denying the challenges that come with working or living on a reservation. Many shows and movies depict the reservation as crime ridden and entrenched in poverty. The reality is that this depiction is not the full story and is often exaggerated. What is never portrayed is the hardworking, heroic, and good people who make up the reservation. On many reservations, grassroots sobriety movements are growing greatly in size. Native and non-Native teachers, against all odds, teach Native children their Native language and culture. And around Mille Lacs Lake in Central Minnesota, people as well as the author stop to help stranded motorists and give people rides when needed. Employees must look at the positive things happening around them as well. Looking forward and keeping your head up is crucial. Meeting and associating with positive/hardworking people will help managers and tribal employees. Due to the personal nature of working for tribal governments, working on the reservation can be a life-changing and rewarding experience. Employees can expect to make friends that last a lifetime, obtain extensive knowledge about the environment, be proud of community successes, and obtain a new outlook on life. Even though times may be hard, Native people never forget to laugh, and neither should employees. The reservation can be a funny place, and funny stories retold help tell the culture and history as well as get tribal members through hard times. The humor, positivity, and stories of the people who help run the reservation are some things few people are lucky to know.

The story of the reservation is one of American history and is rarely ever told in its entirety. Hopefully, this chapter reaches those who need to learn about indigenous peoples' history and culture. Non-Natives should not shy away from the notions of social/economic reparations or be consumed with guilt from the mistreatment of Native peoples. The conversations that need to be had will cause discomfort for non-Natives, but it is only through these uncomfortable discussions that progress can be made. Serious and immediate action needs to be taken by non-Natives to repair the past damage, so Natives and non-Natives can begin the healing process. If you think you cannot make a difference, you are wrong. Many healing ceremonies and events have already begun in Minnesota and across the country. One such event, which is open to the public, is called Mikwendaagoziwag (We remember them). Here, non-Native collaborators, allies, and students canoe side by side with Natives to remember the Sandy Lake Tragedy, an event in which the US government delayed annuity payments and food, causing over 400 Ojibwe to perish in the winter of 1850 (Ring 2017). In 2016, the ceremony had less than 100 attendees, but organizers kept working hard. In 2019, the ceremony had close to 1000 participants, all which danced and took part in different ceremonies. As the ceremony continues to grow, so does the healing every year. We owe it to our future generations of children and country to begin that healing process. Regardless of your past beliefs, you can become part of the solution. Start listening to what Native Americans tell us and then become a person who changes history for the better.

Acknowledgments This chapter uses well-known literature cited and also contains many personal accounts that helped the author learn over the years. Some of these individuals may have not been cited, but here we give special thanks to all those who helped: Terry Kemper, Kelly Applegate, Harvey Goodsky Jr., Cameron Weous, Natalie Weyaus, Susan Klapel, Larry Smallwood, Lee Staple, Michelle Goose, Tanya Aubid, Colin Cash, Mic Isham Jr., Jim Zorn, Joe Dan Sr., GLIFWC representatives and staff, Katie Draper, Bradley Harrington, and many more countless tribal members who I have spoken to. Without the eagerness of tribal members to help the author learn, the ability to write this chapter would have been impossible.

References

Aaseng N (1992) Navajo code talkers. Walker and Company, New York

Adovasio JM, Gunn JD, Donahue J, Stuckenrath R (1978) Meadowcroft Rockshelter, 1977: an overview. Am Antiq 43(4):632–651

Applegate K (2020) Mille Lacs Band Department of Natural Resources, personal communication

Bernardo DV, Neves WA, Kipnis R (2017) The origins project and the first Americans' controversy. Archaeological and Paleontological Research in Lagoa Santa:137–212

BIA (2021) Frequently asked questions. Bureau of Indian Affairs. Available: https://www.bia.gov/frequently-asked-questions

Bigelow B (1992) Review: once upon a genocide: Christopher Columbus in Children's literature. Soc Justice/Global Opt 19:106–121

Bishop L (2020) Findings of fact, approval of MPCA Certification under Section 401, Line 3. Minnesota Pollution Control Agency. Available: https://www.pca.state.mn.us/sites/default/files/Final-401-Findings-of-Fact-for-EnbridgeLine-3-Project.pdf

Bombay A, Matheson K, Anisman H (2014) The intergenerational effects of Indian residential schools: implications for the concept of historical trauma. Transcult Psychiatry 51(3):320–338

Bougher B (2006) Treaty tribes have helped cut salmon harvest by 84 percent. Northwest Indian Fisheries Commission. Available: https://nwifc.org/treaty-tribes-have-helped-cut-salmonharvest-by-84-percent/

Boyd L (2017) They are remembered at big Sandy Lake. Inaajimowin 2017:13

Brown JJ (1994) Treaty rights: twenty years after the Boldt decision. University of Minnesota Press 10(2):1-16

Brown-Rice K (2013) Examining the theory of historical trauma among native Americans. Professional Counselor 3(3)

Cartwright M (2018) Feudalism. Ancient history encyclopedia. Available: https://www.ancient.eu/Feudalism/. October (2020)

Cave AA (2003) Abuse of power: Andrew Jackson and the Indian removal act of 1830. Historian 65 (6):1330–1353

Chase SM (2016) GLIFWC awarded $10,000 grant for mini documentary series. Ashland Daily Press. Available: https://www.apg-wi.com/ashland_daily_press/news/glifwc-awarded-10000-grant-for-mini-documentary-series/article_9c64e990-a6f0-11e6-b916 3b535a741893.html

Clark WG (1985) Fishing in a sea of court orders: Puget Sound salmon management 10 years after the Boldt decision. N Am J Fish Manag 5(3B):417–434

Coates P (2013) Nature: western attitudes since ancient times. Wiley, Hoboken
COIA (2010) The Indian reorganization act–75 years later: renewing our commitment to restore tribal homelands and promote self-determination. Committee on Indian Affairs United States Senate. Available: https://www.govinfo.gov/content/pkg/CHRG112shrg68389/pdf/CHRG-112shrg68389.pdf
Conder T, Hand D, Olsen D, Lovtang J (2008) Distribution and survival of adult hatchery spring chinook salmon radio-tagged and released upstream of warm springs national fish hatchery in 2008 progress report. United States Fish and Wildlife Service. Available: https://fws.gov/columbiariver/publications/Radio_Telemetry_Paper.pdf
DeCarlo P (2017) Fort Snelling at Bdote: a brief history. Minnesota Historical Society Press, St. Paul
Deloria V, Lyttle C (1983) 1934: President Franklin Roosevelt signs the Indian Reorganization Act. American Indians, American Justice. Available: https://www.nlm.nih.gov/nativevoices/timeline/452.html. December (2020)
Dimitrova-Grajzl V, Grajzl P, Guse JA (2014) Jurisdiction, crime, and development: the impact of public law 280 in Indian country. Law & Society Review 48(1):127–160
Dinnerstein L, Reimers DM (2009) Ethnic Americans: a history of immigration. Columbia University Press, New York
Dolin EJ (2010) Fur, fortune, and empire: the epic history of the fur trade in America. WW Norton & Company, New York
Duoos K (2020) On this day in history: the battle of sugar point. Leech lake band of Ojibwe. Leech Lake News Available: https://www.leechlakenews.com/2018/10/05/onthis-day-in-history-the-battle-of-sugar-point/. December (2020)
Emerson TE, Lewis BR (1991) Cahokia and the hinterlands: middle Mississippian cultures of the Midwest. Illinois Historic Preservation Agency, Chicago
Enger J (2016) Explaining Minnesota's 1837, 1854, and 1855 Ojibwe treaties. Minnesota Public Radio News. Available: https://www.mprnews.org/story/2016/02/01/explainingminnesota-ojibwe-treaties. December (2020)
Erickson S (2003) Bands enjoy fair weather, good fishing at mille Lacs. Mazina'igan: a Chronicle of the Lake Superior Ojibwe. Great Lakes Indian Fish and Wildlife Commission. Summer (2003)
Evans WN, Topoleski JH (2002) The social and economic impact of Native American casinos National Bureau of Economic Research: 91-98
FDL (2020) Fisheries-spearing and netting. Fond Du Lac Band of Lake Superior Chippewa, Fisheries Program. Available: http://www.fdlrez.com/RM/spearnet.htm. December (2020)
Ferguson K (1998) Indian fishing rights: aftermath of the fox decision and the year 2000. American Indian Law Review 23(1):97–154
Gagnon GO (2012) Ojibwe Culture. The Wiley-Blackwell Companion to Religion and Social Justice 62:425–437
George LJ (1997) Why the need for the Indian child welfare act? J Multicultural Soc Work 5 (3-4):165–175
Gilbert J, Rose JD, Luehring M, Ray A, Shultz A (2019) Fishery status update in the Wisconsin treaty ceded waters. U.S. Department of Interior, Bureau of Indian Affairs. Available: http://www.glifwc.org/publications/pdf/FisheryStatus2019.pdf. December (2020)
GLIFWC (2013) Treaty rights: sustaining a way of life. Great Lakes Indian Fish and Wildlife Commission. Available: https://www.glifwc.org/publications/pdf/jointtreatyrightspub.pdf. December (2020)
GLIFWC (2016) Model off-reservation code for the 1837 ceded territory. Great Lake Indian Fish and Wildlife Commission. Available: https://www.glifwc.org/Regulations/MN37ConservationCode.2016.pdf. December (2020)
GLIFWC (2019) 2018 tribal hatchery production. Great Lakes Indian Fish and Wildlife Commission. Available: https://www.glifwc.org/Mazinaigan/Spring2019/Hatcheryinsert.pdf. October (2020)

GLIFWC (2020) Treaty rights recognition and affirmation. Great Lakes Fish and Indian Wildlife Commission. Available at: https://www.glifwc.org/Recognition_Affirmation/.October (2020)
Goodsky Jr H (2020) Mille Lacs Band of Ojibwe Department of Natural Resources, personal communication
Goose M (2020) Fond Du Lac Tribal and Community College, personal communication
Haines MR, Steckel RH (2000) A population history of North America. Cambridge University Press, Cambridge
Henig JR (1989) Privatization in the United States: theory and practice. Political Science Quarterly 104(4):649–670
HHS (2021) Profile: American Indian/Alaska Native. U.S. Department of Health and Human Services Office of Minority Health. Available: https://minorityhealth.hhs.gov/omh/browse.aspx?lvl=3&lvlid=62#:~:text=As%20of%20017%2C%20there%20wereof%20the%20total%20U.S.%20population. January (2021)
Hilton R (1976) March feudalism and the origins of capitalism. Hist Work J:9–25
Hoffmann RC (2005) A brief history of aquatic resource use in medieval Europe. Helgol Mar Res 59(1):22–30
Irwin MH, Roll S (1995) The psychological impact of sexual abuse of native American boarding-school children. J Am Acad Psychoanal 23(3):461–473
Isakson HR, Sproles S (2008) A brief history of Native American land ownership. In: Indigenous peoples and real estate valuation:63–75
Isham Jr MJ, Zorn JE (2015) Tribal sovereignty and natural resource management relationship of people to place. Great Lakes Indian Fish and Wildlife Commission. Available: https://www.fws.gov/sfbpc/doc/Zorn_SFBPC_Winter_2015.pdf. October (2020)
Jarus O (2018) Cahokia: North Americas First City. Live Science. Available:https://www.livescience.com/22737cahokia.html#:~:text=The%20pre%2DColumbian%20settlement%20at,living%20there20at%20its%20peak.&text=Cahokia%20was%20a%20city%20that,many%20European20cities%2C%20including%20London. January (2021)
Jewell S (2014) Order no. 3335: reaffirmation of the Federal Trust Responsibility to federally recognized Indian tribes and individual Indian beneficiaries. US Department of Interior. Available: https://www.doi.gov/sites/doi.gov/files/migrated/news/pressreleases/upload/Signed-SO3335.pdf. December (2020)
Kalk J (2019) Mille Lacs Band of Ojibwe Tribal Harvester, personal communication
Kappler CJ (1904) Treaty with the Chippewa, 1837. Indian Affairs: Laws and Treaties 2:491–492
Kay J (1984) The fur trade and native American population growth. Ethnohistory 31:265–287
Keesing RM (1974) Theories of culture. Annu Rev Anthropol 3:73–97
Kelly LC (1975) The Indian reorganization act: the dream and the reality. Pac Hist Rev 44 (3):291–312
Kladky WP (2013) Federal Reservations. Encyclopedia of American Indian Issues Today 2:444
Koskinen A (2017) Dayton faces protesters' blockade in fishing expedition. Alphanews. Available: https://alphanewsmn.com/dayton-faces-protesters-blockade-fishing-expedition/. October (2020)
Krogseng K (2000) Minnesota v. mille Lacs band of Chippewa Indians. Ecol Law Quart 27:771–797
Kukla J (2003) A wilderness so immense: the Louisiana purchase and the Destiny of America. Anchor Books, New York
Lafond K (2019) Tribal Citizens say harassment affects how they hunt, fish. Michigan Radio. Available: https://www.michiganradio.org/post/tribal-citizens-say-harassment-affects-how-they-hunt-fish. December (2020)
Lazarus E (1991) Black Hills White justice: the Sioux nation versus the United States, 1775 to the present. University of Nebraska Press, Lincoln
LDF (2019) Non Native Fires Gun at Tribal Harvesters on Lac du Flambeau Reservation. LacDu Flambeau Tribal Communications Office. Available: https://www.nativeknot.com/news/

Native-American-News/NonNative-Fires-Gun-at-Tribal-Harvesters-on-Lac-du-Flambeau-Rese.html. December (2020)
Lewis J, Hoover MKD (2017) Mining and environmental health disparities in native American communities. Current Environmental Health Reports 4(2):130–141
Lhamon CE (2018) Broken promises: continuing federal funding shortfall for native Americans. U.S. Commission on Civil Rights. Available: https://www.usccr.gov/pubs/2018/12-20-Broken-Promises.pdf. December (2020)
LLBO (2004) Assessment of double-crested cormorant food habits and predation effects on selected fish species in Leech Lake. Leech Lake Band of Ojibwe Division of Resources Management. Tribal Wildlife Grants Program Application FY 2004
Loew P (1997) Hidden transcripts in the Chippewa treaty rights struggle: a twice told story. Race, resistance, and the politics of power. American Indian Quarterly 21:713–728
Marohn K (2018) Mille Lacs County, Band of Ojibwe settle dispute over policing. Minnesota Public Radio. Available: https://www.mprnews.org/story/2018/09/19/mille-lacs-county-band-of-ojibwe-settle-dispute-over-policing. December (2020)
McCann JM (1999) Before 1492: the making of the pre-Columbian landscape: part 1: the environment. Ecol Restor N Am 17:15–30
McGruther F (1999) Michigan's 1836 treaty fishery guide. Chippewa Ottawa treaty fishery management authority. Available: https://www.glifwc.org/publications/pdf/1836TreatyFisheryGuide.pdf. December (2020)
McLaughlin MR (1996) The Dawes act, or Indian general allotment act of 1887: the continuing burden of allotment. A selective annotated bibliography. Am Indian Cult Res J 20(2):59–105
MIDNR (2021) Northern Inland lakes citizens fishery advisory committee. Michigan Department of Natural Resources. Available: https://www.michigan.gov/dnr/0,4570,7350-79137_79763_79903_85163%2D%2D-,00.html. January (2021)
MNDNR (2020). Mille Lacs lake court decision documents: mille Lacs Band v. State of Minnesota. Stipulation Exhibit D Protocols 1-5. Available: https://www.dnr.state.mn.us/millelacslake/mille-lacs-lake-court-decision documents.html. December (2020)
Mountjoy S (2009) Manifest Destiny: westward expansion. Infobase Publishing, New York
Mulier V (2006) Recognizing the full scope of the right to take fish under the Stevens treaties: the history of fishing rights litigation in the Pacific northwest. Am Ind Law Rev 31(1):41–92
Nairn C (2020) Largest dam removal project in U.S. History restarts with interstate partnership. EcoWatch Environmental News. Available: http://www.glifwc.org/publications/pdf/FisheryStatus2019.pdf. December (2020)
Nayquonabe P (2019) Mille Lacs Band of Ojibwe employee, personal communication
NCAI (2020) Tribal nations and the United States: an introduction. National Congress of American Indians. Washington, D.C.
Nesper L (2002) The walleye war: the struggle for Ojibwe spearfishing and treaty rights. University of Nebraska Press, Lincoln
Neumeyer K (2015) Video: tribe keeps tradition alive with baker sockeye fishery. Northwest Indian Fisheries Commission. Available: https://nwifc.org/video-tribe-keeps-traditionalive-with-baker-sockeye-fishery/. December (2020)
NRCS (2018) Conservation program application. Natural resources conservation service, US Department of Agriculture. Available: file:///C:/Users/keith/Downloads/Blank_EQIP_CCC1200.pdf. December (2020)
NTT (2017) Tribes released more than 34 million Hatchery Salmon. Northwest Treaty Tribes. Available: https://nwtreatytribes.org/tribes-released-more-than-34-million-hatcherysalmon/. October (2020)
Nunnally ML (2007) American Indian wars: a chronology of confrontations between native peoples and settlers and the United States military 1500's-1901. McFarland & Company, Inc., London
NWIFC (2020) About us. Northwest indian fisheries commission. Available: https://nwifc.org/about-us/. December (2020)
O'Brien S (1989) American Indian tribal governments. University of Oklahoma Press, Norman

Oberly JW (2014) Glifwc: the founding and early years of the great lakes Indian fish and wildlife commission. Indigenous Perspectives of North America: A Collection of Studies 32

Ong J, Erdei E, Rubin RL, Miller C, Ducheneaux C, O'Leary M, Pacheco B, Mahler M, Henderson PN, Pollard MK, Lewis JL (2014) Mercury, autoimmunity, and environmental factors on Cheyenne River Sioux tribal lands. Autoimmune Diseases Volume 2014

Ostler J (2010) The Lakota's and the Black Hills: the struggle for sacred ground. The Penguin Group, New York

Pember MA (2019) Death by civilization. The Atlantic. Available: https://www.theatlantic.com/education/archive/2019/03/traumatic-legacy-indianboarding-schools/584293/. January (2021)

Pevar SL (2012) The rights of Indians and tribes, 4th edn. Oxford University Press, New York

Pevar SL (2015) Dollar general's attack on tribal sovereignty. American Civil Liberties Union. Available: https://www.aclu.org/blog/racial-justice/american-indian-rights/dollargenerals-attack-tribal-sovereignty. December (2020)

Philp K (1973) John collier and the crusade to protect Indian religious freedom, 1920-1926. J Ethnic Stud 1(1):22

Pinkerton E (1989) Co-operative Management of Local Fisheries: new directions for improved management and community development. University of British Columbia Press, Vancouver

Pomedli M (2014) Living with animals: Ojibwe Spirit powers. University of Toronto Press, Toronto

Prygoski PJ (1995) From Marshall to Marshall. The Compleat Lawyer

Ring MJ (2017) Sandy Lake tragedy: too long forgotten, a delayed commemoration. University of Wisconsin–Eau Claire, Doctoral Dissertation

Roche P (2019) Native American voting rights: two steps forward, one step Back. Am Univ J Gend Soc Pol Law 28:91

Routel C (2013) Minnesota bounties on Dakota men during the US-Dakota War. William Mitchell Law Review. Available:https://heinonline.org/HOL/LandingPage?handle=hein.journals/wmitch40&iv=5&id=&age=. February (2021)

Savoy A, Proctor RW, Salvendy G (2009) Information retention from PowerPoint™ and traditional lectures. Comput Educ 52(4):858–867

Schmidt PM, Stricker HK (2010) What tradition teaches: indigenous knowledge complements Western wildlife science. USDA National Wildlife Research Center-Staff Publications 1283

Schuster R, Germain RR, Bennett JR, Reo NJ, Arcese P (2019) Vertebrate biodiversity on indigenous-managed lands in Australia, Brazil, and Canada equals that in protected areas. Environ Sci Pol 101:1–6

Sherman E (2020) Indigenous fishing rights in the Pacific Northwest are being threatened. Culture. Available: https://matadornetwork.com/read/indigenous-fishing-rights-pacificnorthwest/. December (2020)

Smith A (2004) Boarding school abuses, human rights, and reparations. Social Justice 31:89–102

Smith BD (2009) Resource resilience, human niche construction, and the long-term sustainability of pre-Columbian subsistence economies in the Mississippi River valley corridor. J Ethnobiol 29 (2):167–183

Spangler GR (1997) Treaty fisheries in the upper Midwest. J Contemporary Water Res Edu 107 (1):11

Stewart M (2007) The Indian removal act: forced relocation. Compass Point Books, Minneapolis

Stricker H, Schmidt PM, Gilbert J, Diana LJD, Crider D, Hoagland S, Kohl MT, Perez CA, VanDaele LJ, Van Daele MB, Dupont D (2020) Chapter 42: managing north American indigenous People's wildlife resources. In: Silvy NJ (ed) The wildlife techniques manual. The John Hopkins University Press, Maryland, pp 288–304

Sturgis AH (2007) The trail of tears and Indian removal. Greenwood Press, Westport

Talbot S (2006) Spiritual genocide: the denial of American Indian religious freedom, from conquest to 1934. Wicazo Sa Review 21(2):7–39

Taylor GD (1980) The new Deal and American Indian tribalism: the Administration of the Indian Reorganization act, 1934-1945. University of Nebraska Press, Lincoln

TCA (2018) Tribes ask PUC to reconsider review of new Enbridge pipeline route, saying cultural study wasn't done. SC Times. Available: https://www.sctimes.com/story/news/2018/01/04/tribes-ask-puc-reconsider-review-newenbridge-pipeline-route-saying-cultural-study-wasnt-done/1005322001/. December (2020)
Thomas DH, Miller J, White R, Nabokov P, Deloria PJ (1993) The native Americans. An Illustrated History. Turner Publishing, Inc., Kansas City
Thornton RA (1987) American Indian holocaust and survival: a population history since1492. University of Oklahoma Press, Norman
Tinker GE (1993) Missionary conquest: the gospel and native American cultural genocide. Fortress Press, Minneapolis
Trafzer CE (2009) American Indians/American presidents: a history. Harper Collins Publishers, New York
Treuer A (2010) Ojibwe In Minnesota. Minnesota Historical Society Press, St. Paul
Treuer A (2012a) Everything you wanted to know about Indians, but were afraid to ask. Minnesota Historical Society Press, St. Paul, MN
Treuer D (2012b) Rez life: an Indians journey through reservation life. Atlantic Monthly Press, New York
Treuer A (2013) Atlas of Indian nations. National Geographic Society, Washington D. C
Treuer A (2015) Warrior nation. Minnesota Historical Society Press, St. Paul
Tuchman BW (1978) A distant Mirror: the calamitous 14th century. Ballantine Books, New York
Turner FJ (1897) New light on the early history of the greater northwest. The manuscript journals of Alexander Henry, fur trader of the northwest company, and David Thompson, official geographer and explorer of the same company, 1792–1814. Am Hist Rev 3(1):157–159
US Census (2017) American Indian and Alaska Native Heritage Month: November 2017. 2017 US Census Newsroom. Available: https://www.census.gov/newsroom/facts-for features/2017/aian month.html. October (2020)
US Census Bureau (2010) The 2006-2010 ACS 5-Year summary file technical documentation. American Community Survey Office. Available:https://www2.census.gov/acs2010_5yr/summaryfile/ACS_2006-2010_SF_Tech_Doc.pdf (January 2021)
USFWS (2013) Summary of projects supported by the U.S. fish and wildlife service. United States Fish and Wildlife Service. Available:https://www.fws.gov/NativeAmerican/pdf/twg-projects-summary-2007-2012.pdf. October (2020)
USFWS (2021) Advisory committees. United States Fish and Wildlife Service. Available: https://www.fws.gov/pdm/faqsac.html. January (2021)
Venables RW (2004) American history: five centuries of conflict and coexistence, conquest of a continent 1492-1783, vol 1. Clear Light Publishing, Santa Fe
Washburn WE (1984) A fifty-year perspective on the Indian reorganization act. Am Anthropol 86 (2):279–289
Wedll D (1985) Against the tide of American history: the story of the Mille Lacs Anishinabe. The Minnesota Chippewa Tribe, Cass Lake
White L (1967) The historical roots of our ecologic crisis. Science 155(3767):1203–1207
Wilkinson CF (1991) To feel the summer in the spring: the treaty fishing rights of the Wisconsin Chippewa. Wis Law Rev 375
Wilkinson CF (2000) Messages from Frank's landing: a story of salmon, treaties, and the Indian way. University of Washington Press, Seattle
Wilkinson CF, Biggs ER (1977) The evolution of the termination policy. Am Ind Law Rev 5 (1):139–184
Wishart DJ (2007) Encyclopedia of the Great Plains Indians. University of Nebraska Press, Lincoln

Can You Hear Me Now? Design Considerations for Large Lake, Multispecies Telemetry Projects

Aaron Shultz, Carl A. Klimah, Jocelyn Curtis-Quick, Rachel Claussen, Jalyn LaBine, and Adam Ray

Abstract Passive acoustic telemetry is more frequently being used by resource managers and researchers to understand the movements, distribution, and interactions of aquatic organisms. Here, we discuss the approaches used to design a study to assess the thermal niche of juvenile and adult Walleye (*Stizostedion vitreum*) across time and space in Mille Lacs Lake in the 1837 Ceded Territory in Minnesota. The objectives of our pilot study were to (1) evaluate the range of different acoustic tag types on two power settings in different habitat types, (2) determine the optimal spacing of receivers in a gridded array and evaluate the performance of tags in this virtual array, and (3) determine if acoustic tags implanted in juvenile Walleye affected the fish. To do this, we conducted range tests using Vemco V7, V13, and V16 acoustic tags in four habitat types: sand, boulder, boulder and cobble, and silt. Next, we used range test data to conduct a random swim simulation through a virtual gridded receiver array to determine how frequently we would hear from each tag and power combination. Lastly, we evaluated the survival, condition, swimming performance, and wound healing of juvenile Walleye that had Vemco V7 tags surgically implanted into their coelomic cavity compared to fish that did not receive a tag. Findings indicate that (1) detection range is variable across habitat types, with the largest detection range observed over sand and the smallest over silt; (2) the spacing of receivers based on the largest detection range in a habitat resulted in a gridded array with receivers spaced apart by 3 km (minimal/no overlap in the detection range around each receiver); (3) signals from larger tags (e.g., V16) traveled farther than the signal from smaller tags (e.g., V7); (4) setting the tag on low or high power did not considerably increase signal transmission distance, but the high power setting did shorten the life span of the battery; and (5) the performance of juvenile Walleye (>180 mm) implanted with a V7 acoustic tag was not altered. Together, these projects optimized the design of the telemetry array, guided the tag selection for

A. Shultz (✉) · J. Curtis-Quick · R. Claussen · J. LaBine · A. Ray
Great Lakes Indian Fish and Wildlife Commission, Odanah, WI, USA

C. A. Klimah
Mille Lacs Band of Ojibwe, Onamia, MN, USA

J. C. Bruner, R. L. DeBruyne (eds.), *Yellow Perch, Walleye, and Sauger: Aspects of Ecology, Management, and Culture*, Fish & Fisheries Series 41,
https://doi.org/10.1007/978-3-030-80678-1_10

adult and juvenile Walleye, and provided insight on the performance of larger tags that could be used in other species (e.g., Muskellunge (*Esox masquinongy*). We strongly encourage researchers and resource managers to consider similar pilot projects prior to initiating a whole-lake telemetry study.

Keywords Telemetry · Range test · Incision healing · Swimming performance · Transmitter performance · Gridded array

1 Introduction

Passive acoustic telemetry is increasingly being used by resource managers and researchers to understand the movements, distribution, and interactions of aquatic organisms (Hussey et al. 2015). At a basic level, passive telemetry requires a tag to be fixed on an aquatic organism and receivers or listening stations to be placed in a water body. An acoustic tag transmits information, either continually or intermittently, on a preprogrammed delay, and the information is recorded by a receiver when the aquatic organism is within range (Kessel et al. 2014). Tags transmit a unique identification code, time, and date and can also be outfitted with temperature and/or pressure sensors, which allows the researcher to collect additional information on the organism. The duration of acoustic telemetry studies can be from days to years depending on study objectives, tag size, delay setting, and additional sensors. Research topics range from broad and fine-scale movement, ecosystem and habitat use, fish passage and survival, and how perturbations in the environment may disrupt the movement of aquatic organisms (Hussey et al. 2015; Crossin et al. 2017). Here, we discuss the methods used to design a study to examine the movement and habitat usage of juvenile and adult Walleye (*Stizostedion vitreum*) and potentially other species.

Designing an acoustic telemetry project can be challenging given the multitude of receiver and tag options and limited information on telemetry equipment performance in a new environment. The performance of acoustic tags in different habitats (i.e., detection range) should be considered in any new study prior to deployment (Kessel et al. 2014). Several approaches have been used to determine the detection range, defined here as the distance at which 80% of transmissions are detected (Starr et al. 2000; Lindholm et al. 2007). These approaches may include moving a tag at known distances away from a receiver, fixing transmitters and receivers at set distances apart, attaching sentinel tags near receivers for the duration of the telemetry study, or a combination of these approaches. Range tests are necessary to understand the most effective areas to place acoustic receivers at the study site and to help with the interpretation of detections from tagged study animals (Payne et al. 2010; Kessel et al. 2014). The spacing of receivers in an array can be based on receiver detection range throughout the study area and how often fish will be detected in that range (Heupel et al. 2006). Detection range can be influenced by biotic and abiotic factors, as well as the type of acoustic tag implanted in the fish (Clements et al. 2005).

There is minimal guidance in the literature on how to arrange receivers in large lake ecosystems to meet the stated study objectives and also plan for the inclusion of other species in the future. A common approach to evaluating large-scale movements is to arrange receivers in a linear way to create "gates/curtains" with overlapping detection ranges to ensure that fish are detected by at least one receiver (Kessel et al. 2014). For studies interested in home ranges or seasonal habitat use, an alternative approach is to arrange receivers to maximize coverage in the habitats of interest, often in a two-dimensional gridded array (How and de Lestang 2012; Kessel et al. 2014). Advances in modeling can help determine if this type of arrangement will allow researchers to address research objectives such as determining how frequently fish will be detected on the array given the tag type, power level, spacing of receivers, and detection range of the receivers (Holbrook et al. 2017; Kraus et al. 2018).

Lastly, little information exists on how Walleye respond to the surgical implantation of tags. Tag validation experiments are performed as a cautionary step to evaluate whether fish performance is affected by the implanted transmitter (Jepsen et al. 2001). If the acoustic tag does affect fish performance, then it may lead to unnatural movement or premature death (Koeck et al. 2013). Early guidelines used the "2% rule," where implanted tags should not exceed 2% of a fish's body weight because it may affect fish performance and health (Jepsen et al. 2005). For adult Walleye, several previous studies have successfully used Vemco Inc. (Halifax, NS) V13 and V16 tags to track movements of fish in lentic and lotic ecosystems (Peat et al. 2015; Struthers et al. 2017; Brooks et al. 2019). For smaller fish, tag weight often exceeds 2% of the fish's body weight (Chittenden et al. 2009), there is less space in the coelomic cavity, and inserting a tag may crowd vital organs. Guidance is lacking on which tag size and type to use in juvenile Walleye and how those tags may affect recovery and performance. In juvenile Chinook Salmon (*Oncorhynchus tshawytscha*), Anglea et al. (2004) found that surgically implanted acoustic tags weighing up to 6.7% of tagged fish body weight did not affect tagged salmon health, swim performance, or predator avoidance. Conversely, other studies on freshwater fish found that tags weighing between 1 and 4% of tagged fish body weight affected fish health (Caputo et al. 2009). Specifically, there was documentation of tissue damage and necrosis of the surgery site from pressure on the coelomic cavity caused by implanted tags. Collectively, these studies suggest that fish response to intercoelomic transmitter implantation is species and life-stage specific; therefore, evaluating the response of juvenile Walleye to transmitter implantation is an unmet knowledge gap.

The purpose of this pilot study was to demonstrate how to (1) evaluate the range of different acoustic tag types on two power settings in different habitat types, (2) determine the optimal spacing of receivers in a gridded array, and (3) determine if acoustic tags implanted in juvenile Walleye affected the fish. Ultimately, this study can be used as a blueprint for resource managers in the design of acoustic telemetry studies across a wide range of habitats and species.

2 Methods

2.1 Study Location

This study was conducted on Mille Lacs Lake, a polymictic, mesotrophic lake in the 1837 Ceded Territory in central Minnesota (Fig. 1). Mille Lacs Lake has a maximum depth of 13.1 m and an average depth of 6.4 m (MAGE 2020). Mille Lacs Lake is the third largest lake in the state, covering 536 square kilometers, and is a popular spot for recreational and artisanal fishing (Minnesota Department of Natural Resources 2020). It is composed of multiple bottom types, including aquatic vegetation, cobble, boulder, silt, and sand bottoms. Each habitat type included an array of depths ranging from 0.3 to 12.8 m (Minnesota Department of Natural Resources 2020).

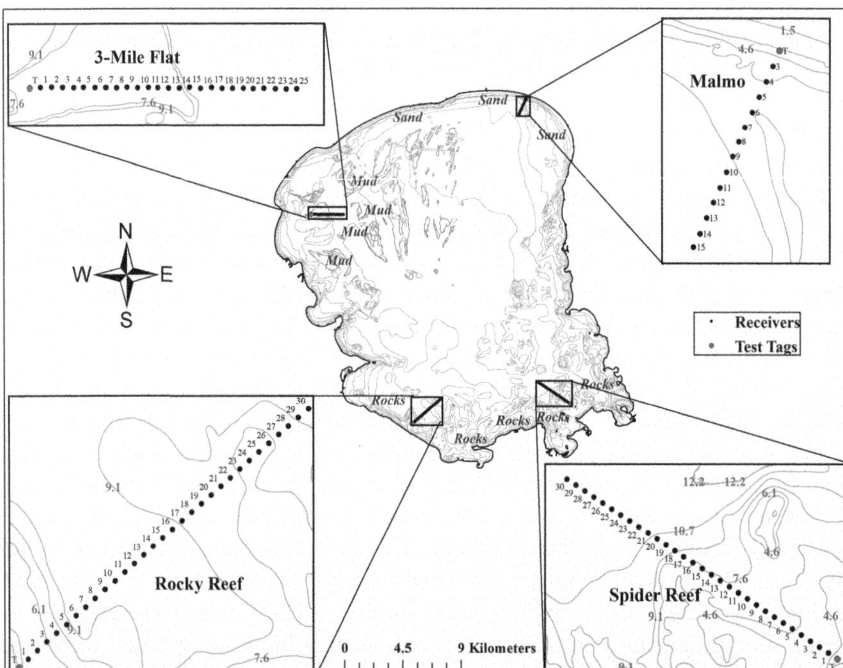

Fig. 1 Image of Mille Lacs Lake with receivers and range test tags noted in four habitat types. Tags were positioned at the following distances from the shoreline: 3 Mile Flat—2.3 km, Rock Reef—0.6 km, Spider Reef—1.3 km, and Malmo—0.3 km. Contour lines are in meters

2.2 Range Test

To conduct the range tests, habitats were broadly defined as sand, cobble, boulder, or silt. "Malmo" was an area that was predominantly sand that had a mixture of silt and sand at depths greater than 3 m. "Rocky Reef" contained cobble and boulders, and "Spider Reef" was predominantly boulders. Finally, "3-Mile Mud Flat" was an area that had nearly 100% silt bottom (Fig. 1).

In May of 2018, VR2W acoustic receivers and acoustic tags were separately moored to 32-kg concrete anchors. The receivers were inserted into a polyvinyl chloride (PVC) pipe at the top of an anchor and fixed in that position with bolts. Acoustic tags were attached to a rope that was suspended in the water column between a subsurface buoy and the top of an anchor. Three types of acoustic tags with temperature and pressure sensors (denoted as TP after the tag name) were used: V7-TP, V13-TP, and V16-TP. Acoustic tags were evaluated at different power levels with a fixed tag delay of 90 seconds (Vemco Inc., Halifax, NS). The V7-TP tags were the smallest (7 mm diameter, 2 g, estimated battery life: 255 days) and are primarily used for juvenile fish (Loher et al. 2017). The V13-TP (13 mm diameter, 6.3 g, battery life: 940 days) tags have a larger size and detection range than the V7-TP but smaller than the V16-TP. The V16 are the largest of the study tags (16 mm diameter, 8.5 g, battery life: 10 years) and emit the strongest detection signal (Oceans Research 2020). Each V13 and V16 tag alternated every 30 seconds between low and high power, whereas the V7 tags switched from low power to high power after 2 days. The low- and high-power rankings determine the decibel strength of the emitted signal, but the higher powered tag leads to lower battery life (Oceans Research 2020). Receivers (13–30 per habitat) were placed 100 meters apart in a straight line in each habitat (Fig. 1). A mooring with the acoustic tags was placed near the shore in each habitat (distance between tags and shoreline 1.1, 0.6–2.3 kilometers; mean, range). The closest receiver was positioned at 100 meters from the acoustic tags, and the farthest receiver was positioned at 3000 meters. Receivers and tags stayed in the water for 4–5 days before retrieval to determine detection probabilities at each receiver up to 3000 m away from the mooring with the tags.

2.3 Range Test Analysis

The range test analysis was used to determine the tag-power level and optimal spacing of receivers around the different habitats of the lake. Data offloading and analysis were completed using VUE® and RangeTest® (Vemco Inc., Halifax, NS) software, respectively. After receivers were retrieved and data downloaded, we analyzed the detection percentage (# of received detections/# expected detections * 100) of each receiver and acoustic tag on low and high power for each habitat type around the lake. We considered the highest detection percentage at 100 m to be the maximum amount because acoustic collisions would occur at this distance, whereas collisions at greater distances were less likely because signals from smaller, lower power tags (e.g., V7 on low power) were not likely to travel as far. We determined the effective range of receivers for each tag and power combination in each habitat type by multiplying the maximum amount of detections at 100 m by 80% (Starr et al. 2000; Lindholm et al. 2007). For example, if a receiver at 100 m received 1000 detections, then the effective range would be the distance at which 800 or more detections were received. In other words, the distance prior to the detections falling below 80% was considered then the effective range for a specific tag and power level in each habitat type.

Simulations of tagged fish were run using the GLATOS package in R (Holbrook et al. 2017) to compare the efficiency and effectiveness of the different tags and powers among the different habitat types in a 3-km grid array of receivers. In R, a probability of detection function was created for each of the 24 observed range test curves. Next, a "tagged" virtual fish would swim a virtual path established as a random walk within the shoreline boundary of Mille Lacs Lake. Tag transmissions were generated along the path using a 10-second burst detection transmission and a random transmission delay of either 120 or 300 seconds. Each habitat, tag, and power combination were simulated 100 times to obtain an average minimum, mean, and maximum time between detections.

2.4 *Juvenile Walleye Collection*

Juvenile Walleye ($n = 40$) between 178 and 279 mm were captured via electrofishing from Mille Lacs Lake. The juvenile Walleye were immediately transferred, in an aerated transport tank, to the Mille Lacs Band of Ojibwe fish hatchery. The collection of juvenile Walleye occurred at night in early June 2018.

Once in the wet lab, the Walleye were placed into one of four sections in a rectangular tank (4451.6 L), with a constant flow of water from Shakopee Lake, a smaller lake connected to Mille Lacs Lake. Water quality was monitored daily using a water quality/fish farming test kit (FF-1A, Hach Company®, Loveland, CO) and a dissolved oxygen meter (Pro 2030, YSI Incorporated®, Yellow Springs, OH). Oxygen levels were maintained by placing oxygen diffusers (airstones) in each section of the tank. A logger (HOBO Pendant® Temperature/Light 64 K, Onset Computer Corporation, Bourne, MA) was used throughout the duration of the experiment to monitor light and temperature changes.

Walleye were acclimated 72 hours before experimentation and fed leeches until satiation (Brauhn and Schoettger 1975). Daily formalin treatments were used to reduce fungal infections on the skin (Rach et al. 1997). After acclimation, Walleye were selected for one of two groups: (1) Walleye ($n = 20$) with the surgically inserted V7 dummy tag (7 mm diameter, 2 g) or (2) Walleye ($n = 20$) that did not receive a dummy tag (i.e., controls).

2.5 Surgery

Before surgeries began, all surgical tools and acoustic transmitters were disinfected using a 10% solution of povidone iodine and then rinsed with deionized water (Hayden et al. 2014). To insert dummy tags, juvenile Walleye were sedated to stage 4 anesthesia using a transcutaneous electrical nerve stimulation (TENS) device (7000 Second Edition Digital TENS Unit, Compass Health Brands, Middleberg Heights, OH) (Summerfelt and Smith 1990). Sedated fish were placed on a V-shaped surgery table, and oxygenated water was supplied to the gills during surgeries.

A size 10 surgical scalpel and a 1.5–3.0-mm cutting tip were used to make an initial 5–10 mm incision posterior to the pelvic girdle on the ventral side along the fish midline (Micro-Unitome Knife by BD, Wagner et al. 2011; Hayden et al. 2014). A V7 dummy tag (7 mm diameter, 2 g) was then inserted into the coelomic cavity and one or two absorbable monofilament sutures (PDS-II, 3–0 and 4–0, Ethicon, Somerville, NJ), and a reverse cutting needle (3/8) was used to close the incision (Dunn and Phillips 2005; Deters et al. 2010). While under anesthesia, Walleye were measured (total length (TL)), were weighed (g), and had a colored external Floy tag implanted below the dorsal fin for future identification (T-Bar Anchor, Floy Tag Inc., Seattle, WA). Both control and tagged Walleye underwent anesthesia for the same duration (5 min) in order to eliminate the possibility of anesthesia being a factor in the experiment.

2.6 Response Metrics

Juvenile Walleye were weighed and measured after surgery (day 1) and at the end of the experiment (day 18). Fulton's condition factor, a measure of body condition and health ($K = [\text{weight (g)}/\text{length (cm)}^3] \times 100$) was calculated for the treatment and control groups (Urbaniak et al. 2016). If Fulton's condition factor (K) is equal to 1, then fish were considered to be in a normal body condition. When values exceed 1, fish were considered to be in a better than normal body condition, while values less than 1 indicate fish in worse body condition than normal. The number of fish that survived in each treatment group was also recorded after 18 days.

A swim test was performed by chasing the fish until exhaustion in a circular swim tank (1068.4 L) on day 18. Fish were chased around a circular piece of plastic that blocks the standpipe, thus creating a swim flume (Szekeres et al. 2014). Oxygen was also maintained by placing an oxygen diffuser in the swim tank and monitoring the dissolved oxygen levels throughout the swim test. A fish was determined to be exhausted when a researcher gently pinched the caudal fin three times and the fish did not swim away (Kieffer 2000). The bottom of the tank had black lines representing quadrants spaced apart by 273 mm. The number of lines the fish crossed while swimming and the time until exhaustion were recorded (Szekeres et al. 2014).

Table 1 Rating scale used to describe the macroscopic appearance of the incision site, reproduced from the incision index of Miller et al. (2014)

Rating	Rating criteria
0	Incision completely closed and healed
1	Incision closed but not healed over
2	Incision held in proximity, but sides of the incision are only partly connected by tissue
3	Incision held in proximity, but sides of the incision are unconnected by tissue
4	Incision less than 50% open
5	More than 50% of the wound open
6	Completely open wound

The wound healing of the incision site was scored (0–6) by using an incision closure index (Table 1, Miller et al. 2014). Briefly, a score of 0 indicates that the incision has completely closed and healed, and a score of 6 indicates a completely open incision. Photographs of the surgery area were taken on day 1, day 7, and day 18 (end of the experiment), and a score was given on those days to each fish in the dummy tag group.

2.7 Analysis

Survival to day 18 was compared across treatment groups by using a chi-squared test. Within a group (control or dummy), Fulton's condition factor K was compared between day 1 and day 18 using a paired t-test. The swimming performance of the control group and dummy tag group was compared using a t-test after log transforming the data to meet the assumptions of normality and equal variances. Lastly, we provided the mean of incision scores on day 1, day 7, and day 18 to assess healing. All statistical analyses were performed using R Studio (3.4.1, R Foundation for Statistical Computing, Vienna, Austria). At the end of the experiment, fish were euthanized, meeting proper animal care standards.

3 Results

3.1 Range Tests

Range tests over a predominately sand habitat (Malmo) yielded the greatest detection range for all tag and power combinations (Fig. 2). Specifically, V7 tags on high and low power had a detection range of 400 m. The V13 tags on low power had a detection range of 1100 m, while a slight increase to 1300 m was noted when it was on high power. The V16 tags on both high and low power had a detection range of

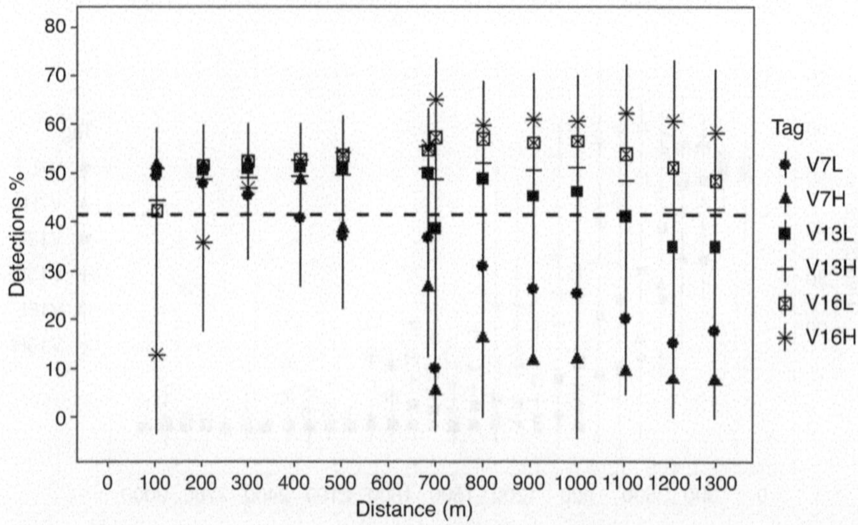

Fig. 2 Percent of detections for three acoustic tags at two power levels (H = high, L = low) at distances up to 1300 m away from the tags over the sandy habitat (Malmo). The dashed horizontal line represents 80% of the maximum detection percentage at 100 m away from the tags. The distance prior to detections falling below 80% was considered the effective range for a specific tag and power level. Error bars on each tag symbol represent ± SD, which may overlap each other

more than 1300 m on sand, but it is worth noting the percent of detections was below the 80% cutoff at 100 m and 200 m for V16 on high power.

The detection range was smaller for all tag and power combinations over the cobble and boulder habitat (Rocky Reef) than the sandy habitat (Malmo; Fig. 3). The V7 tags on low power had the smallest detection range of 200 m. A moderate increase in detection range was observed for V7 tags on high power and V13 tags on low power (400 m). The V13 tags on high power yielded a slight increase in detection range to 500 m. The V16 tags on low power had a detection range of 700 m, and again, the high power setting only increased the detection range by another 100 m.

The habitat composed primarily of boulders (Spider Reef), the most complex habitat, had a smaller detection range for all tag and power combinations than at Rocky Reef (Fig. 4). The V7 tags on high and low power had a detection range of 100 m. The V13 tags on low and high power had a similar detection range of 200 m. There was only a slight gain in detection range for V16 tags on low or high power (300 m), relative to V13 tags on high or low power settings.

Lastly, the silt habitat (3 Mile Flat) had the smallest detection range for most tag and power combinations relative to other habitats (Fig. 5). The V7 tags on high power had a detection range of 100 m, which was similar to the detection range over cobble and boulder habitats; however, V7 tags on low power had a detection range of less than 100 m over the silt habitat. For the V13 tags on low power, the detection range was 100 m, which was the smallest relative to all other habitat types. On the

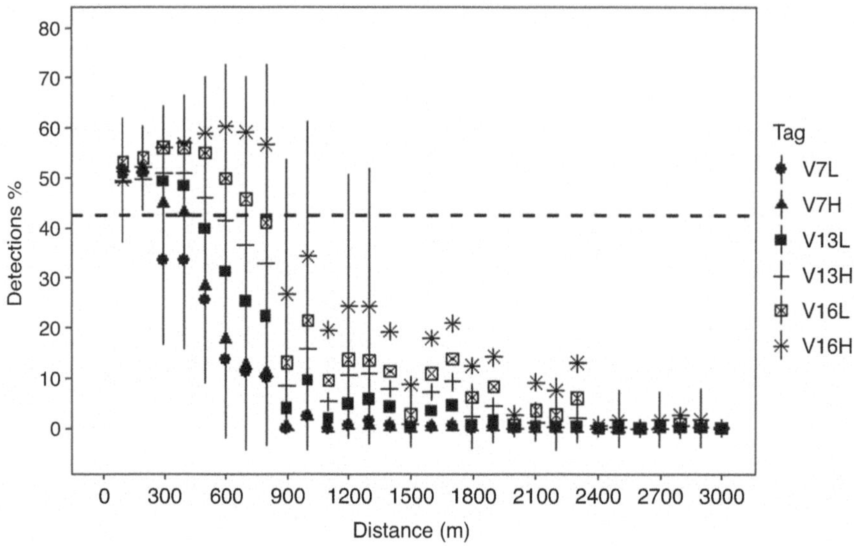

Fig. 3 Percent of detections for three acoustic tags at two power levels (H = high, L = low) at distances up to 3000 m away from the tags over the boulder habitat (Rocky Reef). The dashed horizontal line represents 80% of the maximum detection percentage at 100 m away from the tags. The distance prior to detections falling below 80% was considered the effective range for a specific tag and power level. Error bars on each tag symbol represent ± SD, which may overlap each other

high power setting, the detection range for the V13 tag over silt extended to 200 m, which was similar to the detection range over cobble and boulder. The V16 tag on low had the smallest detection range over silt relative to all other habitats (200 m), while the V16 tag on high power had a similar detection range over silt compared to cobble and boulder.

3.2 *Simulations*

Regardless of habitat, tag type, or power setting, the minimum time between detections ranged from 0 to 2.3 min (Table 2). The sandy habitat (Malmo) had the largest detection range and the shortest average time between detections, ranging from 0.6 min (V16-high) to 26.4 min (V7-high). The cobble and boulder habitat (Rocky Reef) had the second shortest average time between detections, 9.2 min (V16-high) to 41.2 min (V7-low). The boulder habitat (Spider Reef) had shorter average times between detections for the V16 and V13 tags compared to the silt habitat (3 Mile Flat; V16-high: 22.2 and 37.6 min, respectively; V13-low: 109.4 and 184.8 min, respectively). However, the average time between detections for the V7 tag was shorter in the silt habitat (3 Mile Flat) than in the boulder habitat (Spider

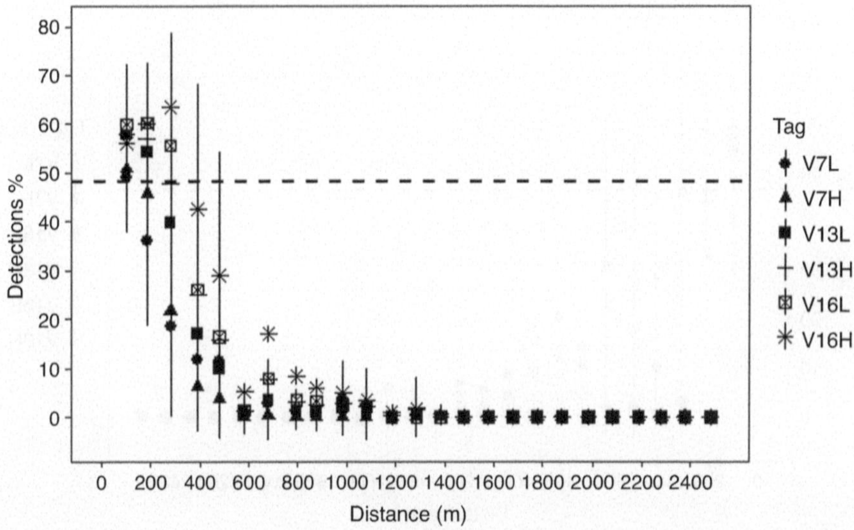

Fig. 4 Percent of detections for three acoustic tags at two power levels (H = high, L = low) at distances up to 2500 m away from the tags over the cobble and boulder habitat (Spider Reef). The dashed horizontal line represents 80% of the maximum detection percentage at 100 m away from the tags. The distance prior to detections falling below 80% was considered the effective range for a specific tag and power level. Error bars on each tag symbol represent ± SD, which may overlap each other

Reef), and there was little difference in times between the high and low power for each site.

3.3 Response Metrics

Fish in the control group were of a similar length and weight as fish that were tagged (Table 3). Tag weight relative to fish body weight averaged 2.6% and had a range of 1.1–4.8%. All fish retained their tags over the 18-day study period. The pH levels in the tank ranged from 7 to 8, and the ammonia content ranged in the tanks from 0.0 ppm to 0.5 ppm. The dissolved oxygen levels were between 6.68 g/mL and 9.48 g/mL, and the temperature ranged from 20.4 °C to 25.2 °C.

Response metrics were similar or slightly improved for juvenile Walleye in the dummy tag group relative to the control group. Specifically, the survival of fish was not significantly different between groups (X-squared = 0.27, $p = 0.60$), and fish that did not survive were some of the largest individuals in each treatment group (Table 3). Fulton's condition factor indicated that both groups had fish that were in a relatively poor condition ($K < 1$), and fish condition did not decline/worsen over the 18-day study period (Fig. 6). Surprisingly, fish in the control group swam slightly slower than fish in the dummy tag group (Fig. 7, $t = -2.29$, $p = 0.03$). All of the

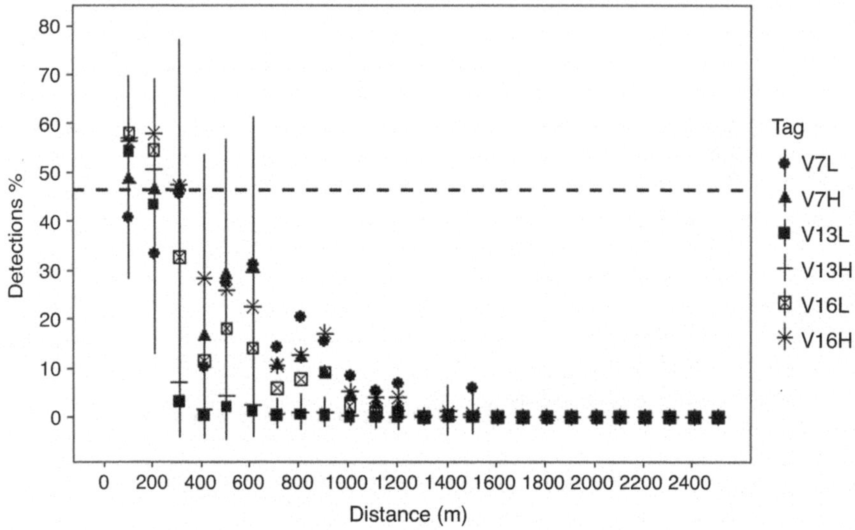

Fig. 5 Percent of detections for three acoustic tags at two power levels (*H* = high, *L* = low) at distances up to 2500 m away from the tags over the silt habitat (3 Mile Flat). The dashed horizontal line represents 80% of the maximum detection percentage at 100 m away from the tags. The distance prior to detections falling below 80% was considered the effective range for a specific tag and power level. Error bars on each tag symbol represent ± SD, which may overlap each other

Table 2 The time between detections in minutes of tags (V7, V13, and V16) on high or low power based on range test results in each habitat. Tag transmissions were generated along the path using a 10-second burst detection transmission and a random transmission delay of either 120 or 300 seconds. Tag and power combinations in each habitat were simulated 100 times to obtain a minimum, mean, and maximum time between detections

Site	Time between detections (min)	V7		V13		V16	
		Low	High	Low	High	Low	High
Malmo (sand and some silt)	Minimum	0.00	2.16	0.00	0.00	0.00	0.00
	Mean	6.20	26.40	3.67	2.55	2.06	0.58
	Maximum	89.05	492.66	69.75	34.91	23.93	6.97
Rocky reef (cobble and boulder)	Minimum	2.21	2.19	2.19	1.59	0.15	0.00
	Mean	41.20	38.61	31.41	14.79	11.17	9.21
	Maximum	565.24	587.07	554.13	313.07	246.20	223.97
Spider reef (boulder)	Minimum	2.27	2.22	2.25	2.21	2.21	1.68
	Mean	103.33	100.92	109.43	52.74	59.63	22.23
	Maximum	911.74	1165.23	1012.52	676.61	819.85	338.22
Three mile (silt)	Minimum	2.07	2.07	2.29	2.27	2.22	2.12
	Mean	36.85	34.62	184.83	147.33	71.01	37.56
	Maximum	488.83	473.90	1504.93	1374.82	913.27	478.06

Table 3 The initial total length and weight of juvenile Walleye, percent of juvenile Walleye that survived, total length of the fish that did and did not survive in each treatment group

Treatment	Initial total length and weight of fish (mean, range)	Initial weight (mean, range)	Survival (%)	Total length of dead fish (*n*, mean, range)
Control (*n* = 20)	214 mm, 185–285 mm	76.4 g, 42.9–201 g	90%	2, 226 mm, 220–232 mm
Dummy (*n* = 20)	216 mm, 180–285 mm	76.9 g, 41.8–180.1 g	75%	5, 242 mm, 217–276 mm

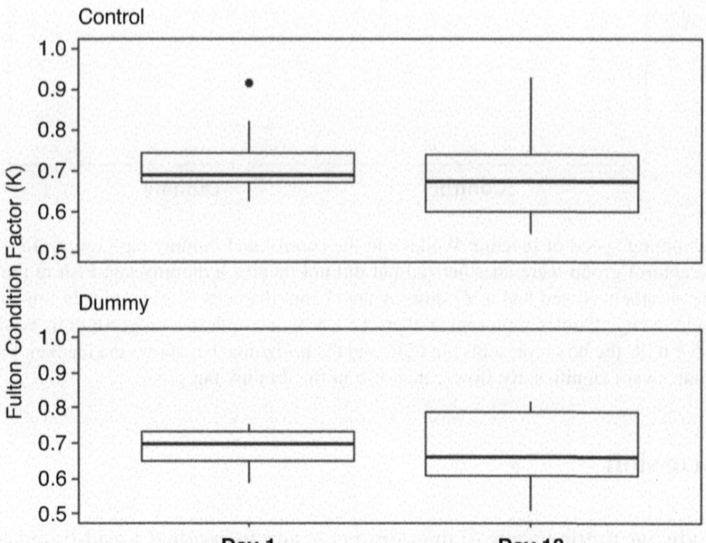

Fig. 6 Fulton's condition factor for juvenile Walleye on day 1 and day 18 in the control and dummy tag treatment groups. Fish in the control group were anesthetized but did not receive a dummy tag. Fish in the dummy group were anesthetized and had a V7 dummy tag (7 mm diameter, 2 g) surgically implanted into the coelomic cavity. Circles represent outliers (>1.5 × interquartile range (IQR)), vertical bars indicate 1.5 * IQR, the box represents the IQR, and the horizontal bar shows the median. No significant differences in condition were detected for fish in either group

juvenile Walleye retained their tags over the 18-day study period. On day 1, the incision healing score (range 0–6) was a mean of 2.3, remained at a mean of 2.3 on day 7, but by day 18, the mean score decreased to 0.2, indicating that most incisions had completely closed and healed (Fig. 8).

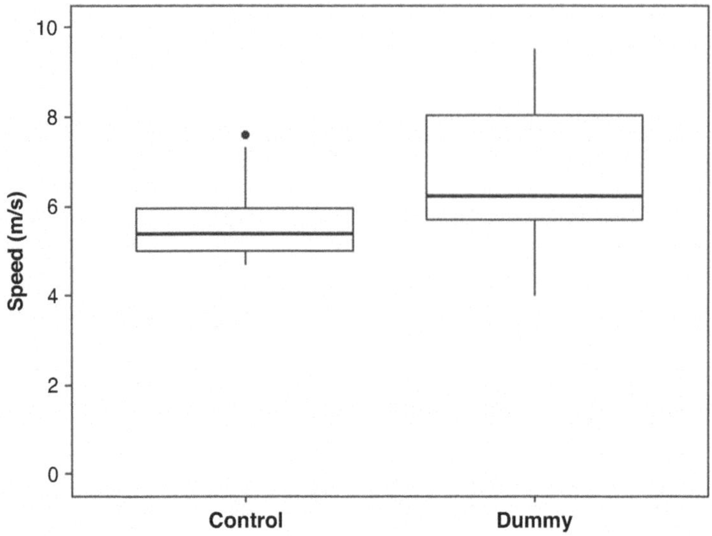

Fig. 7 Swimming speed of juvenile Walleye in the control and dummy tag groups after 18 days. Fish in the control group were anesthetized but did not receive a dummy tag. Fish in the dummy group were anesthetized and had a V7 dummy tag (7 mm diameter, 2 g) surgically implanted into the coelomic cavity. Circles represent outliers (>1.5 × interquartile range (IQR)), vertical bars indicate 1.5 * IQR, the box represents the IQR, and the horizontal bar shows the median. Fish in the control group swam significantly slower than fish in the dummy tag group

4 Discussion

In this study, we outline some of the steps necessary to conduct a grid-based acoustic telemetry study in a large lake. The first step was to understand how different substrate and habitat types can affect transmission and detection. In aquatic environments, physical complexity can impede or deflect transmitter signals, and conversely, open areas with little complexity allow the detection of signals from greater distances. For example, low rugosity hard bottom, high rugosity reef, and mixed hard bottoms were shown to limit the detection range in a marine environment (Shelby et al. 2016). As expected, the detection radius decreased and the time between detections increased as benthic complexity increased in our study, except for the silt habitat, which had the smallest detection range. The limited detection range in the silt habitat may be a result of higher amounts of organic matter in the water column, which has been shown to absorb acoustic energy (Heupel et al. 2006). Indeed, Babin et al. (2019) partially attributed the lower detection range near a hydropower facility to a silty substrate that may have attenuated the acoustic signal relative to the cobble bottom in the reservoir. The detection range in the sandy habitat (Malmo) was undefined for some of the higher decibel tag/power combinations as an inflection point was not detected within the 1300-m test range. This did not alter the grid design as the main concern was identifying the minimum detection

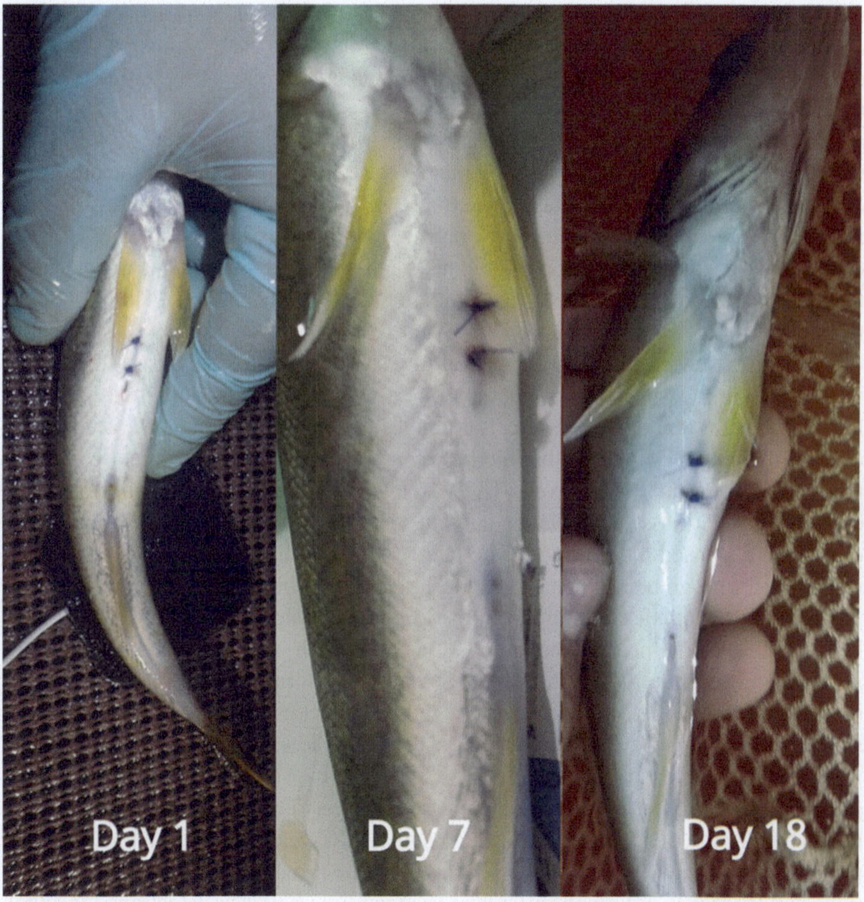

Fig. 8 An average example of the healing of the incision site on juvenile Walleye on day 1, day 7, and day 18. Note the redness and inflammation around the sutures on day 7

range and how tight the grid array needed to be in order to adequately record detections. This unknown range in the sandy habitat may lead to detections of the same fish simultaneously at multiple receivers, which should be considered during data processing. Additionally, detections over the sandy habitat were below the 80% cutoff at 100 m and 200 m for V16 on high power most likely due to close proximity detection interference (CDPI). This phenomenon occurs when hard surfaces such as calm water, hard bottoms, and/or ice cause transmission echoes that interfere with the signal (Kessel et al. 2015). The use of a V16 tag on a high power setting in future telemetry projects in Mille Lacs Lake may be limited because of CDPI and the relatively long distance the signal travels over sandy habitats. Although habitat complexity and substrate are important factors that influence the detection range, there are a multitude of other factors, both biotic (aquatic vegetation) and abiotic (e.g., waves, rain, boating activity), that can change throughout the year and alter the detection range. As recommended by Kessel et al. (2014), sentinel tags can be deployed in the most representative habitats, and a poststudy analysis can be conducted to understand how abiotic and biotic factors may have influenced the detection range over the course of the study.

Before spending resources (time and money) on implementing a grid array of receivers, Kraus et al. (2018) encouraged the use of computer simulations to determine how frequently a tagged animal will be detected and how much time will pass between detections to best optimize the study design. Similarly, Hayden et al. (2016) used a combination of field tests and computer simulations to determine the optimal placement of acoustic receivers in Saginaw Bay and Lake Huron. Unlike the current study, Hayden et al. (2016) used a single tag type (V16), but they observed differences in detection probabilities at the same sites across time. The second step in this study was to conduct random walk simulations in a virtual Mille Lacs Lake using the observed detection ranges of the different habitats. The results of the simulations estimated the time between detections for the different combinations of tag sizes and power levels. The V16 tags provided the shortest time between detections, regardless of power or habitat, yet they are also the largest and most expensive tags used in this study. The increase in detections with the V16 tag does not validate the greater expense and the larger size which may limit the size and species of fish suitable for tagging. In the grid array of this study, the V16 tags may be useful to monitor the movements of large fish (e.g., Muskellunge (*Esox masquinongy*)) in the future; however, given the exploratory nature of these initial studies and the scope of our objectives, these larger tags were not necessary. The simulations provided evidence that the V13 tag on low power would be detected about half as often as on the high power setting; however, the lower power would greatly extend the battery life, allowing for a longer duration of monitoring (estimated battery life of a V13 tag without sensors is 1765 and 760 days on low and high power, respectively). When similar logic was applied to the V7 tags, the simulations indicated there was not much difference in mean time between detections for either power setting, but again the lower power would allow for an extended monitoring period (estimated battery life of a V7 tag without sensors is 404 and 168 days on low and high power, respectively). Random walk simulations are an important part of any telemetry study as they can help set expectations for a given study design or aid in determining ideal receiver placement. For example, Barkley et al. (2019) used random walk simulations to show that the interactions of an animal swim speed, a nominal delay of the tag, and the detection radius of the receiver are critical to optimizing successful detections of tagged sharks and need to be considered when designing the study. In conjunction with laboratory studies, simulations can also provide insight into the appropriate tag and power settings to help meet the objectives of the study, whether it is long-term seasonal trends or more fine-scale movements.

Another important aspect of designing an acoustic telemetry study is to determine the appropriate tag sizes and types for the species/life stages of interest. Given our interest in juvenile Walleye, we used several metrics to determine if juvenile Walleye could carry a surgically implanted tag. We found that after 18 days, survival, condition, and swimming performance were not negatively affected for juvenile Walleye implanted with tags weighing up to 4.8% of their body weight. A study on Brook Trout (*Salvelinus fontinalis*) found no differences in the mortality or swimming performance of tagged fish and suggested tag weights could be up to 7% of Brook Trout weight (Smircich and Kelly 2014). Similarly, Anglea et al. (2004) demonstrated that the swimming performance and predator avoidance of juvenile Chinook Salmon were not affected by the surgical implantation of acoustic transmitters weighing up to 6.7% of tagged fish body weight. In our study, surgeons attempted to tag fish smaller than 180 mm and 41.8 g, but tags would not fit into the coelomic cavity due to limited space, thereby limiting tag weight to 4.8% of the fish's weight. Incisions were nearly healed over the course of 18 days, indicating that tags did not put pressure on the incision site and sutures (i.e., there was enough space for the tag in the body cavity). Formalin treatments may have improved healing at the incision site but were necessary due to several factors: (1) how often fish were handled over 18 days, (2) confinement of fish to tanks, and (3) production of pathogens by the untreated water supply from a natural source. In a similar study on juvenile Muskellunge, growth was not affected by transmitter implantation after 4 months and surgical wounds were healed after 30 days (Walton-Rabideau et al. 2019). By analyzing multiple metrics, we were able to determine that transmitters weighing up to 4.8% of the fish's body weight could be used without affecting their performance.

We conducted this pilot study to aid in the design of a telemetry project that would examine the movement and habitat usage of juvenile and adult Walleye and potentially other species. Kessel et al. (2014) suggested a two-dimensional gridded array for studies interested in home ranges or seasonal habitat use. A two-dimensional gridded array, with receivers spaced 3000 m from each other, was supported by our range tests and modeling results in multiple habitats. The detection range for most of our tags (except for V16-low and -high tags or V13-high tags over sand) was less than 1500 m, which meant only one receiver would hear an individual tag at any given time. Although V16 tags do not appear to be ideal for a gridded array with receivers spaced 3000 m apart, the range tests and modeling results for this tag may be useful in the future if our study objectives change and/or additional species are tagged. Instead, our results indicate that V13 tags would be an optimal tag for adult Walleye in a gridded array with receivers spaced 3000 m apart. V13 transmitters have been successfully used in adult Walleye in large ecosystems (Brooks et al. 2019). As for power selection, we observed a similar detection range across habitats for V13 tags on low and high power, but the estimated battery life was nearly 2.5 times longer on the low power setting. Moreover, the random walk showed similar detection frequencies across habitats on both power settings, indicating to us that we would likely collect more data by selecting the power level (low) with the longest battery life. The same logic was applied to selecting the V7 tag on low power, which our results suggested could be surgically implanted in juvenile Walleye. Specifically, in our quantification of multiple metrics and observance of no negative changes in those metrics, we determined that we could tag juvenile Walleye with V7 tags with temperature and pressure sensors. Collectively, these experiments set the stage for subsequent telemetry studies in Mille Lacs Lake and increase the likelihood of successfully addressing future research objectives. We encourage researchers and resource managers to conduct similar pilot studies prior to implementing telemetry projects so that movement data will adequately address research objectives.

Acknowledgments We thank the team from the Mille Lacs Band of Ojibwe (MLBO) Department of Natural Resources, including George Bigbear and Keith Wiggins, for assisting us with the study design and implementation. We appreciate the logistical support from the Great Lakes Indian Fish and Wildlife Commission (GLIFWC) and the assistance from their intern, Brandon Johnson. Fond du Lac Band, Mole Lake Band, Bad River Band provided letters of support for the proposal. Employees from the Minnesota Department of Natural Resources and the United States Geological Survey commented on the experimental design of these projects. Staff from Vemco quickly responded to technical questions about receivers, transmitters, and software. Financial support for this project came from MLBO, GLIFWC, and a Tribal Wildlife Grant from the Fish and Wildlife Service.

References

Anglea SM, Geist DR, Brown RS, Deters KA, McDonald RD (2004) Effects of acoustic transmitters on swimming performance and predator avoidance of juvenile Chinook salmon. N Am J Fish Manag 24(1):162–170

Babin A, Fitzpatrick L, Linnansaari T, Curry RA (2019) Detection range of acoustic receivers in a largy hydropower reservoir. Aust Fish 4(60)

Barkley AN, Broell F, Pettitt-Wade H, Watanabe YY, Marcoux M, Hussey NE (2019) A framework to estimate the likelihood of species interactions and behavioural responses using animal-borne acoustic telemetry transceivers and accelerometers. J Anim Ecol 89:146–160

Brauhn JL, Schoettger RA (1975) Acquisition and culture of research fish: rainbow trout, fathead minnows, channel catfish, and bluegills. National Environmental Research Center, Office of Research and Development, U.S. Environmental Protection Agency, EPA-660-3-75-011

Brooks JL, Midwood JD, Gutowsky LFG, Boston CM, Doka SE, Hoyle JA, Cooke SJ (2019) Spatial ecology of reintroduced walleye (*Sander vitreus*) in Hamilton harbour of Lake Ontario. J Great Lakes Res 45(1):167–175

Caputo M, O'Connor CM, Hasler CT, Hanson KC, Cooke SJ (2009) Long-term effects of surgically implanted telemetry tags on the nutritional physiology and condition of wild freshwater fish. Dis Aquat Org 84:35–41

Chittenden CM, Butterworth KG, Fiona Cubitt K, Jacobs MC, Ladouceur A, Welch DW, Mckinley RS (2009) Maximum tag to body size ratios for an endangered coho salmon (*O. kisutch*) stock based on physiology and performance. Environ Biol Fish 84:129–140

Clements S, Jepsen D, Karnowski M, Schreck CB (2005) Optimization of an acoustic telemetry array for detecting transmitter-implanted fish. N Am J Fish Manag 25(2):429–436

Crossin GT, Heupel MR, Holbrook CM, Hussey NE, Lowerre-Barbieri SK, Nguyen VM, Raby GD, Cooke SJ (2017) Acoustic telemetry and fisheries management. Ecol Appl 27:1031–1049

Deters KA, Brown RS, Carter KM, Boyd JW, Brad Eppard M, Seaburg AG (2010) Performance assessment of suture type, water temperature, and surgeon skill in juvenile Chinook salmon surgically implanted with acoustic transmitters. Trans Am Fish Soc 139(3):888–899

Dunn DL, Phillips J (2005) Ethicon wound closure manual. Ethicon, Inc., Somerville, NJ

Hayden TA, Holbrook CM, Binder TR, Dettmers JM, Cooke SJ, Vandergoot CS, Krueger CC (2016) Probability of acoustic transmitter detections by receiver lines in Lake Huron: results of multi-year field tests and simulations. Anim Biotelemetry 4(1):19

Hayden TA, Holbrook CM, Fielder DG, Vandergoot CS, Bergstedt RA, Dettmers JM, Krueger CC, Cooke SJ (2014) Acoustic telemetry reveals large-scale migration patterns of walleye in Lake Huron. PLoS One 9(12):e114833

Heupel MR, Semmens JM, Hobday AJ (2006) Automated acoustic tracking of aquatic animals: scales, design and deployment of listening station arrays. Mar Freshw Res 57(1):1–13

Holbrook C, Hayden T, Binder T (2017) GLATOS: A package for the Great Lakes acoustic telemetry observation system. https://rdrr.io/github/jsta/glatos/

How JR, de Lestang S (2012) Acoustic tracking: issues affecting design, analysis and interpretation of data from movement studies. Mar Freshw Res 63:312–324

Hussey NE, Kessel ST, Aarestrup K, Cooke SJ, Cowley PD, Fisk AT, Harcourt RG, Holland KN, Iverson SJ, Kocik JF, Mills Flemming JE, Whoriskey FG (2015) Aquatic animal telemetry: a panoramic window into the underwater world. Science 348(6240):1255642

Jepsen N, Davis LE, Schreck CB, Siddens B (2001) The physiological response of Chinook salmon smolts to two methods of radio-tagging. Trans Am Fish Soc 130(3):495–500

Jepsen N, Schreck C, Clements S, Thorstad EB (2005) A brief discussion on the 2% tag/bodymass rule of thumb. Pages 255–259. In: Spedicato MT, Lembo G, Marmulla G (eds.) Aquatic telemetry: advances and applications. Proceedings of the Fifth Conference on Fish Telemetry held in Europe. Ustica, Italy, 9–13 June 2003

Kessel ST, Cooke SJ, Heupel MR, Hussey NE, Simpfendorfer CA, Vagle S, Fisk AT (2014) A review of detection range testing in aquatic passive acoustic telemetry studies. Rev Fish Biol Fish 24:199–218

Kessel ST, Hussey NE, Webber DM, Gruber SH, Young JM, Smale MJ, Fisk AT (2015) Close proximity detection interference with acoustic telemetry: the importance of considering tag power output in low ambient noise environments. Anim Biotelemetry 3(5)

Kieffer JD (2000) Limits to exhaustive exercise in fish. Comp Biochem Physiol Part A Mol Integr Physiol 126(2):161–179

Koeck B, Gudefin A, Romans P, Loubet J, Lenfant P (2013) Effects of intracoelomic tagging procedure on white seabream (*Diplodus sargus*) behavior and survival. J Exp Mar Biol Ecol 440:1–7

Kraus RT, Holbrook CM, Vandergoot CS, Stewart TR, Faust MD, Watkinson DA, Charles C, Pegg M, Enders EC, Krueger CC (2018) Evaluation of acoustic telemetry grids for determining aquatic animal movement and survival. Methods Ecol Evol 9(6)

Lindholm J, Auster PJ, Knight A (2007) Site fidelity and movement of adult Atlantic cod *Gadus morhua* at deep boulder reefs in the western gulf of Maine, USA. Mar Ecol Prog Ser 342:239–247

Loher T, Webster RA, Carlile D (2017) A test of detection range of acoustic transmitters and receivers deployed in deep waters of Southeast Alaska, USA. Anim Biotelemetry 5(27)

Miller EA, Froehlich HE, Cocherell DE, Thomas MJ, Cech JJ, Klimley AP, Fangue NA (2014) Effects of acoustic tagging on juvenile green sturgeon incision healing, swimming performance, and growth. Environ Biol Fish 97(6):647–658

Minnesota Alliance for Geographic Education (MAGE) (2020) Lake mille Lacs: formation of Landscape & Site Characteristics. Web. https://magemn.com/go-minnesota/lake-mille-lacs-main/

Minnesota Department of Natural Resources (2020) Mille Lacs Lake. https://www.dnr.state.mn.us/millelacslake/index.html

Oceans Research (2020) V7 to V16–69 kHz Transmitters. https://www.oceans-research.com/products/transmitters/coded-tags/v7-v16-69khz/

Payne NL, Gillanders BM, Webber DM, Semmens JM (2010) Interpreting diel activity patterns from acoustic telemetry: the need for controls. Mar Ecol Prog Ser 419:295–301

Peat TB, Hayden TA, Gutowsky LFG, Vandergoot CS, Fielder DG, Madenjian CP, Murchie KJ, Dettmers JM, Krueger CC, Cooke SJ (2015) Seasonal thermal ecology of adult walleye (*Sander vitreus*) in Lake Huron and Lake Erie. J Therm Biol 53:98–106

Rach JJ, Howe GE, Schreier TM (1997) Safety of formalin treatments on warm- and Coldwater fish eggs. Aquac 149(3–4):183–191

Shelby TH, Hart KM, Fujisaki I, Smith BJ, Pollock CJ, Hillis-Starr Z, Lundgren I, Oli MK (2016) Can you hear me now? Range-testing a submerged passive acoustic receiver array in a Caribbean coral reef habitat. Ecol Evol 6(14): 4823–4835.

Smircich MG, Kelly JT (2014) Extending the 2% rule: the effects of heavy internal tags on stress physiology, swimming performance, and growth in brook trout. Anim Biotelemetry 2(16)

Starr RM, Heine JN, Johnson KA (2000) Techniques for tagging and tracking Deepwater rockfishes. N Am J Fish Manag 20:597–609

Struthers DP, Gutowsky LFG, Enders EC, Smokorowski KE, Watkinson DA, Silva AT, Cvetkovic M, Bibeau E, Cooke SJ (2017) Factors influencing the spatial ecology of Lake sturgeon and walleye within an impounded reach of the Winnipeg River. Environ Biol Fish 100: 1085–1103

Summerfelt RC, Smith LS (1990) Anesthesia, surgery, and related techniques. In: Schreck CB, Moyle PB (eds) Methods for fish biology. American Fisheries Society, Bethesda, Maryland, pp 213–263

Szekeres P, Brownscombe JW, Cull F, Danylchuk AJ, Shultz AD, Suski CD, Murchie KJ, Cooke SJ (2014) Physiological and behavioural consequences of cold shock on bonefish (*Albula vulpes*) in the Bahamas. J Exp Mar Biol Ecol 459:1–7

Urbaniak TJ, Barnes ME, Davis JL (2016) Acoustic transmitters impact rainbow trout growth in a competitive environment. Open Fish Sci J 9(1):37–44

Wagner GN, Cooke SJ, Brown RS, Deters KA (2011) Surgical implantation techniques for electronic tags in fish. Rev Fish Biol Fish 21:71–81

Walton-Rabideau SE, Newell M, Jeanson AL, Lédée EJI, Farrell JM, Cooke SJ (2019) Evaluation of tag retention, healing, growth, and behavior in age-0 muskellunge following acoustic transmitter implantation. N Am J Fish Manag 39:652–663

Part III
Sauger (*Stizostedion canadense*)

Sauger Restoration in the Upper Allegheny River Watershed, New York

Justin R. Brewer, Jeffrey J. Loukmas, and Michael Clancy

Abstract Sauger (*Stizostedion canadense*) was historically common throughout several watersheds in New York State but is now considered critically imperiled and in need of reintroduction. Sauger occur in the lower Allegheny River in Pennsylvania, but upstream expansion into New York is blocked by the Kinzua Dam. Thus, in 2014, the New York State Department of Environmental Conservation (NYSDEC) began a stocking program to establish a self-sustaining Sauger population in the upper Allegheny River watershed. From 2014 to 2020, over 37,000 fingerlings and 700,000 fry were stocked in the Allegheny Reservoir and upper river. The annual NYSDEC late summer boat electrofishing catch rates indicate good survival of multiple-year classes. In addition, growth rates were high, with Sauger reaching 380 mm by age 2 and 490 mm by age 5. The survival of stocked Sauger combined with rapid growth has established an adult population likely capable of supporting reproduction. The occurrence of hybrid saugeye (*Stizostedion canadense x Stizostedion vitreum*) during 2019 surveys provides the first evidence that Sauger are attempting to spawn. Despite challenges with hatchery production, unpredictable sampling conditions, and multijurisdictional management implications, preliminary results are encouraging, and stocking is expected to continue through 2023.

Keywords Sauger · Sauger restoration · Allegheny River · New York Sauger · Sauger stocking

---

J. R. Brewer (✉) · J. J. Loukmas · M. Clancy
Division of Fish, Wildlife, and Marine Resources, New York State Department of Environmental Conservation, Allegany, NY, USA
e-mail: justin.brewer@dec.ny.gov; jeffrey.loukmas@dec.ny.gov; michael.clancy@dec.ny.gov

J. C. Bruner, R. L. DeBruyne (eds.), *Yellow Perch, Walleye, and Sauger: Aspects of Ecology, Management, and Culture*, Fish & Fisheries Series 41,
https://doi.org/10.1007/978-3-030-80678-1_11

1 Introduction

Sauger (*Stizostedion canadense*) is one of the most critically imperiled fish species in New York State and is listed as a High Priority Species of Greatest Conservation Need (NYSDEC 2015). A Conservation Management Plan was recently adopted to aid the recovery of the species (Loukmas 2013). The goal of this plan is to establish and maintain self-sustaining Sauger populations in all suitable waters of native watersheds, including the Allegheny River watershed. Sauger have been recorded in the Allegheny River as far north as Warren, Pennsylvania, in the early 1900s (Fowler 1919) but never recorded in the New York portion of the watershed. Widespread pollution beginning in the mid-1800s (Koryak et al. 2009) likely caused their extirpation from the upper portion of the watershed before systematic fish surveys were conducted by the New York Conservation Department in the 1930s (Loukmas 2013). However, this does not mean that Sauger never occurred there as pollution levels of the late nineteenth and much of the twentieth centuries led to the depletion, and sometimes extirpation, of many fish species, particularly those associated with large river habitats, such as Sauger (Eaton et al. 1982; Koryak et al. 2009). The Allegheny River is now in a state of recovery (Koryak et al. 2009), and a popular Sauger fishery currently exists in the lower Pennsylvania section of the river. Until recently, Sauger in the lower river have been prevented from accessing the New York portion of the watershed by dams. Conewango Creek, a tributary that enters the Allegheny River in Pennsylvania below the Kinzua Dam, was recently made accessible to Sauger with the removal of a low head dam at the mouth. While Sauger remain undetected in the creek, there are no longer any obstructions to their movement into this part of the watershed in New York. Sauger are, however, still blocked from entering the New York portion of the main stem Allegheny River by the Kinzua Dam, in Warren County, Pennsylvania. This dam, constructed in 1965, created the 40-km-long Allegheny Reservoir (Fig. 1).

The Allegheny Reservoir and river system upstream contain a diverse and complex habitat with an abundant forage base that is likely capable of supporting a Sauger population. Sauger typically occur in large turbid rivers and lakes (Becker 1983) and prefer diverse natural river channels and associated habitats over relatively simple, uniform channelized segments (Groen and Schmulbach 1978; Seigwarth et al. 1993; Hesse 1994). As evidenced by the excellent recreational Walleye (*Stizostedion vitreum*) fishery that exists there, the New York portion of the Allegheny River system contains suitable spawning habitat and conditions to support riverine percid populations, which provided a promising opportunity for a Sauger restoration program. To establish a self-sustaining population in the river and reservoir section above the dam, a stocking program was developed, initiated in 2014, and will run through 2023. An update on the status of this program is presented here.

2 Methods

2.1 *Study Area*

2.2 *Propagation and Stocking*

The stocking program relies on the generous contributions of fry from Ohio River broodstock from the West Virginia Division of Natural Resources and the Kentucky Department of Fish and Wildlife. Broodstock populations were part of ongoing propagation programs and were annually tested for diseases and met New York State Department of Environmental Conservation (NYSDEC) fish health certification requirements. The original intent was to annually produce 300,000 intensively raised two-week-old fry and 50,000 pond fingerlings at NYSDEC's Chautauqua Fish Hatchery in Mayville, New York, for stocking in the Allegheny River watershed.

In total, 1,251,400 Sauger fry were received from 2014 to 2020 (Table 1). Annual contributions varied widely, ranging from 15,000 in 2017 to 610,000 in 2018, which influenced the methods used to raise them from year to year. When the supply of fry was limited (2014, 2015, 2017), the entire number was raised in ponds for about

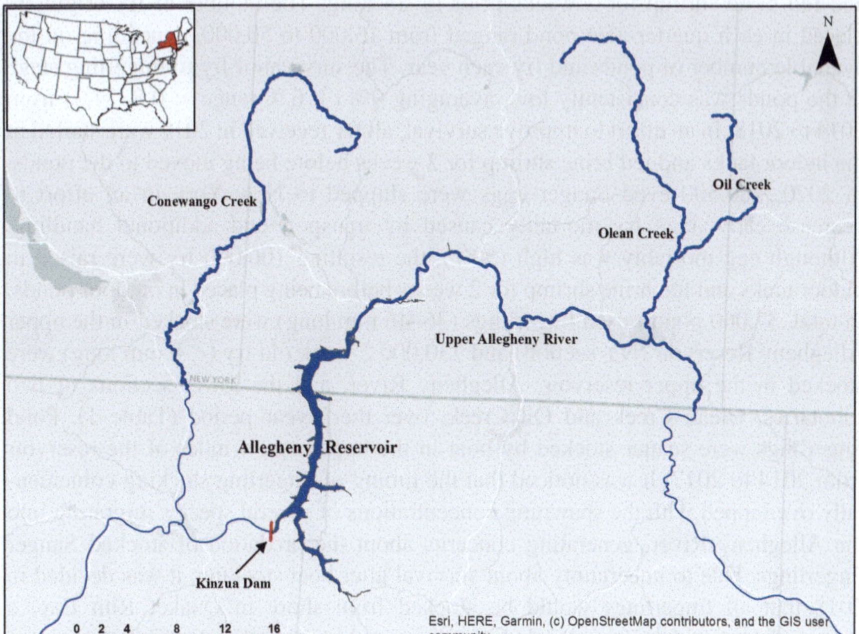

Fig. 1 Sauger restoration area, including the Allegheny River, Allegheny Reservoir, Kinzua Dam, Conewango Creek, Olean Creek, and Oil Creek

Table 1 Number and type of Sauger received and subsequently stocked in the Allegheny Reservoir and upper Allegheny River by life stage, 2014–2020. Fingerlings were stocked into the Allegheny Reservoir, and fry were stocked into the Allegheny Reservoir, upper Allegheny River, and/or tributaries

Year	Agency source of fry	No. of fry received	No. of pond fingerlings stocked (36–46 mm)	No. of fry stocked (< 8 mm)
2014	WVDNR[a]	33,000	5700	0
2015	WVDNR	43,400	5800	0
2016	KDFW[b]	350,000	3260	250,000
2017	WVDNR	15,000	1580	0
2018	WVDNR	200,000	1100	480,000
	KDFW	410,000		
2019	WVDNR	20,000	3020	0
	KDFW	180,000		
2020	WVDNR	106,000[c]	16,600	0
Total		**1,251,400**	**37,060**	**730,000**

[a] West Virginia Division of Natural Resources
[b] Kentucky Department of Fish and Wildlife
[c] Number of surviving fry from 260,500 eyed eggs received

50 days before being stocked. In years with larger numbers of fry (2016 and 2018), about 100,000 were placed in ponds, while the remainder was raised in indoor tanks and fed brine shrimp for 2 weeks prior to stocking. The number of fry originally placed in each quarter-acre pond ranged from 15,000 to 50,000, depending on the available number of ponds and fry each year. The survival of fry to fingerling stage in the ponds was consistently low, averaging 9% (± 6.9, range = 0.8–17.3) from 2014 to 2018. In an effort to improve survival, all fry received in 2019 were started in the indoor tanks and fed brine shrimp for 2 weeks before being moved to the ponds. In 2020, 260,500 eyed Sauger eggs were shipped to New York in an effort to decrease early-stage fry mortality caused by transport and additional handling. Although egg mortality was high (59%), the resulting 106,000 fry were raised in indoor tanks and fed brine shrimp for 2 weeks before being placed in outdoor ponds. In total, 37,060 pond-raised fingerlings (36–46 mm long) were stocked in the upper Allegheny Reservoir (NY section) and 730,000 2-week old fry (< 8 mm long) were stocked in the upper reservoir, Allegheny River, and the lower sections of two tributaries, Olean Creek and Oil Creek, over the 7-year period (Table 1). Pond fingerlings were scatter stocked by boat in the uppermost 4 miles of the reservoir from 2014 to 2017. It was noticed that the timing of fingerling stocking coincidentally overlapped with the spawning concentrations of several species migrating into the Allegheny River, generating concerns about the predation of stocked Sauger fingerlings. Due to uncertainty about survival after boat stocking, it was decided in 2018 that all fingerlings would be stocked from shore in Quaker Run Bay, a protected bay with a complex habitat that would serve as a better nursery area (Fig. 2).

2.3 Population Assessment

To check the status of the stocked Sauger, NYSDEC Region 9 Fisheries staff annually conducted boat electrofishing surveys in the late summer (late August to mid-September) in the stocked section of the upper Allegheny Reservoir. Water-level fluctuations and algal bloom severity were variable and inconsistent from year to year, making the standardization of electrofishing effort difficult. In addition, sampling effort and locations in 2020 were adjusted due to Covid-19 restrictions, preventing access to waters on the Seneca Nation territory where sampling occurred during previous years. From 2014 to 2019, sampling runs were conducted on the uppermost 4 miles of the reservoir to target the immediate stocking area, while in 2020, sampling runs were conducted in two isolated bays in close proximity to past sampling sites (Fig. 2). Surface water temperatures at the time of sampling ranged from 21 to 25 °C. The total duration of sampling runs ranged from 1.6 to 4.8 hours per year (Table 2). Surveys were run at night in all years except 2016, when daytime shocking was used due to unusually low water levels. In contrast, water levels were near flood stage during sampling in 2018 and were closer to normal during all other years (2014, 2015, 2017, 2019, 2020). Total lengths (mm) were recorded for all Sauger collected. Weights (g) and scales for aging were taken from most of the Sauger collected starting in 2016. Walleye were also collected during these surveys

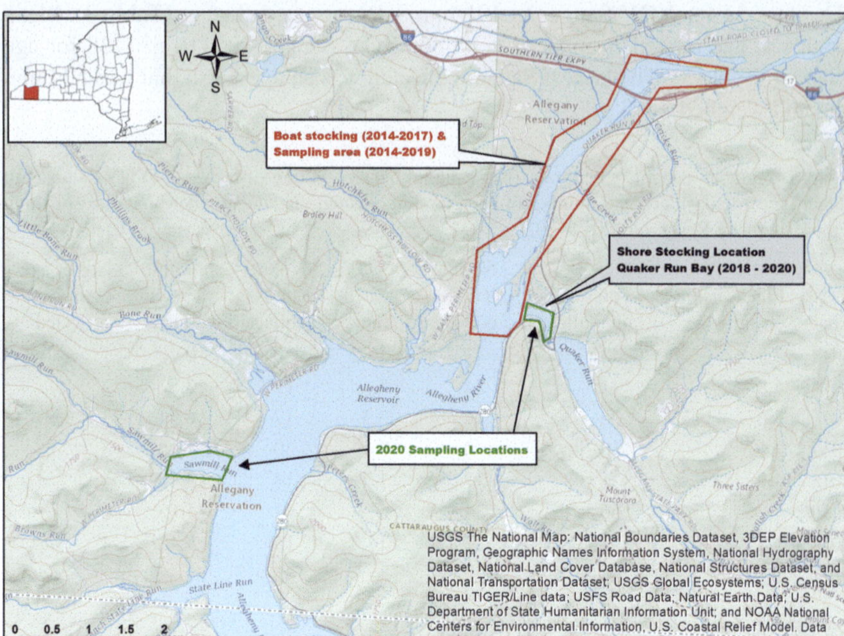

Fig. 2 Map of stocking and sampling locations on the Allegheny Reservoir, New York, 2014–2020

primarily to obtain data for aiding in future management decisions; therefore, only a limited discussion on Walleye is reported in this chapter.

3 Results

A total of 264 Sauger were collected from the upper Allegheny Reservoir from 2014 to 2020, with an overall catch rate of 11.9/h (Table 2). The collection consisted of relatively equal numbers of young of the year (YOY) and age 1 and older fish (129 and 135, respectively). Annual catch rates for all Sauger ranged from 5.1/h in 2018 to 23.1/h in 2020. Young-of-the-year Sauger were collected every year except 2017, a year when relatively few fingerlings, and no fry, were stocked. The highest YOY catch rate (17.2/h) occurred in the initial year of stocking (2014), and the 2020 survey produced the second-highest YOY catch rate (13.8/h). Age 1 and older (adult) Sauger were collected every year that they might have been available (2015–2020), and the adult catch rate during these years ranged from 3.4/h in 2018 to 11.5/h in 2017. Six different age classes of Sauger were collected in 2019 (age 0–age 5), which likely represented all available stocked age classes (Table 2). Only four age classes of Sauger were collected in 2020 (age 0–age 3), likely due to reduced and adjusted sampling effort (Fig. 3). Mean lengths at age ranged from 173 mm ±24 for YOY Sauger to 491 mm ±36 for age 5 Sauger (Table 3). Mean weights at age ranged from 63 g ± 11 for YOY Sauger to 1110 g ± 28 for age 5 Sauger (Table 3). The largest Sauger collected was an age 5 fish that was 534 mm long and weighed 1493 g.

Table 2 Number caught, catch rate (no. caught/hour) of young-of-the-year (YOY) and adult Sauger (age 1 and older), and number of age classes collected via late summer boat electrofishing in the Allegheny Reservoir, 2014–2020

Year	Electrofishing on-time (h)	Total no. collected	No. of YOY collected	YOY catch rate	No. of adults collected	Adult catch rate	No. of age classes collected
2014	2.5	43	43	17.2	0	0	1
2015	4.8	41	24	5	17	3.5	2
2016	2.6	26	9	3.5	17	6.5	3
2017	2.6	30	0	0	30	11.5	3
2018	4.1	21	7	1.7	14	3.4	4
2019	4	66	24	6	42	10.5	6
2020	1.6	37	22	13.8	15	9.4	4
Total	22.2	264	129		135		

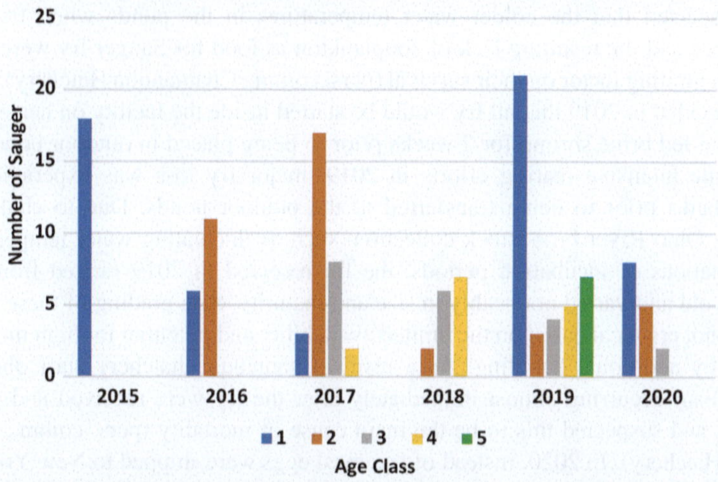

Fig. 3 Age-frequency graph for adult Sauger (age 1 and older) collected via late summer boat electrofishing in the Allegheny Reservoir, 2015–2020

Table 3 Mean lengths (mm) and weights (g) at age (±SD and range) of Sauger collected during boat electrofishing in the Allegheny Reservoir, 2014-2019[a]

Age	No. measured for length	Mean length	No. weighed	Mean weight
0	88	173 ± 24 (126–223)	18	63 ± 11 (45–86)
1	48	310 ± 25 (231–373)	29	237 ± 48 (154–335)
2	33	382 ± 40 (314–446)	26	445 ± 156 (257–761)
3	18	431 ± 26 (381–475)	18	721 ± 143 (483–968)
4	14	468 ± 32 (422–512)	13	916 ± 227 (635–1318)
5	5	491 ± 36 (458–534)	5	1110 ± 281 (872–1493)

[a]Age data from the 2020 survey were not included in this table

4 Discussion

4.1 Propagation and Stocking

To meet the objective of establishing a self-sustaining Sauger population in the New York portion of the Allegheny River watershed, a series of management, monitoring, and outreach actions were detailed in the Conservation Management Plan (Loukmas 2013). The primary management action was the development and implementation of the stocking program that was initially designed to run from 2014 to 2018. Lower than desired numbers of both fry and fingerlings that were available for stocking through that time period resulted in extending the program to a 10-year effort, and thus it is scheduled to continue annually through 2023. Although insect predation was noted as a possible cause of mortality in the outdoor ponds, hatchery staff suspected that the colder water temperatures in the ponds when fry were introduced and the resulting lack of zooplankton as food for Sauger fry were likely the main limiting factor on their survival (pers. comm., Chautauqua Hatchery). Thus, it was decided in 2019 that all fry would be started inside the facility on heated well water and fed brine shrimp for 2 weeks prior to being placed in outdoor ponds.

Despite intensive rearing efforts in 2019, major fry loss was experienced in indoor tanks prior to being transferred to the outdoor ponds. Due to challenges with the Ohio River broodstock collection, such as fluctuating water temperatures and variations in incubation periods, the fry received in 2019 ranged from 1 to 10 days old and varied noticeably in size and maturity. Size grading of these young fry was not practical based on the limited availability and potential for high mortality caused by additional handling. As a result, Chautauqua hatchery staff observed cannibalism occurring almost immediately after the fry were received and placed in tanks and suspected this to be the main cause of mortality (pers. comm., Chautauqua Hatchery). In 2020, instead of fry, eyed eggs were shipped to New York and hatched and reared at Chautauqua Hatchery. The lot of newly hatched fry in 2020 were more uniform in size and age, resulting in pond fingerling production almost three times higher than the previous peak in 2015, suggesting that shipping eyed eggs should be the preferred method for stocking efforts when feasible.

The absence of a propagation program and logistical difficulties in sampling prevented the collection of broodstock from the lower Allegheny River in Pennsylvania. However, Ohio River strain Sauger are collected annually for propagation programs in West Virginia and Kentucky. The Allegheny River is a major tributary of the Ohio River, and these broodstock were identified as a genetically suitable source for stocking in New York (Loukmas 2013). In the future, options to collect and propagate Sauger from the Allegheny River watershed in New York will be explored but are contingent on the successful development of a spawning population that is vulnerable to capture in the spring.

4.2 Population Assessment

Annual late summer boat electrofishing in the Allegheny Reservoir has proven to be valuable in evaluating the success of this program. Sampling has occurred primarily in the upper reservoir, where fingerling Sauger were stocked. The consistently high catch rates and the sustained presence of at least four stocked age classes suggest that many are staying in this area and a relatively high percentage of them are surviving. Although variations in sampling protocols from year to year make statistical comparison with other studies difficult, the mean fall catch rate for total Sauger in New York (11.9/h, range 5.1/h–23.1/h) was more than double the catch rate found on the Kentucky River (4.7/h, range 1.6/h–10.2/h) during a similar 5-year stocking program evaluation, even though the number of fingerlings stocked was more than 12 times higher (Herrala 2014). Also, the growth of Sauger in the reservoir appears to be exceptionally fast, with age 3 fish averaging over 400 mm in total length

Fig. 4 Sauger fingerlings at stocking in mid-June (left) and aquatic biologist Justin Brewer with a 445-mm, age 3 Sauger (right) collected from the Allegheny Reservoir via late summer boat electrofishing in 2019

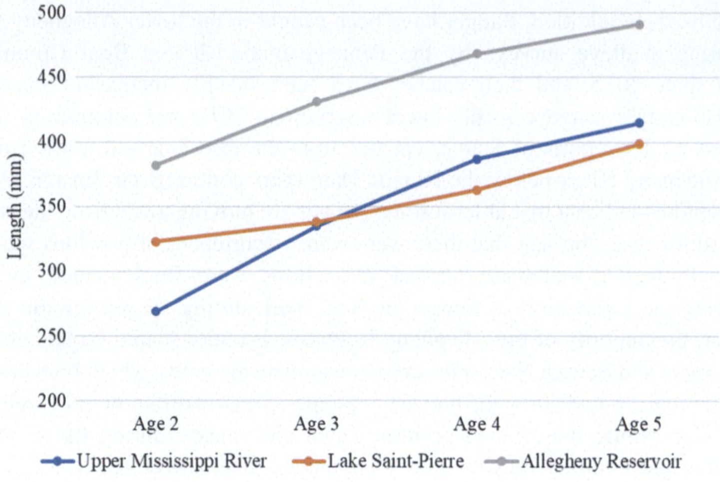

Fig. 5 Length at age distribution of Allegheny Reservoir, Lake Saint-Pierre (Loukmas 2018), and upper Mississippi River Sauger (Mammoliti 2007)

(Fig. 4). Length at age of Sauger collected during this study was consistently and substantially higher for ages 2–5 than in the upper Mississippi River (Mammoliti 2007) and Lake Saint-Pierre (Loukmas 2018) (Fig. 5).

The excellent growth is likely related to the abundant and diverse forage base in the upper reservoir. Specific forage indices were not available; however, late summer electrified bottom trawling surveys in the stocked section of the reservoir collected a total of 37 identifiable species, with percids being the most prevalent family group, followed by cyprinids (Loukmas et al. 2015). Growth also tends to be faster in reservoirs than it is in streams (Preigel 1963), perhaps due to differences in energy demands between lotic and lentic systems.

For Sauger, age at maturity varies widely by location, but, typically, males mature at age 2 and females mature at age 4 (Pitlo Jr. et al. 2004). Because Sauger in the Allegheny Reservoir are growing so well, they may have begun to spawn as early as 2016 and were likely to be spawning by 2018 (when females reached age 4). Early maturity may be critical to the establishment of a self-sustaining population because the average life expectancy of Sauger is only 7 years (Preigel 1969; Gebkin and Wright 1972). Thus far, spawning aggregations of Sauger have not been identified, but in the spring of 2018, two gravid adult female Sauger (both approx. 450 mm) were captured in the section of the Allegheny River, where it enters the reservoir during a spring collection of Walleye for a Seneca Nation of Indians propagation program. This provided the first evidence that the stocked population was ready and able to spawn. Also, in 2019, three saugeyes (Walleye/Sauger hybrids) were caught in the reservoir during the late summer boat electrofishing survey. The saugeye ranged from 279 mm to 330 mm in length and were age 1. This provides evidence that adult Sauger made an effort to spawn in 2018 and used habitats that spawning Walleye used. In addition, Sauger have been caught in the lower Allegheny Reservoir during Walleye surveys by the Pennsylvania Fish and Boat Commission (PFBC) since 2015, and their catches have been steadily increasing since then. They also caught saugeye in the lower reservoir in 2019 and continue to receive numerous angler reports of Sauger catches from the reservoir and a few from the lower Allegheny River below the Kinzua Dam (pers comm. Brian Ensign, PFBC). These captures indicate that at least some Sauger are moving away from the stocked section of the reservoir and that there were some occurrences of previous spawning activity. Protective statewide harvest restrictions were implemented in 2014, prohibiting the possession of Sauger in New York during the restoration efforts. However, the majority of the Allegheny Reservoir is under shared jurisdiction with Pennsylvania and Seneca Nation fishery management agencies, which both currently have open seasons and allow the harvest of Sauger. Despite differences in regulation, we feel that public outreach to promote catch and release during the restoration efforts has had a strong influence on local anglers and, therefore, is not likely to affect the overall Sauger population.

It is expected that Walleye and Sauger spawning activity and habitat utilization will overlap and that hybridization will occur to some degree, but the extent is unknown. The incidence of hybridization tends to increase if the two species have not evolved together in the same system (Billington and Sloss 2011). It has also been noted that in turbid conditions, such as those occurring in the upper reservoir, hybridization can occur presumably due to a failure to recognize a conspecific mate (Billington et al. 1997), a possible scenario in the Allegheny. The current

Walleye population in the Allegheny Reservoir and upper river is thriving and is largely supported by natural reproduction (Loukmas et al. 2015; Ensign 2014). Although the Seneca Nation of Indians and PFBC annually stock Walleye fry in the reservoir, the contribution of stocked fish to the overall population is largely unknown. Based on historical records and the continuing observation of major annual spawning aggregations, natural reproduction contributes substantially to the Walleye population and suggests that an abundance of suitable spawning and rearing habitat is available. Sauger are notably more tolerant of turbid waters and silted bottoms than Walleye (Trautman 1957), characteristics that frequently occur in the upper Allegheny Reservoir, Allegheny River, and several large tributaries. Although habitat and resource competition between Sauger and Walleye are potential concerns, differences in life-history strategies and diet should lessen these concerns (Bozek et al. 2011), and the abundance and complexity of available habitat in the Allegheny River and Reservoir are expected to be sufficient to support both species. Identifying and monitoring spawning and nursery locations as the population develops will provide crucial information for the protection and future management of Sauger in the Allegheny River watershed.

For the purposes of this program, a self-sustaining population is defined as one where at least three naturally produced year classes are present in the population (Loukmas 2013). A broader survey of the Sauger population will be conducted throughout the watershed in 2024 to determine if the objective of establishing a self-sustaining population was met. This survey will include efforts to conduct a population estimate, further investigation into specific spawning habitat and reproduction, and potentially an angler creel survey. In addition, the survey will include exploratory sampling in Conewango Creek to determine if Sauger have naturally migrated into that portion of the Allegheny watershed following the removal of the low head dam at the creek's mouth in Pennsylvania in 2009. If Sauger are not documented in Conewango Creek or its tributaries, a stocking program may be developed to facilitate their recovery in this part of the watershed.

5 Conclusions

Overall, the Sauger stocking program has established an adult population in the Allegheny Reservoir that is likely capable of supporting natural reproduction. Fall surveys in the reservoir have shown increasing trends in both the total and adult Sauger catch rates since 2017, suggesting that good survival over the past 5 years is boosting the potential spawning population. Adult Sauger over 500 mm are now being caught by fishery surveys and anglers in the reservoir and should be in a prime condition to spawn. Surveys in the upper river and tributaries are challenging and have yet to document natural reproduction. However, the catch of saugeye in 2019 and gravid female Sauger in 2018 are encouraging signs that Sauger are attempting to reproduce in the system. Additional stocking should only help to increase the chances of creating a self-sustaining population. Efforts to identify Sauger spawning aggregations and movements will take place over the next few years in hopes to document natural reproduction at the discontinuation of stocking in 2023. The upper Allegheny Reservoir provides an abundance of excellent natural habitat for Sauger and is supplemented by the installation and maintenance of fishery habitat improvement structures (Ensign 2014). Based on the current habitat availability, along with the good survival and exceptional growth of Sauger, no habitat restoration efforts are planned at this time. The reservoir will continue to be the focus of the restoration stocking efforts through 2023, and projections for success appear promising.

Acknowledgments We thank the following individuals and agencies who have contributed to the success of this program: Chris O'Bara, Jim Hedrick, and Ryan Boserman from the West Virginia Department of Natural Resources and Josh Pendleton from the Pfeifer fish hatchery in Kentucky, who enthusiastically and generously provided Sauger fry and eggs for this program; Larry King, Eric Defries, Brad Gruber, Ron Preston, and Jim Rambuski for assistance with Sauger culture at the Chautauqua Hatchery; Shane Titus, Larry Becelia, and Will Miller (Seneca Nation of Indians) for cooperation and assistance with stocking and follow-up monitoring; Brian Ensign, Area 2 Fisheries biologist (Pennsylvania Fish and Boat Commission) for contribution of fishery survey results and information to benefit the program; all NYSDEC Region 9 Fisheries biologists and technicians (James Zanett, Tobias Widger, Amanda Wagner, Cody Whipple, Christina Edwards, Logan Stratton, Joseph Zuppelli, Alec Zerbian, Brad Britt, Alex Smythe, Rob Roth, Kyle Keys, and Eric Maxwell) who have assisted with stocking and follow-up surveys.

References

Becker GC (1983) Fishes of Wisconsin. The University of Wisconsin Press. Madison, WI, 1052 pp

Billington N, Brooks RC, Heidinger RC (1997) Frequency of natural hybridization between saugers and walleyes in the Peoria Pool of the Illinois River determined by morphological and electrophoretic criteria. N Am J Fish Manag 17:220–224

Billington N, Sloss BL (2011) Molecular systematics of sander, and hybridization between walleye and Sauger. In: Barton BA (ed) Biology, management, and culture of walleye and Sauger. American Fisheries Society, Bethesda, Maryland, pp 85–99

Bozek MA, Baccante DA, Lester NP (2011) Walleye and Sauger life history. In: Barton BA (ed) Biology, management, and culture of walleye and Sauger. American Fisheries Society, Bethesda, Maryland, pp 233–301

Eaton SW, Nemecek RJ, Kozubowski MM (1982) Fishes of the Allegheny River above the Kinzua dam. N.Y. Fish and Game J 29(2):189–198

Ensign B (2014) Allegheny reservoir, warren county, Pennsylvania: spring 2014 gill net, trap net and boat electrofishing surveys. Pennsylvania Fish and Boat Commission, Area 2 Biologist Report. 15pp

Fowler HW (1919) A list of fishes of Pennsylvania. Proc Biol Soc Wash 32:49–74

Gebkin DF, Wright KJ (1972) Walleye and Sauger spawning areas study, Pool 7, Mississippi River 1960-1970. Wisconsin Department of Natural Resources Management Report No. 60. 25pp

Groen CL, Schmulbach JC (1978) The sport fishery of the unchannelized and channelized middle Missouri River. Trans Am Fish Soc 107:412–418

Herrala J (2014) Evaluation of a Sauger stocking program on the Kentucky River. Kentucky Department of Fish and Wildlife Resources, Fisheries Bulletin No. 110. 18pp

Hesse LW (1994) The status of Nebraska fishes in the Missouri River. 6. Sauger (Percidae: *Stizostedion canadense*). Transactions of the Nebraska Academy of Sciences 21:109–121

Koryak M, Bonislawsky P, Locy DD, Porter BA (2009) Typical channel fish assemblage of the recovering lower Allegheny River navigation system, Pennsylvania USA. J Freshw Ecol 24 (3):509–517

Loukmas J (2013) Conservation management plan for Sauger in New York state, 2013–2020. Bureau of Fisheries, New York State Department of Environmental Conservation, Albany, New York

Loukmas J (2018) Assessment of Lake Saint-Pierre Sauger for use as a broodstock for Lake Champlain restoration. Bureau of Fisheries, New York State Department of Environmental Conservation, Albany, New York

Loukmas J, Clancy M, Brewer J (2015) Status report on the Allegheny River watershed Sauger management program. Bureau of Fisheries, New York State Department of Environmental Conservation, Albany, New York

Mammoliti CS (2007) Mississippi interstate cooperative resource association: Sauger management investigation. Report to the MICRA gamefish committee. Watershed Institute, Inc., Topeka, KS

NYSDEC (2015) New York state wildlife action plan. New York State Department of Environmental Conservation. Division of Fish and Wildlife, Albany, New York

Pitlo Jr. J, Brecka B, Stopyro M, Brummett K, Jones G (2004) Sauger (*Stizostedion canadense*). Pages 187–197 In: Pitlo J, Rasmussen J (eds) UMRCC Fisheries Compendium, 3rd edition

Preigel GR (1963) Food of walleye and Sauger in Lake Winnebago, Wisconsin. Trans Am Fish Soc 92:312–313

Preigel GR (1969) The Lake Winnebago Sauger-age growth, reproduction, food habits, and early life history. Wisconsin Department of Natural Resources Technical Bulletin No. 43. 63 pp.

Seigwarth GL, Pitlo J, Willis DW, Thompson AL (1993) Walleye and Sauger spawning habitat survey in pool 16 of the upper Mississippi River. Project Completion report. South Dakota State University-South Dakota Cooperative Fish and Wildlife Research Unit, Brookings, SD. 44 pp.

Trautman MB (1957) The fishes of Ohio. Ohio State University Press, Columbus, Ohio, 683 pp

Part IV
Comparison of North American and European Percid Fisheries

International Importance of Percids: Summary and Looking Forward

Robin L. DeBruyne and Edward F. Roseman

Abstract Research presented in the preceding chapters emphasizes recent advancements in the research, management, and aquaculture of Walleye, Sauger, and Yellow Perch in North America. These percid fishes, along with the European Perch and Pikeperch, are economically and ecologically important fishes in their native geographic range. Advances in techniques to evaluate current habitat and predict future habitat conditions provide managers with detailed baseline information and biophysical models useful for evaluating adaptive management practices. Current habitat use and movement assessments have improved substantially with technological advancements in acoustic tags and extensive receiver array networks, which, combined with genetic and genomic tools, are improving percid stock assessments and management. Advances in percid aquaculture techniques have improved growth, survival, and disease resistance, enhancing percid stocking efforts and the production of marketable fish. The exchange of information between researchers and managers will continue to advance techniques of percid management for commercial and recreational exploitation and improve aquaculture practices to provide a lucrative commercial aquaculture industry.

Keywords Yellow Perch · Walleye · European Perch · Pikeperch · Sauger · Management · Aquaculture

R. L. DeBruyne (✉)
Department of Environmental Sciences, Lake Erie Center, University of Toledo, Oregon, OH, USA
e-mail: robin.debruyne@utoledo.edu

E. F. Roseman
U.S. Geological Survey Great Lakes Science Center, Ann Arbor, MI, USA

J. C. Bruner, R. L. DeBruyne (eds.), *Yellow Perch, Walleye, and Sauger: Aspects of Ecology, Management, and Culture*, Fish & Fisheries Series 41,
https://doi.org/10.1007/978-3-030-80678-1_12

1 Introduction

The chapters and research put forth in this book were part of the symposium "Biology, Management, and Culture of Walleye, Sauger, and Yellow Perch: Status and Needs" at the 150th Annual Meeting of the American Fisheries Society. They represent the range of topics presented and emphasize recent advancements in the aquaculture, management, and research of Walleye (*Stizostedion vitreum*), Sauger (*S. canadense*), and Yellow Perch (*Perca flavescens*). These species, as well as the European Perch (also known as Eurasian Perch, (*P. fluviatilis*)) and Pikeperch (also known as Zander (*S. lucioperca*)), provide valuable commercial, recreational, and subsistence fisheries, also serving important ecological roles. Collectively known as pikeperches (*Stizostedion* spp.) and perch (*Perca* spp.; both family Percidae), they are native to North America and Eurasia but have expanded beyond their natural circumpolar range through introductions to habitable systems. Recreationally, they are prized sport fishes pursued for challenge as well as table fare and were introduced to waters beyond their native ranges (including cross-continental transfers) to meet stakeholder desires. Pikeperches and perches have also been introduced to different continents for aquaculture (USFWS 2014; 2019), and lucrative commercial fisheries exist for them across their ranges (Craig 2008; Roseman et al. 2010). In this concluding chapter, we synthesize recent information from topics covered in this book and provide context on the current state and future challenges of percid research and management.

2 Habitat and Climate

Habitat degradation has occurred in the waters inhabited by perches and pikeperches. This includes the degradation of spawning habitat, nursery habitat, overwintering habitat, thermal refuges, and the general in-water environmental conditions. Historical and contemporary alterations (e.g., hydrology, land-use and shoreline changes, non-native species establishment) to aquatic systems often prevent habitat or ecosystem restoration to conditions of even decades prior. Properly remediating degraded habitats does not guarantee improved fishery conditions if the other required habitats or in-water conditions are no longer conducive to the target species. Additionally, other factors may influence habitat quality, availability, or use by any of these percid species (Snickars et al. 2010). As global temperatures predictably rise, percids are a species group that could expand their range to higher latitudes (Lehtonen 1996), increasing yields in lakes currently at their northern limits (Lappalainen et al. 2005). However, the current locations of valuable fisheries, especially in southern latitudes, may experience declines in the survival of species that rely on cool temperatures to meet their life-history strategies (Hokanson 1977; Henderson et al. 1996; Lehtonen 1996; Collingsworth et al. 2017; Hansen et al. 2017). Systems with long-term monitoring will be at the forefront of distinguishing effects from climate and other environmental variables on percid populations (e.g., Oneida Lake, New York; Rudstam et al. 2016).

Percid reproductive success for early-life stages is dependent on habitat conditions, and consistent recruitment into advanced stages often depends on repeatable and stable environmental conditions, as well as available food resources. Maintaining consistent conditions is important during a spawning season, especially in systems where manipulated hydrology fluctuates water levels, possibly exposing eggs and reducing reproductive success (e.g., Draštík et al. 2008; Westrelin et al. 2018; Matt et al. 2021). Even in systems where water levels are not manipulated and remain hydrologically stable, other abiotic conditions can impact percid spawning and larval abundances (e.g., temperature: Snickars et al. 2010; Kallasvuo et al. 2017; DeBruyne et al. 2021; wind and water currents: Beletsky et al. 2007; Zhao et al. 2009; Kallasvuo et al. 2017; turbidity: Wellington et al. 2010; Veneranta et al. 2011; Manning et al. 2014). Spawning and nonspawning habitats in the Baltic Sea for European Perch and Pikeperch are expected to shift as water levels and other abiotic variables change (Čech et al. 2012; Bergström et al. 2013), further complicating the development of long-term population and habitat management plans (Sundblad and Bergström 2014). Research on mechanisms affecting habitat and climate-induced variability across all percid life stages and waterbody types applicable to other areas and possibly across species remains a continued need for these ecologically and economically important species (Neuman et al. 1996; Roseman 2004).

3 Management and Harvest

Managing percid populations across a landscape is inherently complex due to variable fish recruitment, mixtures of genetically distinct stocks, diverse stakeholder desires, and often interjurisdictional management authorities. Examples of large-scale management planning are as follows: Clapp et al. (2021) brought together a case study with statewide fishery survey, creel, and social survey information, along with input from stakeholders, to recommend a standardized statewide bag limit for Yellow Perch in Michigan and provide a framework for similar regulation changes in the future. Lake Erie Walleye and Yellow Perch are managed internationally in the Laurentian Great Lakes using a quota system that allocates portions of a total allowable catch to each jurisdiction (New York, Pennsylvania, Ohio, Michigan, and Ontario) based on population projections derived from statistical catch-at-age models (Hatch et al. 1990; Roseman et al. 2012; Belore et al. 2021). Additionally, much of the Laurentian Great Lakes and other inland US treaty waters are cooperatively managed with First Nations, ensuring that traditional ecological knowledge and harvest methods are incorporated into management strategies and regulations (Gaden et al. 2008; GLIFWC 2020; Klimah 2021). Baltic Sea Pikeperch and European Perch populations are also fished and managed across multiple jurisdictions; however, there is a recognized need for greater interjurisdictional coordination for these mixed stock fisheries (Saulamo and Thoresson 2005). Many European inland fisheries are managed as individual waters with discrete populations using a combination of size and harvest limits (Wysujack et al. 2002; Vainikka et al. 2017). Effective management of these percid species and their habitats could be improved by incorporating multiple sources of information and evaluating additional factors (predator abundance, abiotic conditions, available habitat) unique to that system (Kallasvuo et al. 2017; Skov et al. 2017; Donadi et al. 2020; Eklöf et al. 2020). Long-term monitoring programs, where they persist, routinely measure population-level trends and environmental factors that have proven invaluable in directing research programs addressing specific questions and informing science-based management decisions (Radinger et al. 2019).

Molecular genetics and genomic techniques continue to provide information critical to percid aquaculture and management, including identifying stock structure, stock contributions to fisheries, and stock mixing over large landscapes (Euclide et al. 2020, 2021; Houston et al. 2020; Toomey et al. 2020; You et al. 2020). Euclide et al. (2021) has a detailed account of recent advances in the use of genomic techniques for Walleye management in the Laurentian Great Lakes, as well as providing insight for future research and management applications of these techniques. Genomic techniques have been used to aid the management and aquaculture of Walleye in other areas (Thorstensen et al. 2020; Zhao et al. 2020), as well as of Yellow Perch (Feron et al. 2020), European Perch (Kalous et al. 2017; Ozerov et al. 2018), and Pikeperch (Nguinkal et al. 2019; Schäfer et al. 2021).

4 Movement

Traditional tagging and radio transmitter techniques remain valuable tools for monitoring fish migration patterns, growth, harvest, and mortality rates. Recent advances in technology, namely the use of long-lived electronic transmitter-receiver networks (e.g., GLATOS 2021), have vastly improved the ability and capacity to monitor fish movements and population vital rates across broad spatial ranges and for extended time periods (Krueger et al. 2018). Numerous studies are measuring habitat use, migration patterns, and population vital rates in the Laurentian Great Lakes (GLATOS 2021) to include Walleye (Hayden et al. 2018, 2019; Faust et al. 2019; Fielder et al. 2020) and Yellow Perch (Halfyard et al. 2017; Rous et al. 2017). On a smaller scale, similar marking and telemetry studies have identified spawning areas, movement patterns, exploitation rates, and sex-based distribution and movement patterns (Shultz et al. 2021; Smith et al. 2021). Tagging and movement studies of Pikeperch (Huuskonen et al. 2019) and European Perch (Monk and Arlinghaus 2017; Nakayama et al. 2018; Westrelin et al. 2018) also reveal multiscale movement patterns, habitat use, ontogenetic diet shifts, and growth and survival rates.

5 Aquaculture and Stocking

Percids are intensively reared in aquaculture settings to stock into fishable waters (mainly *Stizostedion* spp.; Specziár and Turcsányi 2017; Raabe et al. 2020) or to raise to marketable size as commercial products (Kestemont et al. 2015; Steenfeldt et al. 2015; Policar et al. 2019). Harvestable aquaculture is common for percids in Europe, and facilities incorporate advanced techniques to promote fast growth and high survival to produce marketable fish economically. Advances in fish-rearing protocols and practices are aimed to improve growing conditions, increase yields, combat disease, and reduce predation from external predators (e.g., European otter (*Lutra lutra*) and Great Cormorant (*Phalacrocorax carbo*): Manikowska-Ślepowrońska et al. 2016; Kestemont et al. 2015). Even though growing to harvestable size is common in Europe, it is not as common in North America where aquaculture facilities often focus on raising Walleye, Sauger, saugeye (*S. vitreum* × *S. canadense*), and, to a lesser extent, Yellow Perch to fry or advanced fingerling size. These fish are stocked into natural rivers and lakes within and outside of their native ranges. Stocking is typically done to sustain or enhance recreational fisheries due to reduced recruitment for a variety of reasons, such as insufficient habitat or environmental conditions to support early-life stages (Ellison and Franzin 1992; Fenton et al. 1996; Raabe et al. 2020; Brewer et al. 2021).

Aquaculture facilities in North America and Eurasia face similar challenges in maintaining the maximum output of stockable fish, emphasizing the importance of sharing technology and solutions for percid species grown in aquaculture systems. Collaborative endeavors between researchers and fishery managers to improve the fish-rearing practices and performance of stocked products through disease prevention, hatchery and pond system engineering, and food selection were shared in this book (Doyle et al. 2021; Eroh et al. 2021; Johnson et al. 2021) and in the review by Policar et al. (2019). Continued improvements in broodstock management and selection, gamete collection and storage, larval propagation and survival, and new evaluation techniques would advance the efficiency and understanding of percid aquaculture (Overton et al. 2015; Policar et al. 2019). The propagation of these valuable species continues to improve and evolve with new techniques as they become available and more widespread across the landscapes.

6 Future Considerations

Continued analytical and technological advances are improving fishery management and aquaculture for pikeperches and perches. However, continued investments in research and new technology will be needed to address anticipated and unforeseen challenges, such as climate change (Hansen et al. 2017) and changing stakeholder desires for fisheries. Variability in water quality, water levels, connectivity of habitat patches, land use, combined with the anticipated increased variability of temperature and precipitation associated with climate change will affect individual stock (Madenjian 2015; Christensen et al. 2020) and broader percid population levels (Lappalainen et al. 2005; Bergström et al. 2013; Hansen et al. 2017). Uncertain responses to invasive species will likely further exacerbate percid recruitment variability, including the threat of percids as invaders (USFWS 2014, 2019). Further, unpredictable changes in the availability of percid prey resources are likely to occur, causing increased variability in percid production for both naturally reproduced and stocked populations (Heino et al. 2009). The growth and recruitment of percids are generally a function of prey availability, and most percids have a broad diet range that includes food resources from across the food web, possibly reducing the effects of dramatic shifts in food availability. Percids also generally tolerate a wide range of thermal and other environmental conditions, as evidenced by their persistence through periods of severe environmental perturbations (e.g., Lake Erie Walleye and Yellow Perch: Koonce et al. 1996; Caspian Sea Pikeperch: Falahatkar et al. 2018). However, little information exists that identifies when populations have reached their stress tolerance limits under the unprecedented range of biotic and abiotic conditions possible, given the forecasted climate scenarios and continued exploitation pressure (Sullivan 2003; Monk and Arlinghaus 2017; Lyach and Remr 2019). These anticipated, yet unpredictable, changes in the fish communities stress the need for continued investment in standardized long-term monitoring and research throughout the percid ranges (Heino et al. 2009).

Finally, continued integration of information and how it is applied to management will be vital to sustaining these valuable recreational, commercial, and subsistence fisheries. Numerous symposia bringing together percid researchers to share information and experiences occur regularly at regional and international venues (e.g., PERCIS, Schmidt et al. 2017; https://www.percis-v.eu/). Results of scientific research and management experiments presented during these conferences and in the scholarly literature continue to be used as the foundation for contemporary fishery management and aquaculture programs across the family's range. Going forward, continued efforts to examine the long-term effects of global climate change (Kao et al. 2020), eutrophication, aquaculture, and invasive species on percid populations will be paramount to the development of fishery research and management programs aimed to conserve percid populations and provide sustainable fisheries and aquaculture.

Acknowledgements The authors are grateful for the reviewers' comments that improved this manuscript. This work was supported by the Great Lakes Restoration Initiative and the U.S. Geological Survey Great Lakes Science Center. Any use of trade, product, or firm names is for descriptive purposes only and does not imply endorsement by the U.S. government.

References

Beletsky D, Mason DM, Schwab DJ, Rutherford ES, Janssen J, Clapp DF, Dettmers JM (2007) Biophysical model of larval yellow perch advection and settlement in Lake Michigan. J Great Lakes Res 33:842–866. https://doi.org/10.3394/0380-1330(2007)33[842:BMOLYP]2.0.CO;2

Belore M, Briggs A, Cook A, DuFour M, Faust M, Heerschap M, Hosack M, Knight C, Schmidt B, Thomas S, Wilkins P, Zhao Y (2021) Report of the Lake Erie yellow perch task group. Great Lakes Fishery Commission, Ann Arbor, 37pp

Bergström U, Sundblad G, Downie A-L, Snickars M, Boström C, Lindegarth M (2013) Evaluating eutrophication management scenarios in the Baltic Sea using species distribution modelling. J Appl Ecol 50:680–690. https://doi.org/10.1111/1365-2664.12083

Brewer JR, Loukmas JJ, Clancy M (2021) Sauger restoration in the upper Allegheny River watershed, New York. In: Bruner JC, DeBruyne RL (eds) Ecology, management, and culture of yellow perch. Walleye and Sauger. Springer, Cham, Switzerland, pp 293–305

Čech M, Peterka J, Říha M, Vejřík L, Jůza T, Kratochvíl M, Draštík V, Muška M, Znachor P, Kubečka J (2012) Extremely shallow spawning of perch (*Perca fluviatilis* L.): the roles of sheltered bays, dense semi-terrestrial vegetation and low visibility in deeper water. Knowledge and Management of Aquatic Ecosystems 406:09

Christensen EA, Svendsen MB, Steffensen JF (2020) The combined effect of body size and temperature on oxygen consumption rates and the size-dependency of preferred temperature in European perch *Perca fluviatilis*. J Fish Biol 97:794–803

Clapp DF, Briggs AS, Claramunt RM, Fielder DG, Zorn TG (2021) Evaluation of a statewide yellow perch bag limit for Michigan. In: Bruner JC, DeBruyne RL (eds) Ecology, management, and culture of yellow perch. Walleye and Sauger. Springer, Cham, Switzerland, pp 55–88

Collingsworth PD, Bunnell DB, Murray MW, Kao Y-C, Feiner ZS, Claramunt RM, Lofgren BM, Höök TO, Ludsin SA (2017) Climate change as a long-term stressor for the fisheries of the Laurentian Great Lakes of North America. Rev Fish Biol Fisheries 27:363–391. https://doi.org/10.1007/s11160-017-9480-3

Craig JF (2008) Percid fishes: systematics, ecology and exploitation. Wiley, Dunscore, Scotland

DeBruyne RL, Tucker TR, Lloyd C, Briggs AS, Belore M, Roseman EF (2021) Distribution and abundance of pelagic larval yellow perch in Lake St. Clair (US/Canada) and adjoining waters. In: Bruner JC, DeBruyne RL (eds) Ecology, management, and culture of yellow perch. Walleye and Sauger. Springer, Cham, Switzerland, pp 89–111

Donadi S, Bergström L, Bertil Berglund JM, Anette B, Mikkola R, Saarinen A, Bergström U (2020) Perch and pike recruitment in coastal bays limited by stickleback predation and environmental forcing. Estuar Coast Shelf Sci 246:107052. https://doi.org/10.1016/j.ecss.2020.107052

Doyle CM, Culver DA, Pugh ME, Filbrun JE (2021) A comparison of aquaculture production methods for optimizing production of fingerling yellow perch (*Perca flavescens*). In: Bruner JC, DeBruyne RL (eds) Ecology, management, and culture of yellow perch. Walleye and Sauger. Springer, Cham, Switzerland, pp 33–54

Draštík V, Kubečka J, Tušer M, Čech M, Frouzová J, Jarolím O, Prchalová M (2008) The effect of hydropower on fish stocks: comparison between cascade and non-cascade reservoirs. Hydrobiologia 609:25–36

Eklöf JS, Sundblad G, Erlandsson M, Donadi S, Hansen JP, Eriksson BK, Bergström U (2020) A spatial regime shift from predator to prey dominance in a large coastal ecosystem. Commun Biol 3:459. https://doi.org/10.1038/s42003-020-01180-0

Ellison DG, Franzin WG (1992) Overview of the symposium on walleye stocks and stocking. N Am J Fish Manag 12:271–275

Eroh GD, Bringolf RB, Camus AC, Williams-Woodward JL, Jennings CA (2021) Effects of parasiticidal hydrogen peroxide treatments on walleye hatching success in a recirculating system. In: Bruner JC, DeBruyne RL (eds) Ecology, management, and culture of yellow perch. Walleye and Sauger. Springer, Cham, Switzerland, pp 191–208

Euclide PT, MacDougall T, Robinson JM, Faust MD, Wilson CC, Chen K-Y, Marschall EA, Larson W, Ludsin SA (2020) Mixed-stock analysis in the age of genomics: rapture genotyping enables evaluation of stock-specific exploitation in a freshwater fish population with weak genetic structure. Genomics. https://doi.org/10.1101/2020.11.10.376350

Euclide PT, Robinson J, Faust M, Ludsin SA, MacDougall TM, Marschall EA, Chen K-Y, Wilson C, Bootsma M, Stott W, Scribner KT, Larson WA (2021) Using genomic data to guide walleye management in the Great Lakes. In: Bruner JC, DeBruyne RL (eds) Ecology, management, and culture of yellow perch. Walleye and Sauger. Springer, Cham, Switzerland, pp 115–139

Falahatkar B, Efatpanah I, Kestemont P (2018) Pikeperch *Sander lucioperca* production in the south part of the Caspian Sea: technical notes. Aquac Int 26:391–401

Faust MD, Vandergoot CS, Brenden TO, Kraus RT, Hartman T, Krueger CC (2019) Acoustic telemetry as a potential tool for mixed-stock analysis of fishery harvest: a feasibility study using Lake Erie walleye. Can J Fish Aquat Sci 76:1019–1030. https://doi.org/10.1139/cjfas-2017-0522

Fenton R, Mathias JA, Moodie GEE (1996) Recent and future demand for walleye in North America. Fisheries 21(1):6–12

Feron R, Zahm M, Cabau C, Klopp C, Roques C, Bouchez O, Eché C, Valière S, Donnadieu C, Haffray P, Bestin A, Morvezen R, Acloque H, Euclide PT, Wen M, Jouano E, Schartl M, Postlethwait JH, Schraidt C, Christie MR, Larson WA, Herpin A, Guiguen Y (2020) Characterization of a Y-specific duplication/insertion of the anti-Mullerian hormone type II receptor gene based on a chromosome-scale genome assembly of yellow perch, *Perca flavescens*. Mol Ecol Resour 20:531–543. https://doi.org/10.1111/1755-0998.13133

Fielder DG, Hayden TA, Vandergoot CS, Krueger CC (2020) Large-scale fish movement affects metrics of management importance as indicated by quantitative stock assessment. J Great Lakes Res 46:633–642

Gaden M, Krueger C, Goddard C, Barnhart G (2008) A joint strategic plan for Management of Great Lakes fisheries: a cooperative regime in a multi-jurisdictional setting. Aquat Ecosyst Health Manage 11:50–60. https://doi.org/10.1080/14634980701877043

Great Lakes Acoustic Observation System (GLATOS) (2021) https://glatos.glos.us/

Great Lakes Indian Fish and Wildlife Commission (GLIFWC) (2020) Treaty rights recognition and affirmation. Great lakes fish and indian wildlife commission. Available at: https://www.glifwc.org/Recognition_Affirmation/

Halfyard EA, Webber D, Del Papa J, Leadley T, Kessel ST, Colborne SF, Fisk AT (2017) Evaluation of an acoustic telemetry transmitter designed to identify predation events. Methods Ecol Evol 8:1063–1071

Hansen GJA, Read JS, Hansen JF, Winslow LA (2017) Projected shifts in fish species dominance in Wisconsin lakes under climate change. Glob Change Biol 23:1463–1476. https://doi.org/10.1111/gcb.13462

Hatch RW, Nepszy SJ, Rawson MR (1990) Management of percids in Lake Erie, North America. In: von Densen WLT, Steinmetz B, Hughes RH (eds) Management of freshwater fisheries. Wageningen, Podoc, pp 624–636

Hayden TA, Binder TR, Holbrook CM, Vandergoot CS, Fielder DG, Cooke SJ, Dettmers JM, Krueger CC (2018) Spawning site fidelity and apparent annual survival of walleye (*Sander vitreus*) differ between a Lake Huron and Lake Erie tributary. Ecol Freshw Fish 27:339–349

Hayden TA, Vandergoot CS, Fielder DG, Cooke SJ, Dettmers JM, Krueger CC (2019) Telemetry reveals limited exchange of walleye between Lake Erie and Lake Huron: movement of two populations through the Huron-Erie corridor. J Great Lakes Res 45:1241–1250

Heino J, Virkkala R, Toivonen H (2009) Climate change and freshwater biodiversity: detected patterns, future trends and adaptations in northern regions. Biol Rev 84:39–54. https://doi.org/10.1111/j.1469-185X.2008.00060.x

Henderson BA, Wong JL, Nepszy SJ (1996) Reproduction of walleye in Lake Erie: allocation of energy. Can J Fish Aquat Sci 53:127–133

Hokanson KEF (1977) Temperature requirements of some percids and adaptations to the seasonal temperature cycle. J Fish Res Bd Can 34:1524–1550

Houston RD, Bean TP, Macqueen DJ, Gundappa MK, Jin YH, Jenkins TL, Selly SLC, Martin SAM, Stevens JR, Santos EM, Davie A, Robledo D (2020) Harnessing genomics to fast-track genetic improvement in aquaculture. Nat Rev Genet 21:389–409. https://doi.org/10.1038/s41576-020-0227-y

Huuskonen H, Piironen J, Syväranta J, Eronen R, Biasi C, Kiiskinen P, Kortet R, Vainikka A (2019) Diet and movements of pikeperch (*Sander lucioperca*) in a large oligotrophic lake with an exceptionally high pikeperch yield. Ecol Freshw Fish 28:533–543

Johnson JA, Kelsey K, Summerfelt R (2021) Walleye larviculture in water reuse aquaculture systems. In: Bruner JC, DeBruyne RL (eds) Ecology, management, and culture of yellow perch. Walleye and Sauger. Springer, Cham, Switzerland, pp 141–190

Kallasvuo M, Vanhatalo J, Veneranta L (2017) Modeling the spatial distribution of larval fish abundance provides essential information for management. Can J Fish Aquat Sci 74:636–649. https://doi.org/10.1139/cjfas-2016-0008

Kalous L, Kuříková P, Kohout J, Rylková K, Petrtýl M, Čech M (2017) Differences in spatial communities of European perch (*Perca fluviatilis* Linnaeus, 1758) fry in a canyon-shaped reservoir are not attributable to genetics. J Appl Ichthyol 33:306–313. https://doi.org/10.1111/jai.13305

Kao Y-C, Rogers MW, Bunnell DB, Cowx IG, Qian SS, Anneville O, Beard TD, Brinker A, Britton JR, Chura-Cruz R, Gownaris NJ, Jackson JR, Kangur K, Kolding J, Lukin AA, Lynch AJ, Mercado-Silva N, Moncayo-Estrada R, Njaya FJ, Ostrovsky I, Rudstam LG, Sandström ALE, Sato Y, Siguayro-Mamani H, Thorpe A, van Zwieten PAM, Volta P, Wang Y, Weiperth A, Weyl OLF, Young JD (2020) Effects of climate and land-use changes on fish catches across lakes at a global scale. Nat Commun 11:2526. https://doi.org/10.1038/s41467-020-14624-2

Kestemont P, Dabrowski K, Summerfelt RC (eds) (2015) Biology and culture of percid fishes. Springer, Dordrecht, Netherlands, 901pp

Klimah CA (2021) Managing tribal fisheries and employees on the reservation. In: Bruner JC, DeBruyne RL (eds) Ecology, management, and culture of yellow perch. Walleye and Sauger. Springer, Cham, Switzerland, pp 239–270

Koonce JF, Busch WN, Czapla T (1996) Restoration of Lake Erie: contribution of water quality and natural resource management. Can J Fish Aquat Sci 53(S1):105–112

Krueger CC, Holbrook CM, Binder TR, Vandergoot CS, Hayden TA, Hondorp DW, Nate N, Paige K, Riley SC, Fisk AT, Cooke SJ (2018) Acoustic telemetry observation systems: challenges encountered and overcome in the Laurentian Great Lakes. Can J Fish Aquat Sci 75:1755–1763. dx.doi.org. https://doi.org/10.1139/cjfas-2017-0406

Lappalainen J, Malinen T, Rahikainen M, Vinni M, Nyberg K, Ruuhijarvi J, Salminen M (2005) Temperature dependent growth and yield of pikeperch, *Sander lucioperca*, in Finnish lakes. Fish Manag 12:27–35. https://doi.org/10.1111/j.1365-2400.2004.00416.x

Lehtonen H (1996) Potential effects of global warming on northern European freshwater fish and fisheries. Fish Manag Ecol 3:59–71. https://doi.org/10.1111/j.1365-2400.1996.tb00130.x

Lyach R, Remr J (2019) The effects of environmental factors and fisheries management on recreational catches of perch *Perca fluviatilis* in the Czech Republic. Aquat Living Resour 32: 15. https://doi.org/10.1051/alr/2019013

Madenjian CP (2015) Bioenergetics modeling of percid fishes. In: Kestemont P, Dabrowski K, Summerfelt RC (eds) Biology and culture of percid fishes. Springer Netherlands, Dordrecht, pp 369–397

Manikowska-Ślepowrońska B, Szydzik B, Jakubas D (2016) Determinants of the presence of conflict bird and mammal species at pond fisheries in western Poland. Aquat Ecol 50:87–95

Manning NF, Bossenbroek JM, Mayer CM, Bunnell DB, Tyson JT, Rudstam LG, Jackson JR (2014) Modeling turbidity type and intensity effects on the growth and starvation mortality of age-0 yellow perch. Can J Fish Aquat Sci 71:1544–1553

Matt KJ, Welsh SA, Smith DM (2021) Spawning characteristics of yellow perch during periods of water level fluctuations in a hydropower reservoir. In: Bruner JC, DeBruyne RL (eds) Ecology, management, and culture of yellow perch. Walleye and Sauger. Springer, Cham, Switzerland, pp 3–32

Monk CT, Arlinghaus R (2017) Eurasian perch, *Perca fluviatilis*, spatial behaviour determines vulnerability independent of angler skill in a whole-lake reality mining experiment. Can J Fish Aquat Sci 75:417–428

Nakayama S, Doering-Arjes P, Linzmaier S, Briege J, Klefoth T, Pieterek T, Arlinghaus R (2018) Fine-scale movement ecology of a freshwater top predator, Eurasian perch (*Perca fluviatilis*), in response to the abiotic environment over the course of a year. Ecol Freshw Fish 27:798–812

Neuman E, Roseman E, Lehtonen H (1996) Determination of year-class strength in percid fishes. Ann Zool Fenn 33(3/4):315–319

Nguinkal JA, Brunner RM, Verleih M, Rebl A, de los Ríos-Pérez L, Schäfer N, Hadlich F, Stüeken M, Wittenburg D, Goldammer T (2019) The first highly contiguous genome assembly of pikeperch (*Sander lucioperca*), an emerging aquaculture species in Europe. Genes 10:708. https://doi.org/10.3390/genes10090708

Overton JL, Toner D, Policar T, Kucharczyk D (2015) Commercial production: factors for success and limitations in European percid fish culture. In: Kestemont P, Dabrowski K, Summerfelt RC (eds) Biology and culture of percid fishes. Springer, Dordrecht, pp 881–890

Ozerov MY, Ahmad F, Gross R, Pukk L, Kahar S, Kisand V, Vasemägi A (2018) Highly continuous genome assembly of eurasian perch (*Perca fluviatilis*) using linked-read sequencing. G3 g3.200768.2018. https://doi.org/10.1534/g3.118.200768

Policar T, Schaefer FJ, Panana E, Meyer S, Teerlinck S, Toner D, Żarski D (2019) Recent progress in European percid fish culture production technology—tackling bottlenecks. Aquacult Int 27: 1151–1174. https://doi.org/10.1007/s10499-019-00433-y

Raabe JK, VanDeHey JA, Zentner DL, Cross TK, Sass GG (2020) Walleye inland lake habitat: considerations for successful natural recruitment and stocking in north Central North America. Lake Reservoir Manag 36:335–359

Radinger J, Britton JR, Carlson SM, Magurran AE, Alcaraz-Hernández JD, Almodóvar A, Benejam L, Fernández-Delgado C, Nicola GG, Oliva-Paterna FJ, Torralva M (2019) Effective monitoring of freshwater fish. Fish Fish 20:729–747

Roseman EF (2004) Percid management: Current status and future research needs. In: JA BTPM (ed) Proceedings of Percis III: The Third International Percid Fish Symposium. WISCU-W-03-001. University of Wisconsin Sea Grant, Madison, pp 7–8

Roseman EF, Drouin R, Gaden M, Knight R, Tyson J, Zhao Y (2012) Managing inherent complexity for sustainable walleye fisheries in Lake Erie. In: Taylor WW, Lynch AJ, Leonard NJ (eds) Great Lakes fisheries policy and management: a binational perspective, 2nd edn. Michigan State University Press, East Lansing, pp 475–494

Roseman EF, Kocovsky P, Vandergoot C (eds) (2010) Status of walleye in the great lakes: proceedings of the 2006 Symposium. Great lakes fishery commission, Ann Arbor, MI. Technical Report 69. 29pp

Rous AM, Midwood JD, Gutowsky LF, Lapointe NW, Portiss R, Sciscione T, Wells MG, Doka SE, Cooke SJ (2017) Telemetry-determined habitat use informs multi-species habitat management in an urban harbour. Environ Manag 59:118–128

Rudstam LG, Mills EL, Jackson JR, Stewart DJ (eds) (2016) Oneida Lake: long-term dynamics of a managed ecosystem and its fishery. American Fisheries Society, Bethesda, Maryland, 541pp

Saulamo K, Thoresson G (2005) Management of Pikeperch migrating over management areas in a Baltic archipelago area. AMBIO J Hum Environ 34:120–124

Schäfer N, Kaya Y, Rebl H, Stüeken M, Rebl A, Nguinkal JA, Franz GP, Brunner RM, Goldammer T, Grunow B, Verleih M (2021) Insights into early ontogenesis: characterization of stress and development key genes of pikeperch (*Sander lucioperca*) in vivo and in vitro. Fish Physiol Biochem. https://doi.org/10.1007/s10695-021-00929-6

Schmidt B, Castle D, Roseman E, Saat T, Lehtonen H (2017) Synopsis of the fourth international percid fish symposium. Fisheries 42(11):567–570

Shultz A, Klimah C, Curtis-Quick J, Claussen R, LaBine J, Ray A (2021) Can you hear me now? Design considerations for large lake, multi-species telemetry projects. In: Bruner JC, DeBruyne RL (eds) Ecology, management, and culture of yellow perch. Walleye and Sauger. Springer, Cham, Switzerland, pp 271–290

Skov C, Jansen T, Arlinghaus R (2017) 62 years of population dynamics of European perch (*Perca fluviatilis*) in a mesotrophic lake tracked using angler diaries: the role of commercial fishing, predation and temperature. Fish Res 195:71–79. https://doi.org/10.1016/j.fishres.2017.06.016

Smith DM, Welsh SA, Hilling CD (2021) Seasonal movement patterns and distribution of walleye in a central Appalachian hydropower reservoir. In: Bruner JC, DeBruyne RL (eds) Ecology, management, and culture of yellow perch, walleye and Sauger. Springer, Cham, Switzerland, pp 209–238

Snickars M, Sundblad G, Sandström A, Ljunggren L, Bergström U, Johansson G, Mattila J (2010) Habitat selectivity of substrate-spawning fish: modelling requirements for the Eurasian perch *Perca fluviatilis*. Mar Ecol Prog Ser 398:235–243. https://doi.org/10.3354/meps08313

Specziár A, Turcsányi B (2017) Management of Pikeperch stocking in Lake Balaton: effect of season, area, fish size and method of release on the rate and distribution of recaptures. Knowledge & Management of Aquatic Ecosystems 418:52

Steenfeldt S, Fontaine P, Overton JL, Policar T, Toner D, Falahatkar B, Horváth Á, Khemis IB, Hamza N, Mhetli M (2015) Current status of Eurasian percid fishes aquaculture. In: Kestemont P, Dabrowski K, Summerfelt RC (eds) Biology and culture of percid fishes. Springer Netherlands, Dordrecht, pp 817–841

Sullivan MG (2003) Active management of walleye fisheries in Alberta: dilemmas of managing recovering fisheries. N Am J Fish Manag 23:1343–1358

Sundblad G, Bergström U (2014) Shoreline development and degradation of coastal fish reproduction habitats. Ambio 43:1020–1028. https://doi.org/10.1007/s13280-014-0522-y

Thorstensen MJ, Wiens LM, Jeffrey JD, Klein GM, Jeffries KM, Treberg JR (2020) Morphology and blood metabolites reflect recent spatial and temporal differences among Lake Winnipeg walleye. *Sander vitreus* J Great Lakes Res. https://doi.org/10.1016/j.jglr.2020.06.015

Toomey L, Dellicour S, Vanina T, Pegg J, Kaczkowski Z, Kouřil J, Teletchea F, Bláha M, Fontaine P, Lecocq T (2020) Getting off on the right foot: integration of spatial distribution of genetic variability for aquaculture development and regulations, the European perch case. Aquaculture 521:734981. https://doi.org/10.1016/j.aquaculture.2020.734981

U.S. Fish and Wildlife Service (USFWS) (2014) Zander (*Sander luciopera*) Ecological risk screening summary. https://www.fws.gov/injuriouswildlife/pdf_files/Sander_lucioperca_WEB_9-15-2014.pdf

U.S. Fish and Wildlife Service (USFWS) (2019) European Perch (*Perca fluviatilis*) Ecological risk screening summary. https://www.fws.gov/fisheries/ans/erss/highrisk/ERSS-Perca-fluviatilis_Final.pdf

Vainikka A, Olin M, Ruuhijärvi J, Huuskonen H, Eronen R, Hyvärinen P (2017) Model-based evaluation of the management of pikeperch (*Sander lucioperca*) stocks using minimum and maximum size limits. Boreal Environ Res 22:187–212

Veneranta L, Urho L, Lappalainen A, Kallasvuo M (2011) Turbidity characterizes the reproduction areas of pikeperch (*Sander lucioperca* (L.)) in the northern Baltic Sea. Estuarine. Coastal Shelf Sci 95:199–206

Wellington CG, Mayer CM, Bossenbroek JM, Stroh NA (2010) Effects of turbidity and prey density on the foraging success of age 0 year yellow perch *Perca flavescens*. J Fish Biol 76:1729–1741

Westrelin S, Roy R, Tissot-Rey L, Bergès L, Argillier C (2018) Habitat use and preference of adult perch (*Perca fluviatilis* L.) in a deep reservoir: variations with seasons, water levels and individuals. Hydrobiologia 809:121–139. https://doi.org/10.1007/s10750-017-3454-2

Wysujack K, Kasprzak P, Laude U, Mehner T (2002) Management of a Pikeperch stock in a long-term biomanipulated stratified lake: efficient predation vs. low recruitment. Hydrobiologia 479: 169–180

You X, Shan X, Shi Q (2020) Research advances in the genomics and applications for molecular breeding of aquaculture animals. Aquaculture 526:735357. https://doi.org/10.1016/j.aquaculture.2020.735357

Zhao Y, Jones ML, Shuter BJ, Roseman EF (2009) A biophysical model of Lake Erie walleye (*Sander vitreus*) explains interannual variations in recruitment. Can J Fish Aquat Sci 66:114–125. https://doi.org/10.1139/F08-188

Zhao H, Silliman K, Lewis M, Johnson S, Kratina G, Rider SJ, Stepien CA, Hallerman EM, Beck B, Fuller A, Peatman E (2020) SNP analyses highlight a unique, imperiled southern walleye (*Sander vitreus*) in the Mobile River basin. Can J Fish Aquat Sci 77:1366–1378

Select Percidae References

Barry TP, Malison JA (eds) (2004) Proceedings of PERCIS III, the third international percid fish symposium. Madison, WI, July 20–24, 2003. University of Wisconsin, Sea Grant Institute, Madison. Available: digital.library.wisc.edu/1711.dl/EcoNatRes.Percis (January 2021)

Barton B (ed) (2011) Biology, management, and culture of Walleye and Sauger American Fisheries Society, Bethesda, MD, 600p

Carlander KD (1997) Handbook of freshwater fishery biology volume three life history data on ichthyopercid and percid fishes of the united states and Canada Iowa State University Press, Ames, IA, 397p

Colby PJ, McNicol RE, Ryder RA (1979) Synopsis of biological data on the walleye *Stizostedion v. vitreum* (Mitchill 1818). Food and agriculture organization of the United Nations, fisheries synopsis series 119, FAO, Rome

Couture P, Pyle G (eds) (2016) Biology of perch. CRC Press, Boca Raton, FL, 307p

Craig JF (1987) The biology of perch and related fish. Timber Press, Portland, OR, 333p

Craig JF (2000) *Percid Fishes: Systematics, Ecology, and Exploitation*. Blackwell Sciences, Oxford

Ebbers MA, Colby PJ, Lewis CA (1988) Walleye – Sauger bibliography Minnesota. Department of Natural Resources Investigational Report No. 396:1–201

Kestemont P, Dabrowski K (eds) (1996) Recent advances in the aquaculture of percid fish. J Appl Ichthyol 12(3–4):137–200

Kestemont P, Dabrowski K, Summerfelt RC (eds) (2015) Biology and culture of percid fishes principles and practices. Springer, Dordrecht, 901p

Kuehne RA, Barbour RW (1983) The American darters University Press of Kentucky, Lexington, KY, 177p

Marsden JE, Gunderson JL, Klepinger MR, Bronte CR (eds) (1998) Special section on an international symposium on Eurasian Ruffe (*Gymnocephalus cernuus*) biology, impacts, and control. J Great Lakes Res 24(2):165–382

Page LM (1983) Handbook of darters T.F.H. Publications, Neptune City, NJ, 271p

Rask MW, Van Densen WLT, Lehtonen H, Rutherford E (eds) (1996) PERCIS II. The second international percid fish symposium, Vassa, FL, August 21–25, 1995. *Annales Zoologici Fennici* 33:303–723. Available: www.sekj.org/anz/anz3334.htm (January 2021)

Schmidt B, Castle D, Roseman E, Saat T, Lehtonen H (2017) Synopsis of the Fourth International Percid Fish Symposium. Fisheries 42(11):567–570

J. C. Bruner, R. L. DeBruyne (eds.), *Yellow Perch, Walleye, and Sauger: Aspects of Ecology, Management, and Culture*, Fish & Fisheries Series 41,
https://doi.org/10.1007/978-3-030-80678-1

Simon TP, Wallus R (2006) Reproductive biology and early life history of fishes in the Ohio river drainage percidae – perch, pikeperch, and darter, vol 4. Taylor & Francis. Boca Raton, FL, 619p

Stevenson JC (ed) (1977) Special issue. Proceedings of the 1976 percid international symposium (PERCIS). J Fish Res Board Canada 34(10):1445–2000

Summerfelt RC (ed) (1996) Walleye culture manual NCRAC culture series 101, North Central Regional Aquaculture Center Publications Office, Iowa State University, Ames, IA, 416p

Thorpe J (1977) Synopsis of biological data on the perch *Perca fluviatilis* Linneaus, 1758 and *Perca flavescens* Mitchill, 1814. Food and agriculture organization of the United Nations, fisheries synopsis series 113, FAO, Rome

Index

J. C. Bruner, R. L. DeBruyne (eds.), *Yellow Perch, Walleye, and Sauger: Aspects of Ecology, Management, and Culture*, Fish & Fisheries Series 41,
https://doi.org/10.1007/978-3-030-80678-1

The manufacturer's authorised representative in the EU is Springer Nature Customer Service Centre GmbH, Europaplatz 3, 69115 Heidelberg, Germany. If you have any concerns regarding our products, please contact ProductSafety@springernature.com

Printed and bound by CPI Group (UK) Ltd, Croydon, CR0 4YY
30/09/2026
02225689-0012